ZINC IN SOILS AND PLANTS

Developments in Plant and Soil Sciences

VOLUME 55

The titles published in this series are listed at the end of this volume

Zinc in Soils and Plants

Proceedings of the International Symposium on
'Zinc in Soils and Plants'
held at The University of Western Australia,
27-28 September, 1993

Edited by

A.D. Robson
Soil Science and Plant Nutrition
School of Agriculture
The University of Western Australia
Perth
Western Australia

KLUWER ACADEMIC PUBLISHERS
DORDRECHT / BOSTON / LONDON

Library of Congress Cataloging in Publication Data

ISBN 0-7923-2631-8

Published by Kluwer Academic Publishers,
P.O. Box 17, 3300 AA Dordrecht, The Netherlands

Kluwer Academic Publishers incorporates
the publishing programmes of Martinus Nijhoff,
Dr W. Junk, D. Reidel, and MTP Press.

Sold and distributed in the U.S.A. and Canada
by Kluwer Academic Publishers
101 Philip Drive, Norwell, MA 02061, U.S.A.

In all other countries, sold and distributed
by Kluwer Academic Publishers Group,
P.O. Box 322, 3300 AH Dordrecht, The Netherlands.

Table of Contents

9. Interactions between Zinc and Other Nutrients Affecting the Growth of Plants

Jack F Loneragan and Michael J. Webb

10. Zinc Phytotoxicity

R.L. Chaney

11. The Distribution and Correction of Zinc Deficiency

P.N. Takkar and Colin D. Walker

12. Diagnosis of Zinc Deficiency

R.F. Brennan, J.D. Armour and D.J Reuter

13. Zinc Concentrations and Forms in Plants for Humans and Animals

Ross M. Welch

14. The Zinc Requirements of Grazing Ruminants

C.L. White

List of Contributors

Armour, J, DPI, PO Box 1054, Mareeba, Queensland 4880, Australia

Barak, P, Soil Science Department, University of Wisconsin, Madison WI 53706, USA

Barrow, N J, CSIRO, Private Mail Bag, Post Office, Wembley WA 6014, Australia

Brennan, R F, Dept of Agriculture, Albany WA 6330, Australia

Brown, P H, Plant Nutrition, Pomology Department, University of California at Davis, USA

Cakmak, I, University Cukurova, Faculty of Agriculture, Soil Science and Plant Nutrition, Adana, Turkey

Chaney, R L, USDA-ARS Environmental Chemistry Lab, Bldg 318 BARC-East, Beltsville Maryland 20705, USA

Gilkes, R J, Soil Science and Plant Nutrition, School of Agriculture, The University of Western Australia, WA 6009, Australia

Graham, R D, Dept of Agronomy, Waite Agricultural Research Institute, Glen Osmond SA 5064, Australia

Helmke, P, Soil Science Department, University of Wisconsin, Madison WI 53706, USA

Kochian, L, US Plant Soil and Nutrition Lab, Tower Road Ithaca NY 14853-0331, USA

Loneragan, J F, Murdoch University, Murdoch WA 6150, Australia

Longnecker, N E, Soil Science and Plant Nutrition, School of Agriculture, The University of Western Australia, WA 6009, Australia

Marschner, H, Universitat Hohenheim, Institut fur Pflanzenernahrung, 7000 Stuttgart 70 Hohenheim, Germany

Mortvedt, J J, Agronomy Dept, Colorado State University, Ft Collins, CO 80523 USA

Rengel, Z, Dept of Agronomy, Waite Agricultural Research Institute, Glen Osmond SA 5064, Australia

Reuter, D J, CSIRO Division of Soils, Glen Osmond SA 5064, Australia

Robson, A D, Soil Science and Plant Nutrition, School of Agriculture, The University of Western Australia, WA 6009, Australia

Takkar, P N, Indian Institute of Soil Science, Z-6 Zone-I Maharana Pratap Nagar, Bhopal 462011 Madhya Pradesh, India

Walker, C D, Dept Environmental and Life Sciences, Murdoch University, Murdoch WA 6150, Australia

Webb, M, Dept of Agronomy, Waite Agricultural Research Institute, Glen Osmond SA 5064, Australia

Welch, R M, US Plant Soil and Nutrition Lab, Tower Road, Ithaca NY 14853-0331, USA

White, C L, CSIRO, Post Office, Private Mail Bag, Wembley WA 6014, Australia

Zhang, Q, Department of Pomology, University of California at Davis, USA

Preface

The symposium on "Zinc in Soils and Plants" is the third in a series which began with "Copper in Soils and Plants" in Perth in 1981 and continued with "Manganese in Soils and Plants" in Adelaide in 1988.

The symposium brings together a series of valuable accounts of many aspects of the reactions of zinc in soils, the uptake, transport and utilization of zinc in plants, the diagnosis and correction of zinc deficiency in plants and the role of zinc in animal and human nutrition.

I am grateful for the financial support provided by Grains Research and Development Corporation, Rural Industries Research and Development Corporation, Wool Research and Development Corporation, Ansett Australia, and Qantas Australian.

I am most appreciative of the willingness of many scientists to act as referees: G S P Ritchie, R J Gilkes, N C Uren, K Tiller, B Leach, H Greenway, N E Longnecker, J F Loneragan, Z Rengel, C A Atkins, J W Gartrell, P J Randall, D G Edwards, R J Hannam, R J Moir, J E Dreosti, N Suttle, C L White, H Marschner, N Wilhelm, M McBride. All provided valuable comments on the manuscripts.

Finally, I thank Mrs M Davison who provided excellent secretarial assistance.

A.D. Robson September 1993

Chapter 1.

The Chemistry of Zinc

PHILLIP BARAK and PHILIP A. HELMKE

"There is another metal, zinckum, which is in general unknown to the [scientific] fraternity...It admits of no mixture...but keeps entirely to itself."
Paracelsus (c. 1490-1541, *Liber Mineralium*, publ. c. 1570)

1. Abstract

Zinc is a metallic element with atomic number 30 and stable isotopes of mass 66, 67, 68, and 70, averaging 65.38 a.m.u. The terrestrial chemistry of Zn is that of Zn (II) rather than Zn(0). The Zn (II) ion has an electron configuration of $1s^2$, $2s^2$, $2p^6$, $3d^{10}$, and therefore lacks unfilled *d* subshells in the well-known oxidation state, the requisite criterion for true transition metals. Zinc(II) has an ionic radius comparable with Mg(II) but a Lewis acidity more like that of the smaller Cu(II) ion.

Numerous zinc minerals-sulfides, sulfates, oxides, carbonates, phosphates, and silicates - have been described, most containing tetrahedral or octahedral coordination polyhedra with either apical S or O. Several minerals contain both tetrahedrally and octahedrally coordinated Zn(II), while others have fivefold coordination polyhedra alternating with octahedrally coordinated Zn(II), reflecting the full and spherical 3*d* subshell which does not favour one coordination over another. The stabilities of aqueous zinc complexes have been measured for hydroxides, chloride, carbonate, sulfate, sulfide, phosphate, simple organic acids, amino acids, and synthetic chelates. Coordination of Zn(II) in aqueous complexes is usually octahedral, although fourfold and fivefold coordination are also known.

Zinc is an essential element for terrestrial life since it is required as either a structural component or reaction site in numerous proteins, the zinc-binding portions of which are highly conserved among species. Zinc sites in proteins consist of Zn polyhedra with apical S, N, or O, associated with cysteine, histidine, glutamic acid, aspartic acid, and water. Coordination numbers for zinc range from four in the case of structural Zn associated with four thiol groups derived from cysteine to six in the case of a number of reactive sites containing O and N as apices.

The total concentration of zinc in soils depends on the composition of the parent material and soil mineralogy, especially the concentration of quartz, which tends to dilute most elements. Only a small fraction of the total zinc is exchangeable or soluble. About one-half of the dissolved zinc exists as the free hydrated cation. The concentration of dissolved complexes of zinc with inorganic ligands can be estimated by computer-implemented models and total concentrations as input. Similar approaches with organic ligands await further research. Most analytical determinations of zinc are made by spectrometric techniques such as atomic absorption spectrophotometry, inductively coupled plasma atomic emission spectroscopy, and inductively coupled plasma mass spectroscopy.

2. Atomic and physical properties

One might be tempted to begin a chapter on zinc chemistry by stating that zinc is a malleable, ductile, bluish-white metal element with atomic number 30 and average atomic mass 65.38. This is not entirely correct. Although the occurrence of zinc as a native element has been reported from several places, the natural occurrence of the metal is doubtful. Very likely the first zinc metal on earth was produced artificially by ancient metallurgists in the Mediterranean basin by 20 A.D., or at the very latest in the thirteenth century by Indian metallurgists. This early interest in zinc was directed toward the manufacture of brass, a copper-zinc alloy, from pure copper, a less desirable metal. Prior to the discoveries of metallurgists, all zinc on earth likely existed solely in various natural zinc compounds. There is no evidence that zinc was recognized in Europe as a metal prior to the beginning of the sixteenth century, at which time Portuguese ships rounding the Cape of Good Hope carried zinc ingots from China. The sixteenth century Swiss physician and alchemist Paracelsus is often credited as the first European to recognize zinc as a distinct metal. Elsewhere, however, Paracelsus wrote that "the metals, then, are seven in number, exclusive of Mercury, namely gold, silver, tin, lead, iron, steel, and copper" (The Economy of Minerals), indicating that full recognition of zinc as an element necessitated the falling away of alchemy in favor of chemistry.

One might also be tempted to describe zinc as one of the transition elements, broadly defined as those elements of atomic number 21-31, 39-49, and 71-81, inclusive. Many modern chemists however prefer a more restricted classification limited to elements 22-28, 40-46, and 72-78, inclusive, all of which have one or more electrons in an unfilled *d* subshell in at least one well-known oxidation state. Elemental zinc has an electronic configuration of $1s^2$, $2s^2$, $2p^6$, $3d^{10}$, $4s^2$ and the only natural oxidation state of zinc, Zn(II), has an electronic configuration of $1s^2$, $2s^2$, $2p^6$, $3d^{10}$ in both the high- and low-spin configurations. Since there are no unfilled *d* subshells of zinc in either Zn(0) or Zn(II), Zn may be considered, together with the other Group IIb elements Cd and Hg and Group Ib elements Cu, Ag, and Au, as a post-transition element.

Zinc metal ranks above hydrogen in the electrochemical series so that it reacts in both acidic and alkaline solutions to form Zn(II) and H_2 gas. Clearly such a metal could not persist even in the most reducing aqueous environment on earth. The metal is not stable under terrestrial oxidizing conditions. Freshly polished zinc metal tarnishes in moist air to give the metal its more typical dull grayish color and can be ignited to give a blue-green flame in air and Zn(II)O smoke.

The terrestrial chemistry of zinc in nature is therefore that of Zn(II). The Zn(II) ion, as mentioned above, lacks unpaired electrons, unfilled electron shells, and variable oxidation states. Without these features, Zn(II) is colorless, unlike the true transition metals which often form brightly colored compounds and complexes. Those compounds and solutions of Zn(II) that are not colorless, or white in powder form, take their color from the other constituents present.

The ionic radius of Zn(II) ranges from 0.68 in fourfold coordination to 0.83Å in sixfold coordination, which places Zn(II) approximately in the size range of Mg(II), Cu(II), and Fe(II) (Whittaker and Muntus, 1970). The divalent ion Ca(II), has a lower atomic number than zinc (Z=20 vs 30) while occupying the same row of the periodic table, yet has an ionic radius of 1.08 Å(CN=6), significantly larger than that of Zn(II). Ca(II) has an empty 3*d* subshell, so that the nuclear charge is shielded effectively by the 1,

2, and 3*s* electrons. In fact, all divalent Row 4 elements containing 3*d* electrons have ionic radii smaller than that of Ca(II). This difference is usually attributed to *d* orbital penetration of core electrons, leading to a general contraction of the core and *d* orbitals, and a resulting decrease in the size of the ion (Purcell and Kotz, 1977). Another explanation given for differences in radii across Row 4 is based on the fact that electrons in *d* orbitals are particularly inefficient in shielding a positive nuclear charge. Since Zn(II) (Z 30, d^{10}) has more unshielded nuclear charge than Ca(II) (Z 20, d^0), the effective ionic charge z_{eff} is higher in Zn(II) than Ca(II), causing decreased ionic radius as well. The difference between formal ionic charge z and z_{eff} leads to the observation that ionic potential, most simply calculated as z/r, when calculated as z_{eff}^{r} is virtually twice as large for Zn(II) as Mg(II), even though both have the same formal charge and similar ionic radii. The charge-to-radius ratio of a cation is related to the Lewis acidity of the cation, i.e., the tendency of the cation to polarize the electron cloud of a donor ligand, leading to covalent bonding. Among the more common divalent bioelements, Zn(II) and Cu(II) seem to be the most effective Lewis acids from the standpoint of z_{eff}^{r} (Clementi and Rainmondi, 1963; Ochiai, 1987) .

3. Zinc minerals

Zinc(II) is found as numerous, named minerals of widely different types, including sulfides (ZnS, sphalerite and wurtzite), sulfates ($ZnSO_4$, zincosite; $ZnSO_4 \cdot 2H_2O$, goslarite), oxides (ZnO, zincite; $ZnFe_2O_4$ franklinite; $ZnAl_2O_4$, gahnite), carbonates (smithsonite, $ZnCO_3$), phosphates ($Zn_3(PO_4)_2 \cdot 4H_2O$, hopeite) and silicates (Zn_2SiO_4, willemite; $Zn_4Si_2O_7(OH)_2 \cdot H_2O$, hemimorphite). The principal zinc ores of commerce are sphalerite, smithsonite, and hemimorphite. Sphalerite and wurtzite are relatively common chalcophilic minerals, and are often found in conjunction with sulfidic minerals of Cu, Fe(II), Mn(II), and Pb. The zinc minerals of the other groups are largely alteration products of the zinc sulfides in response to hydrolysis and sulfide oxidation. Some zinc minerals, such as franklinite, a double Zn-Fe oxide of the spinel group, are known from a single type location and only scattered other reported occurrences..

All of the above-mentioned Zn minerals are compounds of Zn with either S(-II), as in sulfides, or O(-II), as in the sulfates, oxides, carbonates, phosphates, and silicates. The coordination number of Zn minerals is usually either 4 (e.g., sphalerite, wurtzite, zincite, franklinite, gahnite, willemite, hemimorphite) or 6 (e.g., smithsonite, goslarite). The ease with which Zn(II) will accept either tetrahedral coordination (CN=4) or octahedral coordination (CN=6) rests ultimately in the electron structure of Zn(II). Since a completely filled set of *d* orbitals has spherical geometry, there can be no ligand field stabilization when the metal ion has a filled shell configuration (nor a half-filled *d* orbital [Mn(II), $3d^5$] nor empty *d* orbital [Ca(II), $3d^0$]). Crystal field stabilization energy, roughly speaking the measure of stabilization observed for a given geometrical structure over the hypothetical configuration in which ligands are associated with the metal but with no specified geometry, i.e., sphere-to-sphere interactions, is therefore identical for Zn(II) [and Mn(II) and Ca(II)] in tetrahedral and octahedral coordination (Fackler, 1971). Some zinc minerals contain zinc in two different coordination numbers, such as hopeite and hydrozincite [$Zn_5(OH)_6(CO_3)_2$], both of which contain both a tetrahedrally- and an octahedrally-coordinated zinc atom. Fivefold coordination of a central ion does not favor a regular three dimensional structure, yet the zinc minerals adamite (Zn_2OHAsO_4) and tarbuttite (Zn_2OHPO_4) contain one of the two Zn(II) ions in fivefold (bipyramidal) coordination, with the other zinc atom in octahedral coordination (Brehler and Wedepohl, 1978). The fivefold coordination, like the fourfold and sixfold coordinations. is permissible because the crystal field stabilization energy does not favor one coordination over the other because of the full and spherical 3*d* subshell.

Aside from the named zinc minerals, zinc can substitute into a great number of minerals as a minor constituent replacing major constituents without major distortion or charge imbalance. Indeed, zinc is considered to be the 24th most abundant element in the earth's crust (~70 $\mu g\ g^{-1}$ of the lithosphere) largely on the basis of its presence as a minor constituent in common rocks and minerals, as well as soils, rather than the presence of massive, pure deposits of zinc minerals at the earth's surface.

4. Aqueous zinc complexes

Common aqueous complexes of Zn(II) have octahedral coordination, as in the hydrolysis series: $Zn(H_2O)_6^{2+}$, $Zn(OH)(H_2O)_5^{+}$, $Zn(OH)_2(H_2O)_4^{0}$, $Zn(OH)_3(H_2O)_3^{-}$, and $Zn(OH)_4(H_2O)_2^{2-}$. Similarly, Zn chloride complexes may range from $ZnCl(H_2O)_5^{+}$ to $ZnCl_4(H_2O)_2^{2-}$. Other coordination numbers in zinc aqueous complexes are known, such as $Zn(CN)_4^{2-}$ (CN=4) and $[Zn(SCN)tren]^{+}$ (CN=5). As in the mineral structures, because of the completed *d* subshells, the stereochemistry of the complexes is determined by considerations of size, electrostatic forces, and covalent binding forces (Cotton and Wilkinson, 1988).

In addition to complexes with hydroxide and chloride mentioned above, formation constants have been measured for complexes of Zn(II) with numerous other ligands, inorganic and organic, among them: carbonate, sulfate, sulfide, phosphate, simple organic acids, amino acids, and synthetic chelates. Values for such formation constants have been critically compiled by Martell and Smith (1974-1989) and current best values have been incorporated into the databases of several computer-implemented speciation models, including MINTEQA2 (Allison et al., 1991) and GEOCHEM-PC (Parker et al., 1994).

5. Forms and distribution of zinc in solid and solution phases of soils

The concentration of Zn in the upper continental crust is approximately 57 $\mu g\ g^{-1}$ (Brehler and Wedepohl, 1978). The concentration of zinc in soils from 18 countries from all continents ranges from 10 to 300 $\mu g\ g^{-1}$ and can be averaged at about 70 $\mu g\ g^{-1}$ (Swaine, 1955). The concentration of zinc in 1300 surface soils of the United States ranged from 5 to 2900 $\mu g\ g^{-1}$ with a geometric mean of 48 $\mu g\ g^{-1}$ and an arithmetic mean of 60 $\mu g\ g^{-1}$ (Shacklette and Boerngen, 1984). The major factors affecting the concentration of zinc in soils is the concentration of zinc in the soil parent material and the soil concentration of quartz. Quartz in the soil dilutes soil zinc because the reported concentrations of zinc in quartz are very low, ranging from 1.0 $\mu g\ g^{-1}$ (Helmke et al., 1977) to <5 to 8 $\mu g\ g^{-1}$ (Brehler and Wedepohl, 1978). Sillanpää (1982) reports global values for the nutrient status by bioassay and selected extractants of zinc in soils.

In igneous systems, the average concentration of zinc in mafic rocks (100 $\mu g\ g^{-1}$) is greater than that in felsic rocks (40 $\mu g\ g^{-1}$) (Taylor, 1964). This occurs because Zn^{2+} can substitute for Mg^{2+} and Fe^{2+} in pyroxene. The greatest concentration of zinc in common sedimentary rocks occurs in shales (95 $\mu g\ g^{-1}$) while limestone (20 $\mu g\ g^{-1}$) and sandstone

(16 μg g^{-1}) have much lower concentrations (Turekian and Wedepohl, 1961). Brehler and Wedepohl (1978) report the concentration of zinc in numerous minerals and rocks.

The results of decades of research using physical and chemical methods to fractionate soils is that zinc in the solid phase of soils is associated with all of the common soil solid phases; including clay minerals, hydrous oxides, and organic matter. Some of the zinc in soils may substitute for major cations in soil minerals but confirmation of this phenomena awaits direct physical observation. Extraction and isotopic exchange experiments confirm that some of the soil zinc is exchangeable.

No indigenous zinc minerals are found in typical soils other than those occasionally inherited from the soil parent material. All naturally occurring zinc minerals support equilibrium concentrations of dissolved zinc that are higher than the submicromolar concentrations of dissolved zinc found in typical soil water. Franklinite does support an appropriate concentration of dissolved zinc (Lindsay, 1979; Ma and Lindsay, 1990) but there is no evidence that it exists in soils. Sauconite, a trioctahedral smectite containing zinc, appears to be inherited from the parent material and is rarely found in soils (Borchardt, 1977).

The results of isotopic exchange experiments show that only a fraction of the total zinc in soils is exchangeable in a period of several days. Tiller et al. (1972) found that less than 14 percent and more commonly 4 to 6 percent of the total zinc (4 to 183 μg g^{-1}) in acid soils from Australia was isotopically exchangeable. Vale (1982) showed that 0.3 to 5 percent of the total zinc (25 to 180 μg g^{-1}) in acid soils from Wisconsin, USA, and Portugal was isotopically exchangeable in 0.01*M* $Sr(NO_3)_2$. This study also showed that 0.1*M* HCl (1:10 soil solution ratio) and DTPA (diethylenetriaminepentaacetic acid) buffered at pH 7.3 (Lindsay and Norvell, 1978) extracted most of the isotopically exchangeable Zn. It is this fraction of the total soil zinc that is likely to be available to plants but the exact reaction mechanisms by which this zinc equilibrates between the soil solid and solution phases awaits further research.

The concentration of total dissolved zinc in water saturation extracts of soils is usually in the micromolar range or less. A survey of saturation extracts from 68 soil samples from 30 soil series in California found zinc concentrations ranging from 0.01 to 0.40 mg L^{-1}, with a mean of 0.07 mg L^{-1}, and a median of 0.04 mg L^{-1}. Current research has shown that about one-half of the dissolved zinc occurs as the free cation, and that the free cation concentration shows linear ion exchange isotherms with the free cation activity of calcium and magnesium (Kadir and Helmke, 1993). The dependence of the free zinc concentration on the concentration of calcium and magnesium cations shows that calcium and magnesium can displace zinc from solution complexes and from adsorption sites on the soil solids.

There is increasing evidence that the free hydrated cationic form of zinc and other trace metals is the only dissolved form that is actively assimilated by plant roots, although complexed forms of dissolved zinc may be active in transporting zinc to the plant roots (Checkai et al., 1987; Chaney, 1988; Ting, 1989; Bell et al., 1991; Thys et al., 1991). This, and concern about the environmental impacts and fates of trace elements contaminating soils, have prompted the development of computer-implemented speciation models (MINTEQA2, SPECIES, GEOCHEM, and GEOCHEM-PC) that calculate the concentrations of dissolved species of elements using total dissolved concentrations of elements and ligands, formation constants, and mineral equilibria as input (Allison et al., 1991; Barak, 1990; Mattigod and Sposito, 1979; Parker et al., 1994).

These chemical speciation models show that $ZnSO_4^{o}$ and $Zn(OH)^+$ may comprise up to 10 percent of dissolved zinc in most soils under normal conditions. Speciation with other inorganic ligands, such as chloride, nitrate, and phosphate, is usually of lesser importance because of either low concentrations of the ligand or weak complexing strength, or both. The situation for complex formation with organic ligands is much more complicated. Zinc forms complexes with many of the common organic acids potentially found in soils (citric, malic, oxalic, etc.) but these ligands are seldom analyzed in soil solution. Zinc also forms complexes with humic substances such as fulvic acid, humic acid, and dissolved organic (humic) carbon. Formation constants have been determined for such substances but the reported values vary over as many as eight orders of magnitude depending upon pH, type of humic substance, and other variables (Stevenson, 1982). Useful formation constants measured for humic substances will require measurements at ionic strengths, pH conditions, micromolar zinc concentrations, and loading of competing metal ions typical of the soil environment. Computer-implemented speciation models allow the user to revise equilibrium constants and to add new species. Their utility will increase as the sophistication of the techniques to analyze soil solution increases and the quality of the soil-derived database improves.

6. Biochemical aspects of zinc chemistry

The essential nature of zinc to life processes among all terrestrial organisms is certainly related to its bioavailability at the time of the origin of life. Modern seawater has an average pH of 8.2, with a relatively oxygenated surface zone and a zinc content of 0.005-0.014 mg L^{-1} . The overall chemistry of ancient seas was likely much different, with more acidic seawater due to more HCl and H_2SO_4 from early volcanic activity, and more reducing conditions due to an anoxic atmosphere. Since the solution chemistry of Zn is not dominated by hydroxide formation nor by oxidation-reduction reactions, the difference in pH and redox potentials between contemporary and primordial seawater would not be expected to have greatly affected the solubility of zinc in seawater, and Zn would have been as bioavailable in seawater at the time of prokaryote evolution c. 3.5 billion years BP as it is currently (Ochai, 1987).

Zinc has been found to be an essential component of hundreds of enzymes isolated from different species; zinc enzymes have been found in each of the six classes of enzymes-oxidoreductases, transferases, hydrolases, lyases, isomerases, and ligases. The role of Zn(II) in metalloenzymes is inherently different than that of metals capable of redox reactions such as Cu and Fe since Zn(II) does not undergo reduction under any conditions compatible with life on earth. Coordinating groups for Zn sites in enzymes, where identified, have been found to be the residues of the amino acids cysteine, histidine, glutamic acid, and aspartic acid, as well as water molecules. Apparently the other S-bearing amino acid methionine, the 5 other secondary N-bearing amino acids, and the 3 hydroxy- and phenoxy- amino acids were not used during the evolution of biochemical processes to construct Zn sites in enzymes. In enzymes whose crystal structure and amino acid sequence are on record, the amino acid ligands forming the Zn site are highly conserved despite other sequence differences among organisms as diverse as yeast, higher plants, and primates.

The chemistry of Zn in enzymes strikingly follows the chemistry of Zn in pure minerals, with the biochemical materials constructing a single polyhedron of ZnS or ZnO (or ZnN) out of the materials at hand, amino acid residues and solvent (water). For example, the sulfur atom of the amino acid cysteine [$HSCH_2CH(NH_2)COOH$] is a basic component in the construction of a sphalerite-like tetrahedron, with Zn as a central ion surrounded by four S atoms. The thiol side chain of cysteine has a dissociation constant (pK_a) of ~8.5, which is intermediate between the stepwise dissociation constants of H_2S, namely 7.0 and 12.9. Average S-Zn bond lengths for two proteins containing cysteine-Zn interactions for which crystallographic data exist ranged from 2.0 to 2.4, averaging 2.1 Å (Chakrabarti, 1989). Of particular interest is the observation that cysteine thiolates dominate those enzymes containing structural zinc, as opposed to those containing catalytic zinc (Vallee and Auld, 1990). In alcohol dehydrogenase, the Zn centre is close to the surface of the molecule but inaccessible to the solvent, and is part of a lobe projecting out of the catalytic domain; in aspartate transcarbamoylase, Zn holds together two loops of the polypeptide chain. Sulfur(-II) is a bulky anion and the Zn/S cation/anion radius ratio, 0.68/1.56 = 0.44, places ZnS well within the range of the ideal radius ratio for tetrahedral coordination (0.225) and very nearly at the lower limit for octahedral coordination (0.414). The ZnS tetrahedron therefore permits all S anions to contact the central ion but S anions have, at the same time, lost contact with each other to the point that two more S (or smaller) anions could be squeezed into the coordination polyhedron without the anions losing contact with the cation. Typically, structural Zn in enzymes is isolated from the bulk solvent, water, by the structure of the enzyme itself, presumably to preserve the structural site undisturbed by perturbations of the coordination polyhedron (Vallee and Auld, 1990). Interestingly, the peptide structure of alcohol dehydrogenase uses a peptide sequence with as few as 14 amino acids to construct the 4 cysteine coordination site for Zn.

Although a 4 cysteine site corresponds with the sphalerite coordination polyhedron, other variants constructed from cysteine have been identified. Structure refinements of the second Zn site in alcohol dehydrogenase reveal that the site is constructed from 2 cysteine residues, histidine, and H_2O. This second Zn site in alcohol dehydrogenase is a functional site, and this is the only appearance of cysteine in a functional site known to date (Vallee and Auld, 1990). Sequence analysis of the ß' subunit of RNA polymerase has shown 4 cysteine residues in a 31-residue peptide segment, making this segment a strong candidate for a Zn site. Data from an XAFS study of *E. coli* RNA polymerase confirms 4 cysteines coordinated with Zn in the ß' subunit, but also point to two additional coordinating atoms, either O or N, which were indistinguishable by the method used; the two additional atoms could be from carboxylic residues or water, but in any case would indicate that the Zn is in sixfold coordination instead of fourfold (Wu et al., 1992). The same XAFS study indicated that the ß' subunit of RNA polymerase may contain five cysteine ligands with an average distance of 2.38+/- 0.01 Å and an additional coordinated O or N, which the authors thought likely from histidine or tyrosine. As mentioned above, considerations of cation/anion radius ratios permit sixfold coordination of anions the size of S(-II), or smaller, around a central Zn(II) ion without losing contact with the Zn.

The common presence of histidine as a Zn ligand, particularly in functional or catalytic sites, would seem at first glance to have no parallel with any known Zn mineral structure. However, N(-III) is isoelectronic with O(-II), the coordinating anion of all common Zn minerals-oxides, sulfates, carbonates, phosphates, and silicates-except the sulfides. Furthermore, if the crystal radius of NH_4^+ (~1.48 Å) can be used to judge the radius of the heterocyclic N, then N(-III) has a radius closer to that of O(-II) (1.27 Å) than

S(-II). Concentrations of NH_3 are relatively low in terrestrial environments and solubility of NH_3 compounds is high, with the notable exception of NH_4MgPO_4, so there are no opportunities for Zn-NH_3 minerals to exist in nature without the concentrating effect of life processes. A number of Zn enzymes, among them DD carboxypeptidase, ß-lactamase, and carbonic anhydrase I and II, have 3 histidine residues and 1 coordinated H_2O as Zn site ligands. Other zinc enzymes are known to have Zn sites composed of 2 histidines, 1 H_2O, and a dicarboxylic amino acid—either glutamic acid (carboxypeptidase A and B, thermolysin, *B. cereus* neutral protease, or phospholipase C) or aspartic acid (alkaline phosphatase) to complete fourfold coordination (Vallee and Auld, 1990).

A related variant on the zinc ligand scheme is that of Cu, Zn superoxide dismutase (Cu, Zn SOD), an enzyme ubiquitous among aerobic organisms that catalyzes the conversion of the superoxide radical to molecular oxygen and hydrogen peroxide. In SOD, the zinc ligands are three histidine and one aspartic acid residue. Particularly unusual, however, is the fact that one of the histidines is bound through the second heterocyclic N to Cu, which is bound by four histidines in a tetrahedrally-distorted square plane as the active site at the bottom of a funnel-shaped channel (Tainer et al., 1991). The enzyme undergoes reduction of Cu(II) and reoxidation of Cu(I) upon completion of the reaction. In the ground state of Cu, Zn SOD, the same geochemical proximity of Cu and Zn that early smelters discovered among oxides and sulfides as they manufactured brass was created at the dawn of aerobic eukaryotic organisms. (Interestingly, there is also a Fe, Mn SOD, details of which remain to be worked out.) A survey of Zn interactions with sp^2 N-containing heterocycles has shown that Zn prefers a head-on and in-plane approach to the sp^2 lone electron pair. Similarly, the Zn interacts head-on with the O in H_2O and the H-O-Zn angle typically ranges between 0 to 60°, corresponding to sp^2 and sp^3 electron configurations, respectively (Vedani and Huhta, 1990).

Using the same radius ratio criteria as above, the Zn/O radius ratio is 0.68/1.27 = 0.535, which is significantly higher than the critical value for tetrahedrons with all anions contacting both each other and the central cation. As before, if the radius ratio exceeds the critical value for a given coordination type, the anions lose contact with each other and may distort from perfect symmetry and even permit the coordination of addition anions. Evidence has been adduced, particularly for carbonic anhydrases, that groups such as O=C=O may squeeze into the inner sphere of the Zn site to form five-coordinate Zn from the four-coordinate Zn in the ground state. Such geometries are occasionally invoked as an explanation of an "entatic" state, in which an elevated energy of either the ground state due to distorted tetrahedral coordination or to the enzyme-substrate complex in fivefold coordination reduces the activation energy for the completion of the catalyzed reaction (Vallee and Williams, 1968; Lindskog, 1983). Since Zn(II) has a spherical and full 3*d* subshell, crystal field stability energy considerations do not favor one coordination number over another. Fivefold coordination has been observed in both Zn minerals and Zn aqueous complexes, and the ability to distort configuration to accommodate substrates would seem to be an attractive feature in the biological function of zinc.

7. Analytical chemistry of zinc

Most measurements of zinc concentrations in plants, soils, and water are now done by sophisticated spectroscopic methods of analysis because of their superior selectivity, sensitivity, and speed of analysis. The use of colorimetric and wet chemical techniques for the determination of zinc is generally restricted to laboratories that do not have access to modern instrumentation.

For most methods of analysis, samples of soils and plants must be dissolved and converted to the liquid state. Exceptions are X-ray fluorescence analysis, where soils are fused with lithium borate and plant material can be pressed into discs, and neutron activation analysis which requires minimal sample preparation. Ashing of plant material and dissolution of the ash with nitric or other acids or wet ashing with oxidizing acids generally results in quantitative recovery of zinc (Gorsuch, 1970). Soils require a predigestion with oxidizing acids to destroy organic matter followed by treatment with hydrofluoric acid to dissolve silicates, or fusion with an alkaline flux followed by dissolution of the fusion cake in dilute acid (Lim and Jackson, 1982; Hossner, 1994)

Atomic absorption spectrophotometry (AAS) is currently the most common instrumental technique for the analysis of zinc (Baker and Amacher, 1982). The limit of detection with an air-acetylene flame is about 0.005 $\mu g\ mL^{-1}$ under ideal conditions and the optimum working linear range is 0.05 to 2 $\mu g\ mL^{-1}$ with the 213.9 nm line. The limit of detection with the graphite furnace is slightly lower and the graphite furnace has the advantage of requiring sample volumes of only a few tens of microliters. Background correction is often needed, especially with the graphite furnace. Phosphate and some other elements suppress the instrument response to zinc. The methods of standard additions can be used for solutions containing interfering elements. The precision of analysis commonly ranges from 2 to 5 percent relative standard deviation but a skilled analyst under ideal conditions can achieve 0.5 to 2 percent.

Inductively coupled plasma optical emission spectroscopy (ICP-OES) is commonly used for the analysis of zinc and other elements in larger laboratories that analyze large numbers of samples. Modern instruments can analyze for many elements simultaneously with a level of detection of a few ng mL^{-1}. The detection limit for zinc is 2 ng mL^{-1} for the 213.856 nm line. A practical working sensitivity is 10 ng mL^{-1}. Arsenic, copper, iron, nickel, and aluminum potentially interfere at 213.856 nm but usually the corrections are small, and for aluminum very small, and easily made with the multielement capability of modern instruments. Aspiration rates are about one mL min^{-1} and several mL is the minimum volume needed for analysis. Sample throughput is about 20 samples per hour. The precision of analysis for ICP-OES is about the same as for AAS.

X-ray fluorescence (XRF) analysis with commercial instruments is not considered a trace element technique, although it is an excellent trace element technique when used with the intense X-rays from synchrotron light sources. It is seldom used for the analysis of zinc in soils because its limit of detection is only slightly lower than the normal concentrations of zinc in soils. The precision of analysis at soil zinc levels is much inferior to other techniques. Soil samples must be finely ground and fused into a lithium borate disc with lanthanum oxide as a heavy element absorber to standardize matrix absorption of X-rays (Karathanasis and Hajek, 1994). The low atomic number of the dominant elements in plant materials is a highly favorable matrix for XRF and yields a detection limit for zinc of a few $\mu g\ g^{-1}$, which yields acceptable precision for many purposes. Plant samples are commonly ground in a Wiley mill and pressed into a disc with pure boric acid as a backing material to strengthen the disc. Analysis for a single element with modern instruments is fast, about 30 samples per hour.

Neutron activation analysis (NAA) can determine about 25 elements in soils and plant material (Helmke, 1994; Koons and Helmke, 1978). The detection limit depends on the sample mass and the duration and intensity of the neutron irradiation. The detection limit for zinc can be as low as one ng g^{-1} but about one µg g^{-1} is a good working limit. The precision is about one percent relative standard deviation unless limited by counting statistics. The linear working range is unlimited. The technique requires access to a research nuclear reactor and this limits its availability. The cost for modern equipment for radioassay is about the same as that for a modest atomic absorption spectrophotometer. Irradiation times vary from 0.5 to several hours, depending upon sample size and zinc concentrations. Instrumental, also known as nondestructive, NAA is commonly used. The isotope ^{65}Zn, which has a half-life of 243.8 days and an intense gamma line at 1,115 keV, is used for radioassay. The 1,120 keV line of ^{46}Sc is very intense in soil samples and can make placement of the baseline for the zinc peak difficult in samples with a high Sc/Zn ratio. Scandium is very low in clean plant samples and zinc is very easily and precisely determine in such samples, even at sub µg g^{1} levels. Twenty or more samples can be irradiated simultaneously and each radioassay requires 2 to 6 hours after a cooling period of several weeks. The 13.76 hour ^{69}Zn isotope with a 574 keV emission line can also be used but sensitive radioassay of this isotope requires radiochemical separations after irradiation to reduce the Compton background from other short-lived radionuclides. The major advantages of NAA is that sample preparation is minimal; usually all that is involved is grinding the sample to obtain a uniform and representative sample. The technique has relative freedom from contamination and sample blanks because no sample dissolutions are required.

Inductively coupled plasma mass spectroscopy (ICP-MS) is a relatively new technique suitable for about 70 elements that offers incredibly low detection limits (about 0.07 ng mL^{-1} for zinc) and still retains a sample throughput of about 20 samples per hour (Jarvis et al., 1992). It also has the advantage of determining stable isotope ratios. The technique has an almost unlimited linear working range and a precision of 1 to 3 percent relative standard deviation when used in a comparison mode. Zinc has four stable isotopes free of isobaric overlaps. Their mass numbers and abundances are: ^{66}Zn (27.81 percent); ^{67}Zn (4.11 percent); ^{68}Zn (18.57 percent); and ^{70}Zn (0.62 percent). Zinc-66 is commonly used for analysis in the comparison mode because it is the most abundant. The precision of analysis is better than one percent if stable isotope dilution analysis is used. The cost of the most highly enriched (70 to 99 percent) stable isotopes of zinc range from about $2.40 U.S. per mg for ^{66}Zn to $70.00 per mg for ^{70}Zn in 100 mg lots. Separated, enriched isotopes of zinc can be used as tracers in soil, plant, and nutrition studies without the hazards of radioisotopes. The availability of four stable isotopes allows double labeling. There have been very few reports of the use of ICP-MS for the analysis of soils and plants but this situation is likely to change quickly as instruments become more generally available.

Spectrophotometric methods for the determination of zinc in soils and plants have the disadvantages of insensitivity and nonselectivity compared to other instrumental techniques. The dithizon (diphenylthiocarbazone) method has been used for decades (Viets and Boawn, 1965). In this method, zinc and other trace elements are extracted from an alkaline solution by dithizon dissolved in an immiscible organic solvent. Zinc is back extracted into an aqueous mineral acid to separate it from copper. Separation of zinc from lead and cadmium requires an additional alkaline extraction with dithizon and diethyldithiocarbamate, which forms stable complexes with lead and cadmium but does not interfere with the extraction of zinc by dithizon in an organic solvent. The zinc-dithizon complex is violet-red in color, with a detection limit of about 1 μg. The average deviation between multiple determinations is reported to be 1.4% (Cowling and Miller, 1941).

New and improved analytical techniques are needed to increase our understanding of the behavior of zinc and other trace elements in soil-water-plant systems. Extended X-ray adsorption fine structure (EXAFS) and other spectroscopic techniques promise to provide details of specific bonding mechanisms of trace elements to soil solid surfaces. In the solution phase, cation exchange membranes and the principles of Donnan equilibrium have been recently used to measure the activities of free hydrated cations of zinc and other elements at their indigenous levels in water saturation extracts of soils (Fitch and Helmke, 1989; Helmke et al., 1993). The data produced by new techniques can verify computer-implemented speciation models and they have the potential to increase our understanding of the reactions that control dissolved concentrations of zinc in soil.

References

Allison JD, Brown DS and Novo-Gradac KJ 1991 MINTEQA2/PRODEFA2, A geochemical assessment model for environmental systems. U.S. Environmental Protection Agency. Report No. EPA/600/3-91/021.

Barak P 1990 SPECIES: A spreadsheet program for modeling speciation of soil solution. J. Agron. Educ. 19, 44-46.

Baker DE and Amacher MC 1982 Nickel, copper, zinc, and cadmium. *In* Methods of Soil Analysis: Part 2. Chemical and Microbiological Properties. Eds. AL Page, RH Miller and DR Keeney. pp 323-336. American Society of Agronomy-Soil Science Society of America. Madison, WI. USA.

Bell PF, Chaney RL and Angle JS 1991 Free metal activity and total metal concentrations as indices of micronutrient availability to barley [*Hordeum volgare* (L.) 'Klages']. Plant Soil 130, 51-62.

Borchardt GA 1977 Montmorillonite and other smectite minerals. *In* Minerals in Soil Environments. Eds. JB Dixon and SB Weed. pp 293-330. Soil Science Society of America, Madison, WI USA.

Bradford GR, Blair FL and Hunsaker V 1971 Trace and major element contents of soil saturation extracts. Soil Sci. 112, 225-230.

Brehler B and KH Wedepohl 1978 Zinc *In* (ed), Handbook of Geochemistry Vol. II/3. Ed. KH Wedepohl. Springer-Verlag, Berlin. 125 pp.

Chakrabarti P 1989 Geometry of interaction of metal ions with sulfur-containing ligands in protein structures. Biochemistry 28, 6081-6085.

Chaney RL 1988 Metal speciation and interactions among elements affect trace element transfer in agricultural and environmental food-chains. *In* Metal Speciation: Theory, Analysis, and Application. Eds. JR Krammer and HE Allen. pp 219-260. Lewis Publ., Chelsea, MI.

Checkai RT, Corey RB and Helmke PA. 1987 Effects of ionic and complexed metal concentrations on plant uptake of cadmium and micronutrient metals from solution. Plant Soil 99, 335-345.

Clementi E and Rainmondi DL 1963 Atomic screening constants from SCF functions. J. Chem. Phys. 38, 2686-2692.

Cotton FA and Wilkinson G 1988 Advanced Inorganic Chemistry. 5th Ed. Wiley, New York. 1455 p.

Cowling H and Miller EJ 1941 Determination of small amounts of zinc in plant materials. Ind. Eng. Chem. 13, 145-149.

Fackler JP, Jr. 1971 Symmetry in Coordination Chemistry. Academic Press, New York. 139 p.

Fitch A and Helmke PA 1989 Donnan equilibrium/graphite furnace atomic absorption estimates of soil extract complexation capacities. Anal. Chem. 61, 1295-1439.

Gorsuch TT 1970 The destruction of organic matter. Pergamon Press, Oxford. 152 p.

Helmke PA 1994 Neutron activation analysis. *In* Methods of Soil Analysis: Chemical Methods. Eds. DL Sparks, PA Helmke, RH Loeppert and MA Tabatabai. American Society of Agronomy-Soil Science Society of America. Madison, WI. USA. (*In Press*).

Helmke PA, Koons RD, Schomberg PJ and Iskander IK 1977 Determination of trace element contamination of sediments by multielement analysis of the clay-size fraction. Environ. Sci. Tech. 11, 984-989.

Helmke PA, Lampert JK and Li Y 1993 Measurement of trace cation activities by Donnan membrane equilibrium and atomic absorption analysis. Anal. Chem. (*submitted*).

Hossner LR 1994 Dissolution for total elemental analysis. *In* Methods of Soil Analysis: Chemical Methods. Eds. DL Sparks, PA Helmke, RH Loeppert and MA Tabatabai. American Society of Agronomy-Soil Science Society of America. Madison, WI. USA. (*In Press*).

Jarvis KE, Gray AL and Houk RS 1992 Handbook of inductively coupled plasma mass spectroscopy. Blackie and Son Ltd., Glasgow. 380 p.

Kadir A. and Helmke PA 1993 Effect of major cations and pH on indigenous free cation activities of zinc and copper in water extracts of soils. Soil Sci. Soc. Am. J. (*submitted*).

Karathanasis AD and Hajek BF 1994 Element analysis by X-ray fluorescence spectroscopy. *In* Methods of Soil Analysis: Chemical Methods. Eds. DL Sparks, PA Helmke, RH Loeppert and MA Tabatabai. American Society of Agronomy-Soil Science Society of America. Madison, WI. USA. (*In Press*).

Koons RD and Helmke PA 1978 Neutron activation of standard soils. Soil Sci. Soc. Am. J. 42, 237-240.

Lim CH and Jackson ML 1982 Dissolution for total element analysis. *In* Methods of Soil Analysis: Part 2. Chemical and Microbiological Properties. Eds. AL Page, RH Miller and DR Keeney. pp 1-12. American Society of Agronomy-Soil Science Society of America. Madison, WI. USA.

Lindsay WL 1979 Chemical equilibria in soils. Wiley-Interscience, New York. 449 pp.

Lindsay WL and Norvell WA 1978 Development of a DTPA soil test for zinc, iron, manganese, and copper. Soil Sci. Soc. Am. J. 42, 421-428.

Lindskog, S 1983 Carbonic anhydrase. *In* Zinc Enymes. Ed. TG Spiro. pp 77-12. Wiley, New York.

Ma Q and Lindsay WL 1990 Divalent zinc activity in arid zone soils obtained by chelation. Soil Sci. Soc. Am. J. 54, 719-722.

Martell AE and Smith RM 1974-1989 Critical Stability Constants. vols. 1-6. Plenum Press, New York.

Mattigod SV and Sposito G 1979 Chemical modeling of trace metal equilibria in contaminated soil solutions using the computer program GEOCHEM. *In* Chemical Modeling in Aqueous Systems. Ed. EA Jenne. pp 837-856. ACS Symp. Ser. No. 93, Am. Chem. Soc., Washington DC.

Ochai, E-I 1987 General Principles of Biochemistry of the Elements. Plenum Press, New York. 102 p.

Parker DR, Norvell WA, and Chaney RL 1994 GEOCHEM-PC: A chemical speciation program for IBM and compatible personal computers. *In* Chemical equilibrium reaction models. Eds. RH Loeppert, S Goldberg and AP Schwab. American Society of Agromony. Madison, Wisconsin. USA. (In Press).

Purcell KF and Kotz JC 1977 Inorganic Chemistry. Saunders Co., New York. 1116 p.

Sillanpää M 1982 Micronutrients and the nutrient status of soils: a global study. FAO Soils Bulletin 48. Food and Agriculture Organization of the United Nations, Rome. pp 75-82.

Stevenson FJ 1982 Humus chemistry. Wiley, New York. 443 p.

Swaine DJ 1955 The trace element content of soils. Commonwealth Bureau Soil Sci. Tech. Comm. No. 48, England.

Tainer JA, Roberts VA, Fisher CL, Hallewell RA and Getzoff ED 1991 Mechanism and structure of superoxide dismutases. *In* A Study of Enzymes. Vol. 2. Mechanism of Enzyme Action. Ed. SA Kuby. pp 499-538. CRC Press, Boca Raton, Fla.

Taylor SR 1964 Abundance of chemical elements in the continental crust: a new tabulation. Geochim. Cosmochim. Acta 28, 1273-1286.

Thys C, Vanthomme P, Schrevens E, and DeProft M 1991 Interactions of Cd with Zn, Cu, Mn and Fe for lettuce (*Lactuca sativa* L.) in hydroponic culture. Plant, Cell and Environ. 14, 713-717.

Tiller KG, Honeysett JL, and de Vries MPC 1972 Soil zinc and its uptake by plants II. Soil chemistry in relation to prediction of availability. Aust. J. Soil Res. 10, 165-182.

Ting YP, Lawson F, and Prince IG 1989. Uptake of cadmium and zinc by the alga *Chlorella vulgaris*: Part 1. Individual ion species. Biotech. Bioeng. 34, 990-999.

Turekian KK and Wedepohl KH 1961 Distribution of the elements in some major units of the earth's crust. Geol. Soc. Amer. Bull. 72, 175-192.

Vale RDV 1982 Comparison of four trace element extractants by isotope dilution analysis. M.S. Thesis. University of Wisconsin-Madison, USA. 131 pp.

Vallee BL and Auld DS 1990 Zinc coordination, function, and structure of zinc enzymes and other proteins. Biochemistry 29, 5647-5659.

Vallee BL and Williams RJP 1968 Metalloenzymes: The entatic nature of their active site. Proc. Natl. Acad. Sci. USA 59, 498-505.

Vedani A and Huhta DW 1990 A new force field for modeling metalloproteins. J. Am. Chem. Soc. 112, 4759-4767.

Viets FG Jr and Boawn LC 1965 Zinc. *In* Methods of Soil Analysis: Part 2. Chemical and Microbiological Properties. Ed. CA Black. pp 1090-1101. American Society of Agronomy-Soil Science Society of America.

Madison, WI. USA.
Whittaker EJW and Muntus R 1970 Ionic radii for use in geochemistry. Geochim. Cosmochim. Acta 34, 945-956.
Wu F Y-H, Huang W-J, Sinclair RB and Powers L 1992 The structure of the zinc sites of *Escherichia coli* DNA-dependent RNA polymerase. J. Biol. Chem. 267, 25560-25567.

Chapter 2

Mechanisms of Reaction of Zinc with Soil and Soil Components

N. J. BARROW

1. Abstract

Zinc can react with clay minerals, with organic matter and with metal oxides. In all cases, both the ionic species present in solution and the charge on the solid reactant are important in determining the amount of zinc sorption. Reactions with metal oxides have been closely modelled because the charge on the oxide can be modelled. Reaction with oxides, and with soil, may continue for many weeks. This is consistent with diffusion into the particles. Reaction with soil can be closely modelled by taking the diffusive penetration into account and by assuming that the reacting surfaces are not uniform. Fractionation schemes for identifying the forms of zinc in soil have limitations.

2. Introduction

Rudyard Kipling is no longer very popular. He is regarded as far too jingoistic for modern taste. Yet he wrote some wise things – for example: "And what should they know of England who only England know". Jingoistic perhaps; but true of many things – including zinc. If we want to understand zinc, it helps to discuss it in the context of other other metals which behave similarly, such as cadmium, nickel and copper, of metals which behave rather differently, such as mercury, and even of anions such as phosphate and borate which appear to behave very differently. The emphasis in this review is therefore on "compare and contrast".

3. Zinc in solution

Zinc may be added to soil in fertilizer – either deliberately or inadvertently, as a component of sewage sludge, or as part of normal recycling as decaying plant or animal tissue. Whatever the form of addition, zinc must pass through the solution phase before it can react with soil or be taken up by plants. However, it is simplistic to think of "zinc" in solution; we should think about zinc ions. We therefore need to know the species of ions which are present and how they change with changes in conditions.

In solution, divalent zinc ions, like the ions of many other metals, tend to be surrounded by six water molecules arranged in an octahedron. We might write such an ion as $Zn(H_2O)_6^{2+}$. Such ions are multiprotic acids because the water molecules may lose protons. The first step gives $ZnOH(H_2O)_5^+$ the monovalent cation - more simply written $ZnOH^+$. Further steps give the uncharged species, and later the monovalent anion. In the case of zinc the acid is very weak; the pK for the first dissociation step is large. There is, however, some uncertainty about its precise value. Baes and Mesmer (1976) give the value for pK_1 as 8.96. A similar value is used by Sposito and Mattigod (1980) in the model "GEOCHEM". On the other hand, Lindsay (1979) gives the value as 7.69 and this lower value is often used by soil scientists. This poses a dilemma – how can we choose the correct value? The ideal recourse of separate measurement is not available because hydrolysis proceeds to only a small extent before the onset of precipitation thus making measurement difficult (Baes and Mesmer, 1976). I prefer the Baes and Mesmer value because it is in a work specifically devoted to evaluating such measurements and because the source of the data is given.

Calculation of the proportion of the zinc ions present as $ZnOH^+$ is important because these ions may play a role in the reaction of zinc with soil (Tiller et al., 1972). Neglecting for the moment further dissociation, and also the formation of complexes, the proportion present as $ZnOH^+$ ions is given by: $K_1/((H^+)+K_1)$. When (H^+) is much bigger than K_1 – say 10^{-4} compared to $10^{-8.96}$ – then the expression approximates to: $K_1/(H^+)$. In this range, a ten-fold decrease in (H^+) – that is, unit increase in pH – causes a ten-fold increase in the proportion present as $ZnOH^+$. The value assigned to K_1 is therefore important not only for determining the value for the proportion present as $ZnOH^+$ but also for determining the range over which the "ten-fold" relation with pH holds.

Several of the other metals also have pK_1 values above the normal range of soil pH: cobalt, nickel and cadmium are examples. For all of these metals, there is therefore an increase in the proportion present as MOH^+ with the range over which the "ten-fold" rule applies depending on the individual values for pK_1. Mercury is an exception. It is stronger acid with both pK_1 and pK_2 near 3 (Baes and Mesmer, 1976). Because of this second dissociation, the proportion present as $HgOH^+$ *decreases* with increasing pH (Barrow and Cox, 1992).

At say pH 4, only about one zinc ion in 10^5 is present as $ZnOH^+$. It is tempting to argue that such a small proportion could not be important but there is evidence from work on borate that ions present in small proportions can be important. For boric acid the pK_1 is also about 9. Thus, at pH 4 only about one in 10^5 of the molecules is present as the borate ion – the remainder being uncharged boric acid. There is nevertheless an effect of ionic strength of the background solution on adsorption (Barrow, 1989a). This can be readily explained if reaction is with the borate *ions,* despite their low proportions. Thus the properties of the ions also matter. Disturbance of the shell of water molecules around a metal ion by conversion of an H_2O to an OH^- ligand makes it easier to remove the ion from its hydration sheath (James and Healy, 1972) and therefore increases the probability of reaction.

In soil solutions, and in the solutions mixed with soil in laboratory experiments, there may be many ligands present which form complexes with zinc. The outcome depends on the affinity of the ligand for zinc, on the affinity of the zinc-ligand complex for the reacting surface, and on the affinity of the ligand for the surface. Even in the apparently simple case of zinc-chloride complexes, it can be difficult to extrapolate from one sorbing material to another. Thus, for goethite, zinc sorption was greater from chloride solutions than from perchlorate (Bolland, 1970), and greater from chloride than from nitrate (Forbes, 1973; Forbes et al., 1976; Padmanabham, 1983). This would suggest that chloride complexes – such as $ZnCl^+$ - had some affinity for the reacting surface and this assumption was successfully included in models to describe the reaction (Barrow et al., 1981). For soil, however, the opposite result was obtained: at the same pH, sorption was smaller from a chloride solution than from nitrate (Barrow and Ellis, 1986). This suggests that $ZnCl^+$ had little affinity for the reacting surface in the soil. Such apparently conflicting results are not infeasible. Experience with mercury is similar. Mercury forms very strong complexes with chloride and, in most cases, the presence of chloride decreases mercury sorption (Forbes et al., 1974; Kinniburgh and Jackson, 1978; Barrow and Cox, 1992). Even so, Barrow and Cox (1992) had to assume that the $HgCl^+$ ions also participated in the reaction in order to model the data. This is consistent with reports that, under some conditions, chloride increased sorption of mercury by clays (Farrah and Pickering, 1978).

Chairidchai and Ritchie (1990) found that six organic ligands all decreased sorption of zinc by a lateritic soil. In general, the effects were greatest for those ligands which complexed zinc most strongly. They also included a "humate" to simulate soil organic material. It also decreased zinc sorption but it was not known to what extent it complexed zinc. Similarly, Bar-Tal et al. (1988) showed that fulvic acid decreased sorption of zinc by montmorillonite.

4. The soil components which react with zinc

Many studies have shown that zinc can react with clay minerals, with organic matter, and with oxides of metals such as iron and manganese. The reactions with clay minerals have been summarized by Pickering (1980). He emphasizes the differences between three main kinds of clay mineral. Kaolinites have relatively small isomorphous substitution and a large proportion of the charge is due to disruption of the structure at the edges of particles. The illites and montmorillonites have a structure with one aluminate layer sandwiched between two silicate layers. Isomorphous substitution is more marked and the permanent charge is greater than kaolinite. The interlayer spaces may be involved in cation exchange. Exchange tends to be slower for these minerals than for kaolinite and slower in illites than in montmorillonites.

Reactions between zinc and organic matter have also been widely studied because of the frequent occurrence of zinc deficiency on organic soils. Shuman (1980) has summarized several earlier reviews. Although he cited reports which indicated that the functional groups which hold zinc are weak acids, he concluded that more research was needed to characterize the groups and to specify the mechanisms. More recently, McBride (1989) discussed the reaction of metal ions with organic matter in more-modern terms. He pointed out, for example, that at low metal concentrations the relatively few sulfur-containing ligands might be involved in sorption of some metals.

Reactions between metal ions (including zinc) and metal oxides have been intensively studied in recent years. Many of the studies on hydrous ferric oxides have been summarised by Dzombak and Morel (1990) whereas the work of Loganathan et al. (1977) is an example of a study using a hydrous manganese oxide. Before dealing with these studies in some detail, it is desirable to point out different attitudes which have developed towards the three classes of reactants – clay minerals, organic matter, and metal oxides. Studies with clay minerals tend to emphasize cation exchange, studies with organic matter tend to emphasize stability constants (Shuman, 1980), and studies with oxides tend to emphasize specific adsorption and ligand exchange. Yet it is not always easy to define the differences between these terms.

The metal oxides have been taken as examples of variable charge materials. Although there are differing ideas about the precise specification of the reactions involved in generating the charge, the essence of the idea is that metal ions at the edge of a crystal are unable to complete the bond pattern of the bulk of crystal and therefore tend to react with water molecules. Depending on the pH and on the structure of the oxide, these surface water molecules may gain a proton and become positively charged, or lose one and become negatively charged. Consequently, the surface charge is variable. It tends to be positive at low pH and negative at high pH. The edges of crystals of clay minerals also carry a variable charge for analogous reasons. Organic matter is also a variable charge material although the mechanism is different. In this case, we can think of the many functional groups having a wide range of values for the pK so that they dissociate at different pH values. Again the overall effect is that the charge is positive at low pH and negative at high pH. No matter what kind of bond is formed or what functional group is involved, reaction between a surface and zinc involves ions. Positive ions are repelled by positive surfaces and attracted by negative ones. The reaction must therefore be affected by pH. It is desirable to take these effects into account in describing the reaction and much work has been done in this respect for reaction with metal oxides.

5. Models of reaction with metal oxides

The different models which have been published have been reviewed and compared several times – most recently by Sposito (1984), Barrow (1987a) and Goldberg (1992). The models differ in the complexity of the arrangement of the ions near a charged surface and in the way the equations which specify the model are written. All models allocate the various ions which react with a surface to planes and all ions in a given plane are thought to experience the same electric potential. In the simplest model, all the reacting ions are allocated to the one plane. This plane may therefore contain protonated (and deprotonated) water molecules plus specifically adsorbed cations and anions. All other ions are exchangeable and are in the diffuse layer. The work of Dzombak and Morel (1990) is an example of large-scale application of this model. In it, exchangeable ions and specifically adsorbed ions are distinguished by their position. One of the problems of this model is that ions in the diffuse layer are treated mathematically as if they were point charges without dimensions. However, near the surface, the concentration of ions in the double layer is high. The distance of closest approach to the surface is important and the size of the ions and their energy of hydration should be considered. The differences between electrolytes containing Li, Na and K ions can be explained in terms of the hydration status of the cations. K ions are most likely, and Li ions least likely to shed their sheathing water molecules. A greater proportion of the K ions are therefore likely to be in a layer with no water molecules between the ion and the surface (Shainberg and Kemper, 1966). This Stern layer has been included in some models.

Models which include a Stern layer may allocate the ions into three categories: the s-plane which includes protonated (and deprotonated) water molecules; the ß-plane which includes electrolyte ions such as Na^+ and Cl^-; and the diffuse layer which also includes the electrolyte ions. In some cases, specifically adsorbed metal ions have been allocated to the ß-plane. For example, Davis and Leckie (1978) modelled the reaction of lead, cadmium, copper and silver using this assumption. In this case, there is no sharp distinction between specifically adsorbed and exchangeable ions. In other cases, specifically adsorbed ions have been allocated to both the s-plane and the ß-plane. Hayes and Leckie (1987) used this assumption to model the adsorption of lead. In this case, one could argue that the mean plane for adsorption was somewhere between the s and the ß planes. It would therefore seem logical to make the model more general by introducing a further plane – the a-plane – for the specifically adsorbed ions between the s and the ß planes. The precise location of this plane varies with the properties of the ions. For small ions such as fluoride, one would expect it to be close to the s-plane and for larger ions such as phosphate it might be further away. Similarly, for metal ions reacting with a metal oxide, one would expect the ions to fit closely with the structure so that the mean location of the charge was close to the s-plane. For this 4-plane model, there is a clear conceptual difference between specifically adsorbed and exchangeable ions. Nevertheless, there are operational difficulties in distinguishing these categories. In practical soil science, exchangeable cations are taken as those ions displaced by a moderately concentrated solution containing a foreign cation. However, increasing the solution concentration near a negatively charged surface increases the concentration of cations in the ß-plane and in the diffuse layer. The consequent increase in the electric potential in the a-plane will displace some of the specifically adsorbed cations and these would therefore be interpreted as exchangeable.

6. The problem of charge

In simple cation *exchange*, there is no change in surface charge or in the pH of the solution. One cation simply replaces an equivalent amount of another cation. Specific adsorption of cations differs in that some protons are displaced from the surface and therefore the pH is changed. To give an artificial numerical example, imagine that 10 zinc atoms react with a patch of surface carrying a net charge of -100 at a pH of 6. We are not concerned here with the mechanism of the reaction but only with the charge balance. As most of the zinc ions in solution are Zn^{2+}, the total charge of the 10 atoms is close to 20. The outcome might be that the surface charge is increased to -97 and there are 17 protons released. This example is chosen to illustrate two points. One is that, since the charges must balance, one can express the outcome either in terms of the change in surface charge or in terms of the protons released. The other is that simple ratios do not necessarily occur. According to the four-layer model, the choice between the two ways of balancing the charge depends on the location of the a-plane. If this is close to the s-plane, there is a greater probability of displacing protons (Barrow, 1987a). Results suggest that zinc may be located close to the s-plane and calcium somewhat farther away. Thus Kinniburgh (1983) found that sorption of zinc released 1.7 protons per molecule sorbed whereas specifically sorbed calcium released only 0.9. The freedom in locating the a-plane is important in the functioning of the model and the ability of a model to reproduce both the observed changes in charge and the amount of adsorption is an important test of the model.

7. The effects of time on metal adsorption by oxides

In most studies of adsorption of metal ions on oxides, it has been assumed that equilibrium has been reached. Dzombak and Morel (1990) point out that the periods involved have ranged from a few minutes up to several days and that under some conditions there can be a slow continuing reaction. Recently, Brümmer et al. (1988) showed that metal ions continued to react with a sample of goethite for up to 42 days. The rate of the reaction was in the sequence: Ni>Zn>Cd and was increased by increasing the temperature. They concluded that the continuing reaction was due to diffusion of the adsorbed ions into the adsorbing particles. This kind of reaction is of considerable importance to soil science because of its potential effects on effectiveness of zinc fertilizers.

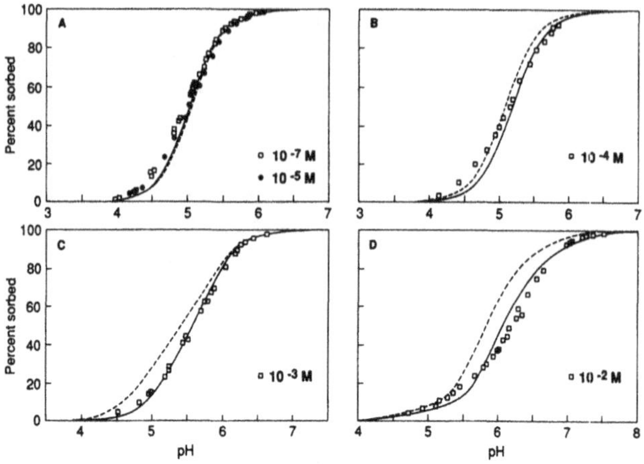

Figure 1. Observed (points) and modelled (lines) for sorption of zinc by an iron hydrous oxide gel. The observations were from Kinniburgh and Jackson (1982) and these were modelled by Dzombak and Morel (1990). The initial Zn concentrations are indicated on the figure. The total iron content was 0.093 *M* and the background electrolyte was 1 *M* $NaNO_3$. The broken lines indicate the fit of the model to all levels of addition of zinc, the solid line indicates the fit to each separate set of data.

8. Modelling of adsorption of metals by oxides

Dzombak and Morel (1990) collected a large amount of published data and fitted the simple model in which ions are in one adsorption plane and are balanced by a simple diffuse layer. They argued that there were advantages in fitting the model separately for each data set – this includes, for example, a separate fit for each level of addition of metal ions. Fig. 1 shows that the separate fits for each level of addition of zinc were good (solid lines). However, the fit to the data as a whole was poor (broken lines). The model was therefore not comprehensive. An important criterion in judging models is that the more comprehensive the data set described, the more confidence one can have in the model. The two subsequent examples given therefore illustrate fits to data for which more than one variable was altered or more than one property was measured.

In the first example (Fig. 2), the identity and concentration of the background electrolyte were varied and both sorption and the change in surface charge as a result of adsorption were measured. The figure shows that a good description of the data was obtained using a model in which metal ions were allocated to the a-plane - separate from the s- and the ß-planes. Because this model described the change in surface charge, it follows that it also described the protons released. In the second example (Fig. 3), there was a range in initial concentration of metal ions, of background electrolyte, of pH, of period of contact and of temperature. Three metal ions were investigated – nickel, zinc and cadmium – but only the results for zinc are presented in Fig. 3. The figure shows that a good fit was obtained using the one set of parameters to the whole of the data set. In fitting the model, it was assumed: that there was an initial adsorption of MOH^+ ions into a separate adsorption plane; that the surface was heterogeneous so that there were a few sites of high affinity, rather more with lower affinity, more still with yet lower affinity, and so on; and that the initial adsorption was followed by diffusive penetration of the surface. Details of the equations and of the parameter values are given in the original

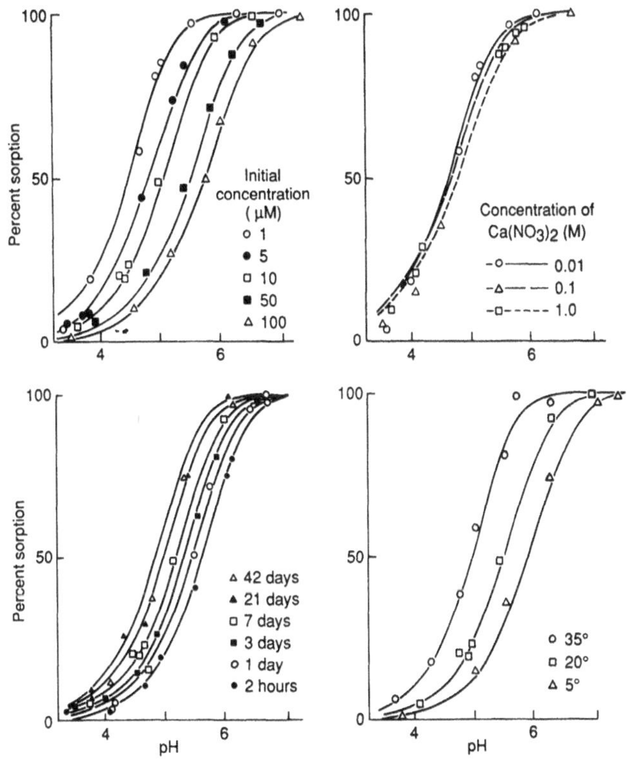

Figure 2. Observed (points) and modelled (lines) values (a) for sorption of copper by goethite, (b) for charge in the absence of copper, and (c) for charge in the presence of copper. The initial copper concentrations and the concentration of electrolyte are indicated on the figure. (From Barrow et al., 1981.)

publication (Barrow et al., 1989).

Fig. 3 shows that there were only small effects of ionic strength of the background electrolyte on sorption of zinc by goethite. According to the model, this arises from the interplay of three factors. First, the value of the practical dissociation constant decreases with increasing concentration. Second the activity coefficient of the reacting ion decreases. Both of these factors decrease adsorption. And third, increasing the ionic strength decreases the absolute value of the electric potential in the plane of adsorption. If adsorption occurs at a sufficiently low pH, as it does for zinc on goethite, the potential is positive. Decreasing its value favours adsorption. The outcome is a small net effect. Cadmium and nickel were adsorbed at a higher pH, the surface was more negative, and there was a decrease in adsorption with increasing ionic strength (Barrow et al., 1989). In the separate study illustrated in Fig. 2, copper was adsorbed at a lower pH, the surface was positive (Fig. 2c), and increasing the ionic strength increased adsorption (Fig. 2a). This again illustrates the value of a comprehensive data set and of "comparing and contrasting".

The model in which the metal ions are adsorbed into an adsorption plane separate from the s- and the ß-planes thus seems to be good at describing the reaction of metal ions with oxide surfaces. It also has advantages when we want to extend the model to other

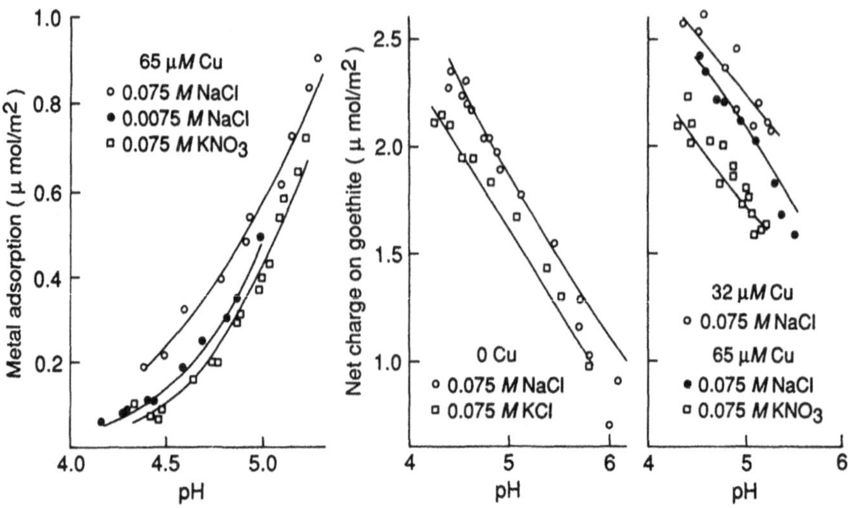

Figure 3. Observed (points) and modelled (lines) values for sorption of zinc by goethite at the indicated initial concentrations, concentration of background electrolyte, period and temperature of reaction. (From Barrow et al., 1989.)

adsorbing materials and to the complex situation existing in soil. To explain this, we need to look more closely into the way the model works.

An important part of the model is the equation relating adsorption (S) to the solution concentration and to the electrical potentials. The equation was derived by Bowden et al. (1977):

$$S = \frac{N K_i c \alpha \gamma_i \exp(-z_i \psi F/R T)}{1 + K_i c \alpha \gamma_i \exp(-z_i \psi F/R T)} \quad (1)$$

where N is the maximum adsorption (in the same units as S), K_i is the equilibrium constant for the ion i, c is the total concentration of the adsorbate, α is the proportion of the adsorbate present as the reacting ion i, γ_i the activity coefficient, z_i the valency (including sign) of the ion i, ψ the electric potential experienced by the ion i in the plane of adsorption and F, R and T have their usual meaning. Equation (1) is a Langmuir equation but with the term: $\alpha \gamma_i \exp(-z_i \psi F/RT)$ added. It is the components of this term which account for the effects of pH. The value of α changes with pH. As it is assumed that it is $ZnOH^+$ which is adsorbed, the concentration of this ion increases 10-fold with unit increase in pH up to about pH 8 and then more slowly. The value of ψ becomes smaller (more negative) with increasing pH but, for cations, the $-z_i$ term reverses the sign so that $\exp(-z_i \psi F/R T)$ becomes larger with increasing pH. The effect of the α term and the $\exp(-z_i \psi F/RT)$ term are therefore in the same direction and there is a very steep increase in sorption with increasing pH. The behaviour of mercury provides a contrast. As indicated earlier, the pK values for mercury are much lower and the concentration of $HgOH^+$ *decreases* 10-fold for each unit decrease in pH. The α term and the $\exp(-z_i \psi F/RT)$ term therefore oppose each other and there is only a small effect of pH on sorption (Barrow and Cox, 1992). Although it is postulated that it is $ZnOH^+$ which reacts with the surface, it does not follow that, at pH 4, reaction is limited to one ion in every 10^5. As the surface reacts with $ZnOH^+$ there will be a further hydrolysis to maintain the ratio of ions in solution appropriate to the particular pH. This is the mechanism by which some of the protons are released to solution. It may be termed a surface-induced hydrolysis.

Thus, within this model, there is a conceptual separation of the effects into two groups – those due to changes in the degree of dissociation (α), and those due to the electrical effects via ψ. It is possible to fit this model to data for oxides because there are effective theories which relate pH to charge and therefore to electric potential. An extension of this approach to organic matter would require appropriate theories to relate pH to charge for this material. An extension to clay minerals would require a partition of the charge into that due to isomorphous replacement and that due to edge effects.

9. Zinc sorption by soils - effects of concentration, pH and time

Soil scientists are always very interested in the partition of a reactant between the solution phase and the solid phase – that is, in sorption curves. This distribution is very important, at least over short times, in controlling the rate of movement of the reactant both to plant roots and in leachate. Much effort has therefore been devoted to describing such curves. In such measurements, reaction of a soil with metal ions always releases protons and therefore decreases the pH. For most metals, the decrease in pH has a feed-back effect and decreases sorption. Unless pH is kept constant, this negates one of the basic requirements of the Langmuir equation. Further, as there are several possible reacting materials in most soils, it seems unlikely that another basic requirement of this equation can be met – reaction with a uniform surface. Nevertheless, Shuman (1980) concluded that "zinc adsorption often follows the Langmuir model" even though he had found (Shuman, 1975) that two Langmuir surfaces were required to describe zinc sorption by acid soils. Two Langmuir surfaces were also used by Elrashidi and O'Connor (1982). The use of two Langmuir surfaces has been criticized on both theoretical and practical grounds. The theoretical grounds were concerned with the reality of the postulated surfaces (Posner and Bowden, 1980; Sposito, 1982). The practical grounds are that it is inefficient because it requires more coefficients to describe the results than other methods and that it has poor statistical properties (Barrow, 1978; Ratkowsky, 1986).
Plots of zinc sorption against concentration have been found to be curved when plotted on a log-log scale (Fig. 4). The curvature visible in Fig. 4 may lead, at very low concentrations, to a slope of unity on a log- log scale - that is, to a linear relation on an arithmetic scale. Thus the Freundlich equation does not apply – although it may be a

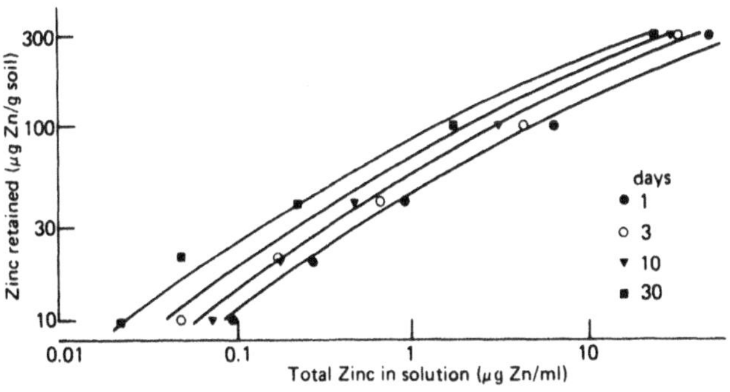

Figure 4. Logarithmic plot of zinc sorption by a soil after periods of reaction from 1 to 30 days. The lines were obtained from the fit of a mechanistic model to the data. (From Barrow, 1986a.)

convenient approximation over limited concentration ranges. At very low concentrations of metal ions, a linear relation between sorption and concentration has been observed (Jarvis, 1981; McLaren et al., 1983; Gerritse and van Driel, 1984). For trace elements in many soils the concentrations may be so low that this is a common situation. A comprehensive description (and explanation) should encompass this behaviour.

The influence of pH on zinc sorption by soils is so large that pH has been called the "master variable" (Msaky and Calvet, 1990). Nevertheless, the effect of pH is much smaller than observed for reaction with iron oxides. As indicated above, for oxide surfaces, the effects could be divided into those due to dissociation plus those due to changes in the electric potential of the variable charge surface. For soils, we can test this model by plotting sorption against the calculated concentration of $ZnOH^+$. Any remaining tendency for sorption to be higher at high pH would then indicate the effects of the electrical component. In some cases, this method of plotting removes most of the effect of pH (Barrow, 1985, 1986b; Chairidchai and Ritchie, 1990). In another case, the outcome varied with the soil used. Thus Msaky and Calvet (1990) found that this method of plotting removed the effect of pH for an oxisol but for a brown silty soil and for a podzol the effects of pH were smaller.

Plotting zinc sorption against $ZnOH^+$ concentration could be a powerful tool for diagnosing which of the possible substrates for reaction with zinc is important in a particular soil. When the overall effect of pH is so small that this method of plotting over-corrects, it may suggest that cation exchange of Zn^{2+} with permanent charge surfaces is important. When it under- corrects, it may indicate that changes in electrical potential with pH are playing a role. When the overall effect is such that points fall close to a common line, it might indicate that there is a nice balance of the above effects. Or, it might indicate that $ZnOH^+$ ions are reacting with a surface which is not variable charge. That is, there is no extra effect due to changes with pH in the charge on the surface of oxides or of organic matter. One way to check this is to study the interaction between ionic strength and pH. By definition, for a variable charge surface there is a pH value at which the mean charge is zero. At this pH, changing the concentration of an indifferent electrolyte has no effect on the electric potential. For some ions, there will therefore be little effect on sorption and so there will be a point at which there is zero salt effect on sorption. For metal cations, it is a little more complex because of the effects on the value of the practical dissociation constant but nevertheless there will be a region of little salt effect. As the pH diverges from this region, there will be an increasingly large effect of ionic strength on sorption. This behaviour has been observed for anions reacting with soil (Barrow, 1989b). However, for zinc and soil, two reports indicate only small interaction between pH and ionic strength (Barrow and Ellis, 1986; Shuman, 1986). These observations also favour the conclusion that zinc did not react with a variable charge surface in these soils.

The possibility that metallic cations do not react with oxides is important. Very large amounts of work are being devoted to the study of the reaction between metal ions and oxide surfaces – see Dzombak and Morel (1990) for a summary. Could it be that this work is irrelevant to soil and to natural waters? Or is there a way to reconcile the effects observed in soils with the observations that these surfaces do indeed react strongly?

Most studies with oxides have used pure samples prepared in the laboratory. The oxides in soil are very impure – they have, after all, been formed in a very dirty environment! Both silicon and phosphorus are found in soil goethites (Norrish, 1975; Norrish and Rosser, 1983). The presence of these anions will make the charge more negative. For manganese oxides, there may also be a replacement of Mn^{4+} ions with Mn^{2+} ions. A more-negative charge could be an advantage in the long-term survival of a given particle in attracting and retaining constituent cations in competition with other particles in the soil. If the negative charge is large enough, then at say pH 5 to 6, the surface sites may be all occupied by protons balancing this charge. Further decreases in pH then cannot make the charge more positive. It is therefore possible for a dirty oxide to be a fixed charge material over part of the pH range (Barrow, 1986c). Because of its negative charge, such a material would also be a preferred site for reaction with zinc.

Studies on the rate of reaction of zinc with soils have produced diverse results. On the one hand, Msaky and Calvet (1990) measured reaction for only 12 hours and concluded that equilibrium seemed to have been reached after 3 hours. On the other hand, Tiller et al. (1984) found that, for a soil clay containing a high proportion of montmorillonite, specific sorption was still marked after 2 weeks. Continuing reaction was less marked for clays dominated by kaolinite or illite. This suggests that there might be marked differences between soils in the rate of reaction. It also indicates the importance of making measurements over an adequate period. After all, zinc fertilizers have many years during which they might continue to react with soil. Fig. 4 shows that reaction between zinc and a soil continued for up to 30 days. In that study (Barrow, 1986b), temperatures of up to 60°C were used to accelerate the reaction and it was found that reaction was continuing, albeit slowly, after the equivalent of almost 1000 days at 25°C (Fig. 5).

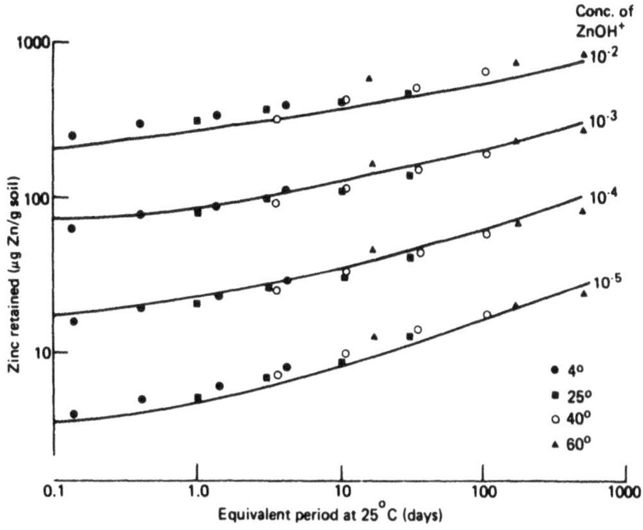

Figure 5. Observed and modelled changes in the amount of zinc retained by a soil through time at the indicated concentrations (µg Zn mL-1) of ZnOH+ in solution. The values at the indicated temperatures have been converted to an equivalent period at 25°C using the fitted values for the activation energy. (From Barrow, 1986a.)

10. Modelling the reaction between zinc and soil

The reaction between zinc and soil has been closely modelled (Barrow, 1986b) using the following assumptions: that ZnOH ions are adsorbed onto a heterogeneous surface; that the initial adsorption reaction is followed by a diffusive penetration of the adsorbing material; and that reaction causes a negative feedback which decreases further adsorption. The feedback effect is caused partly by the decrease in pH which accompanies adsorption and partly by the increase in charge of the surface. The heterogeneity is modelled by assuming that there is a normal distribution of site affinities. The model divides this distribution into a number of uniform slices within which the affinity is taken as that for the midpoint of the slice. The overall behaviour is calculated by summing over the slices. Fig. 4 shows that the model closely reproduced both the shape of the sorption curve and the change in its position through time. The model has the further advantage that extrapolation to lower concentrations predicts a linear relation between sorption and concentration – as is indeed observed. The postulated solid- state diffusion reaction has a large activation energy and therefore a large effect of temperature on rate. Fig. 5 shows that this was also closely modelled. The model has the further advantage that it provides a simple explanation for the reported differences between sorption and desorption curves (Barrow, 1986b). The differences arise because the slow diffusive penetration of the surface takes time to reverse.

11. Competition and cooperation

Natural systems usually contain a mixture of ions. Further, if the pH is low enough, reaction with zinc can release both aluminium and manganese ions. There is therefore interest in the extent to which cations might compete with zinc for sorption and anions might cooperate and increase sorption. Benjamin and Leckie (1981) summarised earlier work on competition and studied competition between metal ions for sorption on iron hydroxide. They found that competition was generally smaller than expected.

Three mechanisms for competition may be discerned: effects on available sites; effects via changes in charge and potential; and effects via changes in pH. The classical view of competition is that reaction with one species decreases the number of sites left for the other species. For competition between ions such as, say, zinc and copper, such effects may seldom be important because the amounts added are small compared to the number of reaction sites and the decrease in the number of vacant sites is trivial. The picture becomes more complex when we consider competition between, say, zinc and calcium because the concentrations of calcium are commonly much larger. There could be a decrease in the number of sites, but there will also be effects on the electrical potential of the surface as a result of the increased concentration. The magnitude of these effects will depend on the charge on the surface. If the charge is small, the effects of changing the calcium concentration will also be small (Fig. 3). If the charge is large, as it often is for soil, the effects are larger. There will also be effects caused by the changes in the charge on the surface as a result of adsorption. However, for metal cations, these effects are commonly small because the charge on the ions is mostly balanced by release of protons. The resulting drop in pH caused by one cation may, however, have been an important factor in competition with another cation in the experiments of Kurdhi and Doner (1983) and of Harter (1992).

Effects via changes in pH may also be important in interactions between ions of opposite charge – for example zinc and phosphate. I showed that sorption of phosphate by a soil could either increase or decrease sorption of zinc depending on the direction of the effect of phosphate sorption on pH (Barrow, 1987b). Possible separation of sorption sites may also play a role in the extent to which phosphate sorption may increase zinc sorption. For reaction with oxides, both phosphate and zinc appear to react with the same surfaces and reaction with phosphate increases reaction with zinc (Stanton and Burger, 1967; 1970:

Bolland et al., 1977; Madrid et al., 1991). That is, decrease in the electric potential as a result of reaction with phosphate makes it easier for zinc to react. In contrast, Madrid et al. (1991) found no effect of phosphate on zinc sorption by montmorillonite and suggested that reaction was with different surfaces. Similarly, in soil, it would be expected that phosphorus would react preferentially with the sites of least negative potential and zinc with those of most negative potential. Overlap would only occur at high levels of addition. This is broadly the picture found by Barrow (1987b).

12. Forms of zinc in soil

We know that zinc *can* react with oxides, with clay minerals and with organic matter. Further we know that it can continue to react for some time. It is therefore reasonable to postulate that zinc is present in soil in four main forms: exchangeable, specifically adsorbed, bound to organic matter, and penetrated into the particles. But how can we judge which fractions are important in a given soil? The answer is probably "only with difficulty". Three approaches seem relevant: interpretation, correlation and fractionation.

By "interpretation" I mean close examination of, and deductions from, the effects of pH, the interaction between pH and ionic strength, and the effects of time and temperature. I have indicated earlier how such observations might be used to infer the kind of reactions involved.

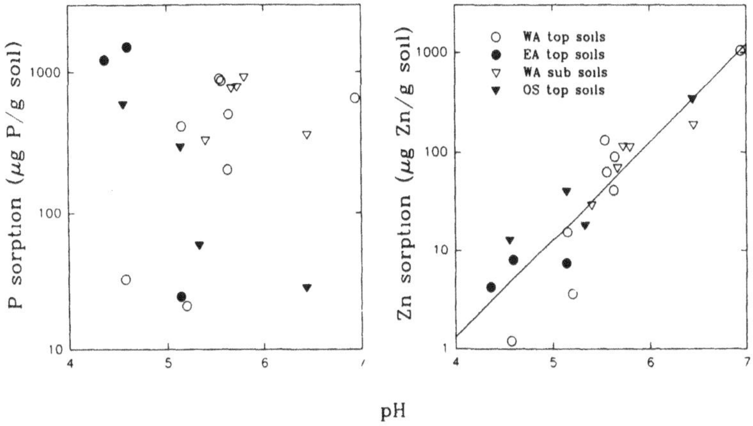

Figure 6. Plots of phosphate sorption and zinc sorption against pH for a group of soils selected so that there was little correlation between phosphate sorption and pH. For both elements, sorption was interpolated at a solution concentration of 1 µg element mL-1. The line indicates the simple regression of zinc sorption on pH. WA indicates western Australia; EA, eastern Australia; OS, non-Australian.

Correlation is a powerful technique which must be used with caution. Shuman (1980) lists seven references reporting a relation between zinc reaction and cation exchange capacity of the soils but, of course, this does not necessarily mean that cation exchange was the mechanism involved. In a previously unpublished study, I tested the idea that both phosphate and zinc sorption was largely on soil oxides and that, after adjustment for the different effects of pH, there would be a correlation between the amounts of phosphate sorbed and the amounts of zinc sorbed. I collected 20 soils and measured phosphate sorption and zinc sorption by mixing the soils with dilute solutions of phosphate or zinc in 0.01 *M* calcium chloride for 24 hours at 25°C. In both cases, sorption was interpolated at a final solution concentration of 1 μg mL^{-1}. The pH value used in regressions was that interpolated at a zinc concentration of 1 μg mL^{-1}. The soils were chosen so that there were large differences in phosphate sorption but little correlation between pH and phosphate sorption (Fig. 6). Despite the large range of soils, pH was easily the most powerful factor with the simple correlation with pH accounting for 81 per cent of the variance (Fig. 6). Adding a term for phosphate sorption significantly increased the variance accounted for to 92 per cent. Adding interaction terms and quadratic terms did not significantly decrease the residual variance. The fitted equation was:

$$\log_{10}(S_{Zn}) = -4.80 + 0.99\ \mathrm{pH} + 0.39 \log_{10}(S_P) \quad (2)$$

where S indicates sorption. The relation between phosphate sorption and zinc sorption was inspected by calculating zinc sorption adjusted for the pH and intercept terms (Fig. 7). This figure shows that there was a strong trend for the adjusted zinc sorption to increase non-linearly as phosphate sorption increased. This figure also shows that, at given values for phosphate sorption, the soils collected from outside Australia tended to have higher values for adjusted zinc sorption than soils collected inside Australia. The significance of this trend was tested by fitting an equation with the same form as Equation (2) separately to Australian and non- Australian soils. There was a significant decrease in the residual variance ($P<0.01$) and the total variance accounted for increased to 97 percent. The four non-Australian soils were a soil derived from rhyolitic ash in New Zealand, a soil from Thailand, a sample of Sutherlin soil and one of Yorkville soil from California. The sample of soils was too small to pursue the investigation further and all that can be said is that differences in the clay mineral content might be responsible for the different behaviour.

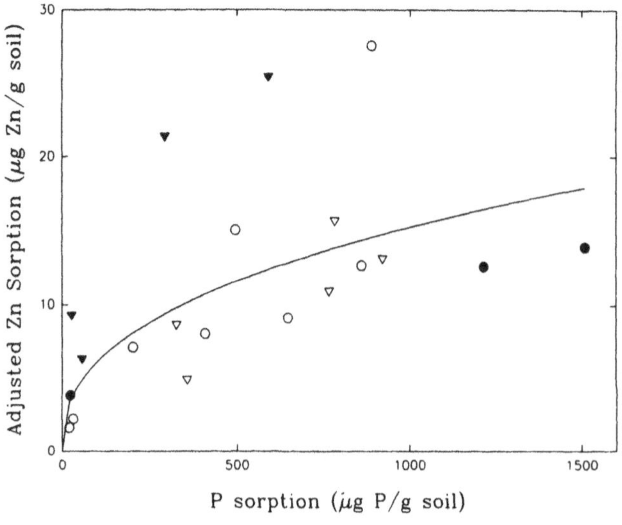

Figure 7. Relation between adjusted zinc sorption and phosphate sorption. Zinc sorption was adjusted for pH by transferring the term for pH to the left side of Equation (2). The position of the intercept was similarly adjusted by transferring the intercept. The symbols are as for Fig. 6.

Uren (1992) pointed out that recent interest in the pollution of soils has led to increased interest in fractionation schemes. He cited four references which point out the shortcomings of existing schemes. Most schemes attempt to identify a soluble fraction. However the relation between the amount of readily desorbable zinc and the solution concentration is a complex one varying with time and with ionic strength. Further, the position on such a curve reached in a given method depends on the solution:soil ratio used and will differ from that of the original soil at field moisture status. Therefore the measured "soluble fraction" is method dependent. As is indicated earlier, it can even be difficult to distinguish between the traditional classes of exchangeable and specifically-adsorbed zinc: increases in the concentration of a salt will displace some zinc from specifically adsorbed forms because of effects on the electric potential of the reacting surface. Attempts to remove zinc from organic forms are also subject to error because of effects on manganese oxides (Uren, 1992). Attempts to dissolve iron using reducing agents will also dissolve the less- reactive manganese oxides remaining after earlier treatments to dissolve organic matter (Uren, 1992). He concluded that "validation of any fractionation scheme is currently impossible because all phases of soil cannot be qualitatively identified and for that reason the schemes will continue to be empirical for some time yet".

References

Baes C F and Mesmer R E 1976 The Hydrolysis of Cations. John Wiley and Sons, New York.489p.

Barrow N J 1978. The description of phosphate adsorption curves. J. Soil Sci. 29, 447-462.

Barrow N J 1985 Reaction of anions and cations with variable-charge soils. Adv. Agron. 38, 183-230.

Barrow N J 1986a Testing a mechanistic model. II. The effects of time and temperature on the reaction of zinc with a soil. J. Soil Sci. 37, 277-286.

Barrow N J 1986b Testing a mechanistic model. IV. Describing the effects of pH on zinc retention by soils. J. Soil Sci. 37, 295-302.

Barrow N J 1986c Testing a mechanistic model. VI. Molecular modelling of the effects of pH on phosphate and on zinc retention by soils. J. Soil Sci. 37, 311-318.

Barrow N J 1987a Reactions with Variable-Charge Soils. Martinus Nijhoff Publishers, Dordrecht 191 p.

Barrow N J 1987b The effects of phosphate on zinc sorption by a soil. J Soil Sci. 38, 453-459.

Barrow N J 1989a Testing a mechanistic model. X. The effect of pH and electrolyte concentration on borate sorption by a soil. J. Soil Sci. 40, 427-435.

Barrow N J 1989b The reaction of plant nutrients and pollutants with soil. Aust. J. Soil Res. 27, 475-492.

Barrow N J, Bowden J W, Posner A M and Quirk J P 1981. Describing the adsorption of copper, zinc and lead on a variable charge mineral surface. Aust. J. Soil Res. 19, 309-321.

Barrow N J and Cox V C 1992 The effects of pH and chloride concentration on mercury sorption. I. By goethite. J Soil Sci. 43, 295-304.

Barrow N J and Ellis A S 1986 Testing a mechanistic model. V. The points of zero salt effect for phosphate retention, for zinc retention and for acid/alkali titration of a soil. J. Soil Sci. 37, 303-310.

Barrow N J, Gerth J and Brümmer G W 1989 Reaction kinetics of the adsorption and desorption of nickel, zinc and cadmium by goethite. II Modelling the extent and rate of reaction. J Soil Sci. 40, 437-450.

Bar-Tal A, Bar-Yosef B and Chen Y 1988 Effects of fulvic acid and pH on zinc sorption on montmorillonite. Soil Sci. 146, 367-373.

Benjamin MM and Leckie J O 1981 Competitive adsorption of Cd, Cu ,Zn, and Pb on amorphous oxyhydroxide. J. Colloid Interface Sci. 83, 410-419.

Bolland M D A 1970 Zinc adsorption by goethite in the absence and presence of phosphate. B Sci Agric (Hons) Thesis, Univ, West Aust.

Bolland M D A, Posner A M and Quirk J P 1977 Zinc adsorption by goethite in the absence and presence of phosphate. Aust. J. Soil Res. 15, 279-286.

Bowden J W, Posner A M and Quirk J P 1977 Ionic adsorption on variable charge mineral surfaces. Theoretical charge development and titration curves Aust. J. Soil Res. 15, 121-136.

Brümmer G W, Gerth J and Tiller K G 1988 Reaction kinetics of the adsorption and desorption of nickel, zinc and cadmium by goethite. II Adsorption and diffusion of metals. J. Soil Sci. 39, 37-52.

Chairidchai P and Ritchie G S P 1990 Zinc adsorption by a lateritic soil in the presence of organic ligands. Soil Sci Soc. Amer. J. 54, 1242-1248.

Davis J A and Leckie J O 1978 Surface ionization and complexation at the oxide/water interface. II Surface properties of amorphous iron hydroxide and adsorption of metal ions. J. Colloid Interface Sci. 67, 90-107.

Dzombak D A and Morel F M M 1990 Surface Complexation and Modelling. John Wiley and Sons, New York. 393 p.

Elrashidi M A and O'Connor G A 1982 Influence of solution composition on sorption of zinc by soils. Soil Sci. Soc. Amer. J. 46, 1153-1157.

Farrah H and Pickering W L 1978 The sorption of mercury species by clay minerals. Water Air and Soil Pollution 9, 403-409.

Forbes E A 1973 The specific adsorption of heavy metal cations on goethite. Ph D. Thesis, Univ, West Aust.

Forbes E A, Posner A M and Quirk J P 1974 The specific adsorption of inorganic Hg(II) species and Co(III) complex ions on goethite. J. Colloid Interface Sci. 49, 403-409.

Forbes E A, Posner A M and Quirk J P 1976 The specific adsorption of divalent Cd, Co, Cu, Pb and Zn on goethite. J. Soil Sci. 27, 154-166.

Gerritse R G and van Driel M 1984 The relationship between adsorption of trace metals, organic matter and pH in temperate soils. J. Environ. Qual. 13, 197-203.

Goldberg S 1992 Use of surface complexation models in soil chemical systems. Adv. Agron. 47, 233-329.

James R O and Healy T W 1972 Adsorption of hydrolysable metal ions at the oxide-water interface. J. Colloid Interface Sci. 40, 65-81.

Jarvis S C 1981 Copper sorption by soils at low concentration in relation to uptake by plants. J. Soil Sci. 32, 257-269.

Harter R D 1992 Competitive sorption of cobalt, copper, and nickel ions by a calcium-saturated soil. Soil Sci. Soc. Amer. J. 56, 444-449.

Hayes K F and Leckie J O 1987 Modelling ionic strength effects on cation adsorption at hydrous oxide/solution interfaces. J. Colloid Interface Sci. 115, 564-572.

Kinniburgh D G 1983 The H^+/M^{2+} exchange stoichiometry of calcium and zinc adsorption by ferrihydrite. J. Soil Sci. 34,759-768.

Kinniburgh D G and Jackson M L 1978 Adsorption of mercury(II) by iron hydroxide gel. Soil Sci. Soc. Amer J. 42, 45-47.

Kinniburgh D G and Jackson M L 1982 Concentration and pH dependence of calcium and zinc adsorption by iron hydrous oxide gel. Soil Sci. Soc. Amer. J. 46, 56-61.

Kurdi F and Doner H E 1983 Zinc and copper sorption and interaction in soils. Soil Sci. Soc. Amer. J. 47, 873-876.

Lindsay W L 1979 Chemical Equilibria in Soils. John Wiley and Sons, New York. 449 p.

Loganathan P, Burau R G and Fuerstenau D W 1977 Influence of pH on the sorption of Co^{2+}, Zn^{2+} and Ca^{2+} by a hydrous manganese oxide. Soil Sci Soc. Amer. J. 41, 57-62.

Madrid L, Diaz-Barrientos E and Contreras M C 1992 Relationships between zinc and phosphate adsorption on montmorillonite and an iron hydroxide. Aust. J. Soil Res. 29, 239-247.

McBride M B 1989 Reactions controlling heavy metal solubility in soil. Adv. in Soil Sci.10, 1-56.

McLaren R G, Williams J G and Swift R S 1983 The adsorption of copper by soil samples from Scotland at low equilibrium solution concentrations. Geoderma 31, 97-106.

Msaky J J and Calvet R 1990 Adsorption behavior of copper and zinc in soils: influence of pH on adsorption characteristics. Soil Sci 150, 513- 522.

Norrish K 1975 Geochemistry and mineralogy of trace elements. *In* Trace Elements in Soil and Plant Systems. Eds. D J D Nicholas and A R Egan. Academic Press. London.

Norrish K and Rosser H 1983 Mineral phosphate. *In* Soils an Australia Viewpoint. Eds. Division of Soils, CSIRO. CSIRO Melbourne/ Academic Press, London.

Padmanabham M 1983 Comparative study of the adsorption-desorption behaviour of copper(II), zinc(II), cobalt(II), and lead(II) at the goethite-solution interface. Aust. J. Soil Res. 21, 515-525.

Pickering W F 1980 Zinc interaction with soil and sediment components. *In* Zinc in the Environment. Ed. J O Nriagu. pp 71-112. John Wiley and Son, New York.

Posner A M and Bowden J W 1980 Adsorption isotherms: should they be split? J. Soil Sci. 31, 1-10.

Ratkowsky D A 1986 A statistical study of seven curves for describing the sorption of phosphate by soil. J. Soil Sci. 37, 183-189.

Shainberg I and Kemper W D 1966 Hydration status of adsorbed cations. Soil Sci. Soc. Amer. J. 30, 707-713.

Shuman L M 1975 The effect of soil properties on zinc adsorption by soils. Soil Sci. Soc. Amer. Proc. 39, 454-458.

Shuman L M 1980 Zinc in soils. *In* Zinc in the Environment. Ed. J O Nriagu. pp 39-69. John Wiley and Son, New York.

Shuman L M 1986 Effect of ionic strength and anions on zinc adsorption by two soils. Soil Sci. Soc. Amer. J. 50, 1438-1442.

Sposito G 1982 On the use of the Langmuir equation in the interpretation of "adsorption" phenomena: II The "two-surface" Langmuir equation. Soil Sci. Soc. Amer J. 46, 1147-1152.

Sposito G 1984 The Surface Chemistry of Soils. Oxford University Press, New York. 234 p.

Sposito G and Mattigod S V 1980 Geochem: a computer model for the calculation of chemical equilibria in soil solutions and other natural water systems. The Kearney Foundation of Soil Science, University of California.

Stanton D A and Burger R Du T. 1967 Availability to plants of zinc sorbed by soil and hydrous iron oxides. Geoderma 1, 13-17.

Stanton D A and Burger R Du T. 1970 Studies on zinc in selected Orange Free State soils: V. Mechanisms for the reaction of zinc with iron and aluminium oxides. Agrochemophysica 2, 65-76.

Tiller K G, Gerth J and Brümmer G 1984 The sorption of Cd, Zn and Ni by soil clay fractions: procedures for partition of bound forms and their interpretation. Geoderma 34, 1-16.

Tiller K G, Honeysett J L and De Vries M P C 1972 Soil zinc and its uptake by plants. Aust. J. Soil Res. 10, 165-182.

Uren N C 1992 Forms, reactions, and availability of nickel in soils. Adv. Agron 48, 141-203.

Chapter 3.

Zinc Fertilizers

J. J. MORTVEDT and R. J. GILKES

1. Abstract

Zinc fertilizers are commonly applied to many crops around the world. The most common sources are $ZnSO_4$ and ZnO, but other inorganic products and sources such as chelates and natural organic complexes also are used. Industrial by-products containing Zn also are being processed and sold as Zn fertilizers. The levels of water-soluble Zn and weak acid-soluble Zn in granular Zn products give a good measure of their effectiveness for crops. Insoluble $ZnNH_4PO_4$ compounds form in ammonium phosphate fertilizers; these reaction products are not very available for crops, especially on sandy, neutral to alkaline soils under dry conditions. Zinc fertilizers are applied to soil mainly with NPK fertilizers, either by incorporating at the factory or bulk blending in granular form with other granular fertilizers. Soluble Zn fertilizers also are applied as foliar sprays to fruit and vegetable crops. Choice of Zn fertilizer depends on the intended method of application, relative agronomic effectiveness, price per unit of Zn, compatibility, and convenience in applicatior ~ither alone or with other fertilizers.

2. Introduction

Numerous Zn fertilizers are being used to correct Zn deficiencies in crops. These fertilizers vary considerably in physical state, chemical reactivity, cost, and availability to plants. Methods of Zn application also differ, depending on the crop, farming system and equipment available. As a result, care must be used in selecting the most economical, effective Zn fertilizer for specific conditions at hand. This paper discusses Zn sources, their methods of production and application, and their behavior in soil, along with chemical evaluation and regulations affecting Zn fertilizers.

3. Zinc sources

There are four main classes of Zn sources: inorganic, synthetic chelates, natural organic complexes, and inorganic complexes (Table 1). These vary considerably in their Zn content, cost and effectiveness for crops.

Inorganic sources include ZnO, $ZnCO_3$, $ZnSO_4$, $Zn(NO_3)_2$ and $ZnCl_2$. $ZnSO_4$ is perhaps the most common source, and it is sold in both crystalline and granular form. While ZnO is also sold as a fine powder or in granular form, its immediate effectiveness for crops is rather low in granular form because it has a very low solubility in water (Mortvedt, 1993). Zinc oxysulfate is produced by partially acidulating ZnO with H_2SO_4. The degree of acidulation dictates the percentage of ZnO which is converted to $ZnSO_4$, the water-soluble form. The percentage conversion is especially important when the resulting product is granulated, because the decreased specific surface of granular fertilizers results in decreased effectiveness of those products with low levels of water-soluble Zn.

Table 1. Commonly used Zn sources

Zn source	Formula or designation	Percent Zn
Zinc sulfate monohydrate	$ZnSO_4.H_2O$	36
Zinc sulfate heptahydrate	$ZnSO_4.7H_2O$	22
Zinc oxysulfate	$xZnSO_4.xZnO$	20-50
Zinc oxide	ZnO	50-80
Zinc carbonate	$ZnCO_3$	50-56
Zinc chloride	$ZnCl_2$	50
Zinc nitrate	$Zn(NO_3).3H_2O$	23
Chelates	$Na_2ZnEDTA$	8-14
	NaZnHEDTA	6-10
	NaZnNTA	9-13
	$Zn_3(C_6H_5O_7)_2.2H_2O$	10-18
Natural organic complexes	---	3-12
Ammoniated zinc sulfate solution	$Zn(NH_3)_4SO_4$	10

Adapted from Mortvedt, 1991

Synthetic chelates generally are formed by combining a chelating agent with a metal ion through coordinate bonding. The stability of the metal-chelate bond affects availability to plants of the chelated metal. An effective chelate is one in which the rate of substitution of the chelated metal for other cations in the soil is quite low, thus maintaining the metal in chelated form. ZnEDTA (Na_2Zn-ethylenediamine tetra-acetate) is the most commonly used Zn chelate. Its stability constant is 17.5, which is much higher than that of CaEDTA (11.6) (Norvell, 1991). Therefore, little chelated Zn is substituted by Ca in neutral and calcareous soils, and ZnEDTA remains effective for plants in these soils.

Synthetic chelates generally are the most effective micronutrient sources, but they usually are the most costly per unit of micronutrient. For example, ZnEDTA may be as much as 2 to 5 times more effective than $ZnSO_4$ for crops, but its cost may be 5 to 10 times higher per unit of Zn. Zinc citrate has a much lower stability constant than ZnEDTA (Norvell, 1991) but Zn citrate also is less effective. However, the cost per unit of Zn also is lower for Zn citrate. Some dealers sell mixtures of the above chelates to reduce the price and yet provide chelated sources.

Natural organic complexes are produced by reacting metallic salts with some organic by-products of the wood pulp industry. Several classes of these complexes are the lignosulfonates, phenols, and polyflavonoids. The type of chemical bonding of metals to the organic components is not well understood. Some bonds may be coordinate, as in the chelates, but other types of chemical bonds also may be present. While Zn complexes are less costly per unit of Zn, they usually are less effective than ZnEDTA (Mortvedt, 1979).

One inorganic complex of Zn is ammoniated $ZnSO_4$ solution. Four NH_3 molecules-may be complexed by each Zn^{2+} cation in solution. However, the Zn^{2+} ions probably do not retain the complexed NH_3 after soil application. Commercial products generally contain 10-15% N, 10% Zn, and 5%S; their main use is in liquid starter fertilizers containing ammonium polyphosphate. Another inorganic complex of Zn is ammoniated $ZnCl_2$ solution.

Most Zn fertilizers have a significant residual effect in soils. Reports have shown that crop response to applied Zn can be demonstrated at least 5 years after application. Because Zn has such a residual effect, levels of available Zn in soils may increase with annual applications of Zn fertilizers. Soil or plant tissue samples should be taken to monitor the available Zn status with time. Zinc rates should decrease or no Zn should be applied if these levels increase to the adequate range.

Table 2. Amounts of some micronutrients sold in the USA

Year	Cu	Fe	Mn	Mo	Zn
			1,000 metric t†		
1967-1968	2.2	3.1	10.5	0.07	13.1
1971-1972	0.6	1.2	11.2	0.09	14.4
1975-1976	0.5	2.3	8.1	0.10	14.7
1979-1980	1.5	5.1	13.2	0.12	40.2
1983-1984	1.1	5.9	15.1	‡	37.3

† Amounts expressed on elemental basis. ‡ Not reported.
Mortvedt, 1991.

4. Production and use

Zinc consumption in the USA appeared to increase dramatically from 1976 to 1980 (Table 2). While actual Zn use did increase somewhat during this period, the major reason for the increase was a change in the reporting procedure to include more producers of Zn sources, especially those making Zn fertilizers from industrial by-products. In 1984, the USDA ceased collecting all fertilizer consumption data. While the National Fertilizer and Environmental Research Center of the Tennessee Valley Authority has continued to collect NPK consumption data, collection of such data on micronutrient usage was discontinued.

A similar situation exists in Australia where detailed recording of Zn fertilizer consumption was discontinued after 1985. During the period 1977-1985, annual consumption of Zn in fertilizers ranged between 900 and 1,700 t (Anon., 1986). Comparable amounts for other micronutrient fertilizers were 2,620 to 3,510 t of Cu, 6 to 39 t of Co, 78 to 223 t of Mo, 917 to 3,080 t of Mn, 267 to 1,060 t of Fe, and 41 to 182 t of B. Data on micronutrient use in other countries is not available, to our knowledge.

5. Methods of application

The most common method of applying Zn is soil application. Recommended application rates are <10 kg of Zn/ha; therefore, uniform application of Zn fertilizers. if applied separately in the field, is difficult. Both granular and fluid NPK fertilizers are used as carriers of Zn, because it allows more uniform distribution with conventional application equipment. Costs also are reduced by eliminating a separate application.

Zinc sources can be uniformly distributed in NPK fertilizers by incorporation during manufacture. Because the Zn source is in contact with compounds of the mixed fertilizers under conditions of high temperature and moisture, chemical reactions may reduce the plant availability of some Zn sources. For example, when ZnEDTA is mixed with phosphoric acid before ammoniation, acid decomposition of the chelate molecule results in decreased availability of Zn (Mortvedt and Cox, 1985). Immediate plant availability of applied Zn decreases with the level of water-soluble Zn incorporated in ammoniated phosphate fertilizers (Mortvedt, 1968).

The main advantage of bulk blending Zn fertilizers with granular NPK fertilizers is that fertilizer grades can be produced which will provide the recommended rates of Zn, N, P, and K for a given field. The popularity of this method of applying Zn has increased in the USA during the past 20 years. There were over 5,000 bulk blending plants in the USA in 1984, with Zn and other micronutrients being added in 73% of these plants (Mortvedt and Cox, 1985).

The main disadvantage of applying Zn with bulk blended fertilizers is that segregation of Zn can occur during the blending operation and with subsequent handling. Segregation results in non-uniform application which is critical with Zn because the application rate is so low. A comparison of micronutrient distribution in bags of a granular NPK fertilizer containing incorporated ZnO which was then bulk blended with powdered MnO just prior to bagging is shown in Table 3. Concentrations of Zn were rather uniform in all bags, but those of Mn were unacceptably variable. Obviously, the powdered MnO had segregated from the granular product during blending and bagging.

Table 3. Micronutrient concentrations in four bags of fertilizer, as affected by incorporating ZnO in granular MAP and blending this product with granular KCl and powdered MnO.

	Test 1		Test 2	
Bag identification	Mn	Zn	Mn	Zn
	%			
A	3.0	0.7	2.9	1.4
B	0.7	0.6	0.9	1.6
C	2.9	0.6	9.8	1.1
D	0.6	0.6	4.9	1.4
Intended concentration	(3.0)	(0.5)	(3.0)	(1.2)

Hignett, 1964.

Segregation can be minimized by carefully matching the particle sizes of all components of the blend.

Gilkes (1975) investigated the distribution of Zn between granules in an ordinary superphosphate (OSP) fertilizer where the Zn had been added as ZnO powder prior to granulation. There was a substantial variation in the Zn concentration of granules (mean value 0.53% Zn, coefficient of variation 33%), but segregation of discrete ZnO grains did not occur.

Coating Zn fertilizers onto granular NPK fertilizers eliminates, for the most part, the possibility of segregation. Fertilizer solutions are preferred as binders, so the fertilizer grade is not reduced as much. The Zn fertilizer should be ground to <0.25 mm (<60 mesh) to adhere to the NPK granules. Agronomic effectiveness of Zn coated onto soluble granular NPK fertilizers should be similar to effectiveness of Zn incorporated during manufacture. Bean (*Phaseolus vulgaris* L.) yields on a Zn-deficient soil were similar with $ZnSO_4$ or ZnO incorporated during manufacture or coated onto a granular NPK fertilizer (Table 4). However, Zn uptake was higher with $ZnSO_4$ than with ZnO using either method of application.

Mixing Zn sources with fluid fertilizers has become a popular method of application, especially in the USA. Clear liquids are commonly used as starter fertilizers, and soluble Zn sources can be easily mixed and applied by this method. Suspensions also can be used as carriers of ZnO because complete solution is not required. Agitation of suspensions is required to maintain uniform consistency during application.

Foliar sprays of Zn are used for certain crops, especially fruits and vegetables. Soluble inorganic salts generally are as effective as synthetic chelates and natural organic complexes as foliar sprays, so the inorganic salts are used because of lower costs. Low residue foliar sprays have been used to correct Zn deficiencies of citrus crops, but sprays which will discolor the fruit should be avoided. Including urea in the foliar spray increases Zn absorption, and including sticker-spreader agents increases efficiency of Zn uptake from foliar sprays.

Table 4. Yield and Zn concentrations in beans, as affected by Zn sources and methods of application with a granular NPK fertilizer.

Zn Source	Method of Application	Yield	Zn Concentration
		kg/ha	mg/kg
—	—	1,230	20
$ZnSO_4$	Blended	1,660	40
$ZnSO_4$	Incorporated	1,640	31
$ZnSO_4$	Coated	1,670	34
ZnO	Incorporated	1,620	30
ZnO	Coated	1,670	26
LSD (0.05)		170	3

Ellis et al., 1965.

6. Agronomic effectiveness of Zn fertilizers

There are various procedures for comparing the relative agronomic effectiveness (RE) of fertilizers for promoting plant growth. Simple comparisons of the plant yield obtained for a single common application rate of each fertilizer are unsatisfactory because the values of RE will depend on the rate of fertilizer applied (Barrow, 1985). This is a consequence of the exponential form of the response function. Comparisons of the Zn concentration (mg Zn/kg plant) or Zn content (mg Zn/plant) of plants receiving a common rate of each fertilizer (i.e. a constant application of Zn expressed as mg Zn/kg soil or kg Zn/ha) provide more sensitive measures of RE. This is because, unlike yield, the Zn concentration and Zn content versus rate of applied Zn response functions do not reach a constant value (plateau) when Zn sufficiency is achieved. However, these values of RE still vary with rate of fertilizer application. The preferred method for determining the RE of fertilizers relates to the substitution values of the fertilizers. That is, how much of fertilizer B is required to produce the same yield as a unit of fertilizer A. Thus, the relative amounts of Zn as different fertilizers required to support a specified yield (typically 90% of maximum yield) provide a measure of the RE of the fertilizer.

If the response curves for all fertilizers under comparison rise to a common value of maximum yield, and if the response curves all follow the same functional form (Mitscherlich function), then values of RE will be independent of the value of yield chosen as a basis for comparison. This procedure is sometimes referred to as a horizontal comparison, whereas calculation of RE based on yields for a single level of application provides a vertical comparison that is dependent on the rate of application of the fertilizer (Palmer et al., 1979). Thus, the substitution values (RE) of Zn fertilizers can only be established by field or glasshouse experiments that generate complete response curves to enable horizontal comparison. Unfortunately, this inevitably requires a large and complex experiment. Furthermore, the considerable experimental variability that is inherent in

Table 5. Response of maize to Zn sources band applied with an ammonium polyphosphate starter fertilizer.

	Zn applied, kg/ha				
Zn source	0	0.11	0.33	1.12	3.36
	Grain yield, t/ha				
-	3.9				
ZnEDTA		8.6	8.7	9.5	8.8
ZnO		8.2	8.2	8.6	9.1
$ZnSO_4$		8.3	8.9	8.7	9.1
$ZnSO_4$-NH_3 complex		7.8	8.6	8.5	8.8
$Zn(NO_3)_2$-UAN		8.2	8.2	8.9	8.8

Hergert et al., 1984.

micronutrient experiments may result in rather imprecise values of RE.

The typical variability of data is illustrated in Table 5 where grain yields of maize (*Zea mays* L.) are compared for four rates of application of five NP-Zn fertilizers. Yields for all fertilized treatments are similar and do not increase systematically with levels of application ranging from 0.11 to 3.36 kg Zn/ha. Thus, for all fertilizers, the lowest rate of application was (almost) sufficient to support maximum yield (about 8.8 t/ha), and no conclusions concerning the RE of the five fertilizers could be made. In order to obtain accurate values of RE, the experiment should have included rates of application lower than 0.11 kg Zn/ha, thus enabling horizontal comparisons to be made in the sensitive ascending part of the response curves.

Ghosh (1990) used the horizontal comparison procedure to compare 21 solid Zn fertilizers, including reagents, commercial fertilizers, experimental fertilizers, and their water-insoluble residues. He used a highly responsive, acid, sandy soil in a greenhouse experiment. Fertilizers were provided in both granular and finely powdered forms, and two crops of wheat (*Triticum* spp.) were grown to provide measurements of both RE and residual RE (RRE).

Some representative results of this work are shown in Table 6 and are typical of results reported by several workers. For the first crop, $ZnSO_4$ solution mixed through the soil was the most effective fertilizer (RE = 1.00), with Zn in powdered OSP, monammonium phosphate (MAP), and diammonium phosphate (DAP) being almost equally effective (RE = 0.78 to 0.95). In marked contrast, Zn applied in granular fertilizers, including granulated water-soluble $ZnSO_4$, was not very effective (RE = 0.03 to 0.12). Presumably, only small volumes of soil adjacent to granules were enriched in Zn, thereby greatly reducing Zn uptake by plants relative to those growing in soils containing mixed powdered fertilizers. For the second crop, the RRE of Zn in $ZnSO_4$ solution and powdered OSP, MAP, and DAP had decreased by about one-third, whereas the RRE of

Table 6. The relative effectiveness (RE) of granular and powdered Zn fertilizers for yield of wheat grown under glasshouse conditions.

Zn Source	RE, Crop 1	RRE*, Crop 2
$ZnSO_4$ solution	1.00	0.64
$ZnSO_4$ granules	0.12	0.61
Superphosphate powder	0.93	0.61
Superphosphate granules	0.09	0.09
MAP powder	0.78	0.68
MAP granules	0.11	0.12
DAP powder	0.95	0.63
DAP granules	0.03	0.06

* RRE = residual relative effectiveness relative to freshly applied $ZnSO_4$ solution for crop 2. Ghosh, 1990.

granular $ZnSO_4$ had increased to about the same value. This was not the case for Zn in granular OSP, MAP, and DAP where RRE values remained low.

Thus, incorporation of Zn into granular NPK fertilizers may greatly reduce the effectiveness of Zn relative to Zn which has been intimately mixed with the soil to promote maximum dissolution and contact with roots. Indeed, for powders, the chemical form of the Zn and NPK carrier do not greatly affect the RE of Zn (RE = 0.78 to 0.95). In marked contrast, the RE of Zn in granulated NPK fertilizers does reflect the nature of the carrier, with much less Zn being provided by DAP relative to MAP and OSP. This sequence relates to the presence of only slightly soluble $Zn(NH_4)PO_4$ and related compounds in ammonium phosphate (AP) fertilizers, and soluble Zn compounds in OSP. Furthermore, the rate and extent of dissolution of Zn is influenced by the alkaline, near-neutral, and acid reactions of DAP, MAP, and OSP, respectively. These observations are similar to those obtained by Terman et al. (1966), Allen and Terman (1966), and Mortvedt (1968) who compared NPK fertilizers as carriers of Zn for maize.

In summary, it appears that for acid soils, diverse Zn compounds are equally effective fertilizers when provided as fine powders and thoroughly mixed through the soil. For alkaline soils, it is advisable to use water-soluble Zn fertilizers. In practice, granulated NPK fertilizers containing Zn are used for all soil types to facilitate uniform application of low rates of Zn per hectare. Under these circumstances, the RE of Zn is influenced by the composition of the NPK fertilizer. In particular, Zn incorporated into DAP-based fertilizers is unlikely to be a fully effective source of Zn, especially in neutral to alkaline soils. However, water-insoluble ZnO and water-soluble $ZnSO_4$ are equally effective when incorporated into granules of acidic OSP (Gartrell and Glencross, 1969).

The amounts of Zn incorporated into NPK fertilizers differ substantially, depending on the requirements of crops and pastures for N, P, K, and Zn. Consequently, a large number of formulations of solid and liquid Zn-NPK fertilizers are available to farmers. For example, Ghosh (1990) identified 50 Australian commercial and experimental solid Zn-NPK fertilizers with total Zn contents of 0.1 to 3.3% Zn and water-soluble Zn contents of 0.0 to 1.7% Zn. It should also be considered that OSP contains appreciable Zn (commonly 0.01 to 0.09%) depending on the source of phosphate rock, whereas AP and K fertilizers contain little Zn. For soils with marginal Zn deficiency, the low concentrations of Zn in OSP may provide an adequate maintenance level of Zn, but a change to AP fertilizers may lead to the development of Zn deficiency in crops and pastures.

7. Reactions of Zn fertilizers in soils

The composition of NPK carriers of Zn considerably influences the dissolution of Zn within granules and the movement of Zn from granules into the surrounding soil. Ghosh (1990) measured the effects of soil pH, soil water content, soil texture, and incubation time on the rate of Zn release from granulated NP fertilizers. In a yellow earth soil maintained at field capacity, the percentages of Zn that had diffused from granules in 21 days were: $ZnSO_4$ - 95%, OSP - 98%, MAP - 90%, DAP - 51%. These differences reflect the occurrence of water-insoluble Zn compounds in AP fertilizers and the alkaline reaction of DAP.

The percentage of Zn lost from granules in 21 days increased linearly with soil water contents up to field capacity. At 40% of field capacity for the same yellow earth soil, the percentages of Zn lost were: $ZnSO_4$ - 60%, OSP - 72%, MAP - 60%, and DAP - 32%.

For slightly soluble forms of Zn, a greater proportion dissolved as soil texture increased, but dissolution of water-soluble Zn compounds was not sensitive to soil texture. Thus, for otherwise similar soils ranging in texture from sand to clay loam and maintained at field capacity for 21 days, the percentage of Zn dissolved increased from 54 to 68% for DAP but remained at 98% for OSP. There was a very strong effect of soil pH on the dissolution of Zn from fertilizer granules for fertilizers such as MAP and DAP that do not generate an acid solution during dissolution. There is little effect of soil pH on dissolution for acidic fertilizers such as OSP. For example, in a yellow earth soil adjusted to pH values of 4.2 to 8.8, the percentages of Zn dissolved in 21 days decreased as follows: $ZnSO_4$ - 100 to 98%, OSP - 100 to 98%, MAP - 97 to 20%, and DAP - 54 to 0%.

The above data clearly show that incorporation of Zn in AP-based fertilizers may greatly reduce the solubility of fertilizer Zn in some soils and hence the availability of this Zn to plants. This inefficient use of fertilizer Zn is likely to be greatest in sandy, alkaline, and seasonally dry soils.

The laboratory measurements of Ghosh (1990), described above, are consistent with analyses of Zn-OSP granules recovered from a lathyritic podzolic soil in the field (Gilkes and Sadleir, 1981). The fertilizer had been drilled into moist soil, and the granules were recovered at various intervals. About 90% of the Zn in the fertilizer was water soluble, and this fraction had diffused from the granule within 7 days. The remaining 10% of water-insoluble Zn remained in the relic granule (anhydrite, apatite, Fe, Al phosphates, etc.) one year after application.

8. Chemical evaluation of solid Zn fertilizers

It has been stressed in this chapter that chemical reactions between Zn additives and some NPK fertilizers may result in products that are not very soluble in soil solutions. This problem is compounded by the influence of soil properties on the dissolution of Zn in solid compound fertilizers. A large variety of Zn-NPK fertilizers is available to farmers, and new combinations are continuously being developed in response to the changing demands of agriculture. Clearly, the effectiveness of all possible Zn-NPK combinations cannot be evaluated under field and glasshouse conditions.

There is a need for a simple and reliable chemical test that can be used to predict the agronomic effectiveness of Zn in NPK fertilizers. Ghosh (1990) investigated seven extractants and ten diverse Zn-NPK fertilizers to identify the relationships between extractable Zn obtained under various conditions and the Zn content of wheat plants grown on acid soil under glasshouse conditions. He concluded that a one-hour extraction at a solid:solution ratio of 1:100 in dilute acids (0.005 *M* HCl, 0.1% citric acid, 0.025 *M* HCl, and 2% citric acid) provided a reasonable prediction of the relative agronomic effectiveness of the fertilizers (i.e., $R^2 = 0.90$, 0.90, 0.88, and 0.83, respectively, for plant Zn content versus Zn soil test value). Water (0.77), *M* HCl (0.66), and DTPA (0.61) were less predictive extractants. One reason for water-soluble Zn as an ineffective predictor of available Zn is that some Zn dissolved from AP fertilizers in water subsequently is precipitated as insoluble $ZnNH_4PO_4$ during the extraction.

When a second crop of wheat was grown on the same fertilized soils, no extractant provided a very good prediction of the Zn content of the plants [2% citric acid ($R^2 = 0.76$) > 0.005 *M* HCl (0.61) > 0.025 *M* HCl (0.56) > 0.1% citric acid (0.50) > *M* HCl (0.49) > water (0.42)]. Presumably much of the fertilizer Zn had dissolved so that agronomic effectiveness of Zn was relatively more dependent on Zn-soil reactions for the second crop than for the first crop.

9. Use of industrial by-products as Zn fertilizers

Industrial by-products now are being used as Zn fertilizers because they are less costly per unit of Zn than products manufactured specifically for that purpose. By-product ZnO flue dusts are commonly used as Zn fertilizers which are incorporated with NPK fertilizers. Because ZnO by-products are dusty and difficult to handle, they may be partially acidulated with H_2SO_4 to form Zn oxysulfates containing 20 to 50% Zn (Mortvedt, 1992a). Such products are granulated during the acidulation process and used for blending with NPK fertilizers.

Some of these industrial by-products also contain heavy metal contaminants such as Cd, Pb, and Ni. Heavy metal concentrations usually are low, and their application rate to soil as Zn fertilizers also would be low because Zn fertilizers are applied at low rates. For instance, application at a rate of 5 kg Zn/ha of a Zn oxysulfate fertilizer containing 40% Zn and 100 mg Cd/kg would result in a Cd application rate of 1.25 g/ha. Since total Cd in the surface 15-cm layer of soil generally ranges from 0.5-1.5 kg/ha, it would take about 1,000 years of annual applications of the above Zn fertilizer to double the Cd concentration in the surface soil.

There have been few reports of the bioavailability of heavy metal contaminants in Zn fertilizers made from industrial by-products. Such availability is related to both soil and plant factors. Availability to plants of some heavy metals usually is greater in coarse- than fine-textured soils and increases with increasing soil acidity. Some leafy vegetable crops tend to accumulate more heavy metals than cereal crops. In a greenhouse study, five Zn fertilizers made from industrial by-products were compared with reagent grade $ZnSO_4$ applied to an acid (pH 5.8) soil. Uptake of Cd, Ni, and Pb each was higher in Swiss chard (*Beta vulgaris* L.) leaves than in maize forage. Uptake of Cd by each crop was highest from those Zn fertilizers containing the highest Cd concentrations (up to 2,165 mg/kg), but uptake of Ni and Pb by either crop was not affected even though these products contained up to 8,950 mg Ni/kg and 52,000 mg Pb/kg (Mortvedt, 1985).

Jarosite waste containing 4.7% Zn and 0.002% Cd from a Zn refining process was evaluated as a fertilizer by Kanabo and Gilkes (1992). When thoroughly mixed with soil, the waste was nearly (80%) as effective as $ZnSO_4$ as a fertilizer for wheat. When extremely large amounts of jarosite waste were applied to soils (2.5% waste), high to excessive levels of Cd accumulated in plant tops [16.5 mg/kg in clover (*Trifolium* spp.), 0.7 mg/kg in ryegrass (*Lolium perenne* L.), 3.1 mg/kg in wheat, and 0.2 mg/kg in rice (*Oryza sativa* L.)].

10. Fertilizer regulations

Regulations concerning the handling and sale of fertilizers vary considerably with each country. Each state in the USA is responsible for such regulations, but attempts have been made for each state to have uniform regulations, especially concerning fertilizer labels which would greatly simplify interstate commerce of fertilizers.

A uniform state fertilizer bill was adopted by the Association of American Plant Food Control Officials (AAPFCO) in 1968. This association, composed mainly of members of the appropriate regulatory bodies in each state, suggested passage of this bill by each state legislature. The bill lists the minimum percentage of each plant nutrient which may be guaranteed on a fertilizer label. The bill also suggested the allowable deficiencies of each plant nutrient from those percentages guaranteed on the fertilizer label. The values for Zn are 0.05% as the minimum percentage guarantee, with an allowable deficiency of 0.005% plus 10% of the amount guaranteed (Mortvedt and Cox, 1985). For example, the allowable deficiency would be 0.205% Zn in a fertilizer grade guaranteeing 2.0% Zn on the label.

In Australia, legislation controlling the quality of fertilizer differs among states, but the total content of Zn must be registered and specified on labels for all states. In New South Wales and Queensland, labels also must indicate the form of Zn added to the fertilizer.

The level of plant available nutrients in fertilizers also is of concern in labelling regulations. Sufficient data should be listed on a label to give adequate information to the consumer. Agronomic effectiveness of Zn fertilizers is partially related to their water solubility. This is especially important with granular fertilizers because the specific surface of a granular fertilizer is significantly decreased from that of powdered fertilizer. Results have shown that at least 40% of the total Zn in granular Zn fertilizers should be in water-soluble form to be fully effective for crops (Mortvedt, 1992b).

More industrial by-products, especially oxides of Zn and Mn, are being at least partially acidulated with H_2SO_4 for use as fertilizers. These products, known as oxysulfates, are sold in granular form. If they have not been sufficiently acidulated to convert a significant portion of the Zn or Mn into the water-soluble sulfate form, they will not be as effective for crops. The state of Florida recently included the requirement of level of water-soluble Mn as well as total Mn on labels for Mn-containing fertilizers (Warren, 1988). Results with Zn described above suggest that the same requirement should be included for granular Zn fertilizers.

References

Allen SE, and Terman GL (1966) Response of maize and sudangrass to Zn in granular macronutrients. In Trans. Comm. II and IV. Int. Soc. Soil Sci. p. 255-266, Aberdeen, Scotland.

Anon. (1966) Chemical Fertilizer in Australia, 8th ed., Dept. Primary Industry, Australian Govt. Publ. Service, Canberra, Australia.

Barrow, NJ (1985) Comparing the effectiveness of fertilizers. Fert. Res. 8, 85-90.

Ellis BG, Davis JF, and Judy WH (1965) Effect of method of incorporation of zinc in fertilizer on zinc uptake and yield of pea beans (*Phaseolus vulgaris* L.). Soil Sci. Soc. Am. Proc. 29,635-636.

Gartrell JW, and Glencross RN (1969) Copper, zinc and molybdenum fertilizers for new land crops and pastures. J. Agric. W. Aust. 9, 517-521.

Ghosh AK (1990) Chemistry and agronomic effectiveness of Zn-enriched fertilizer. Unpublished Ph.D. Thesis, University of Western Australia, Nedlands, W.A.

Gilkes RJ (1975) Factors influencing the release of Cu and Zn additives from granulated superphosphate. J. Soil Sci. 28, 105-111.

Gilkes RJ and Sadleir SB (1981) Dissolution of granulated Cu Zn-superphosphate in soils. Fert. Res. 2, 147-157.

Hergert GW, Rehm GW, and Wiese RA (1984) Field evaluation of zinc sources band applied in ammonium polyphosphate suspension. Soil Sci. Soc. Am. J. 48, 1190-1193.

Hignett TP (1964) Supplying micronutrients in solid bulk-blended fertilizers. Commer. Fert. Plant Food Ind. 108(1), 23-25.

Kanabo IAK, Gilkes RJ (1992) Low-contaminant jarosite waste as a fertilizer amendment. J. Environ. Qual. 21, 679-684.

Mortvedt JJ (1968) Crop response to applied zinc in ammoniated phosphate fertilizers. J. Agr. Food Chem. 16, 241-245.

Mortvedt JJ (1979) Crop response to zinc sources-applied alone or with suspensions. Fert. Solutions 23(3), 64-79.

Mortvedt JJ (1985) Plant uptake of heavy metals in zinc fertilizers made from industrial by-products. J. Environ. Qual. 14, 424-427.

Mortvedt JJ (1991) Micronutrient fertilizer technology. In Micronutrients in Agriculture, 2nd ed., Eds. JJ Mortvedt et al. pp 523-548. Soil Science Society of America, Madison, WI, USA.

Mortvedt JJ (1992a) Use of industrial by-products containing heavy metal contaminants in agriculture. In Residues and Effluents - Processing and Environmental Conditions. Eds. RG Reddy, WP Imrie, and PB Queneau. pp 861-870. The Minerals, Metals and Materials Society, Warrendale, PA, USA.

Mortvedt JJ (1992b) Crop response to level of water-soluble zinc in granular zinc fertilizers. Fert. Res. 33, 249-255.

Mortvedt JJ, Cox FR (1985) Production, marketing, and use of calcium, magnesium, and micronutrient fertilizers. In Fertilizer Technology and Use. Ed. OP Engelstad. pp 455-481. Soil Sci. Soc. Am., Madison, WI, USA.

Norvell WA (1991) Reactions of metal chelates in soils and nutrient solutions. In Micronutrients in Agriculture. 2nd Ed. Eds. JJ Mortvedt, et al. pp 187-227. Soil Sci. Soc. of America, Madison, WI, USA.

Palmer B, Bolland MDA, and Gilkes RJ (1979) A re-evaluation of the effectiveness of calcined Christland Island C-grade rock phosphate. Aust. J. Exp. Agric. Anim. Husb. 19, 605-610.

Terman GL, Allen SE, and Bradford BN (1966) Response of corn to Zn as affected by nitrogen and phosphorus fertilizers. Soil Sci. Soc. Amer. Proc. 30, 119-124.

Warren, JD (1988) Basis for Florida fertilizer rule change. Proc. Soil Crop Sci. Soc. Fla. 47, 1-4.

Chapter 4

Zinc Absorption from Hydroponic Solutions by Plant Roots

LEON V. KOCHIAN

1. Abstract

Zinc is an essential micronutrient that enters the plant primarily via absorption of Zn^{2+} from the soil solution by plant roots. As with the other micronutrients (except iron), there have been relatively few studies in the literature detailing the mechanism(s) and regulation of Zn^{2+} absorption by plant roots. Much of the research in the literature has been based on solution culture techniques; in this paper, the current literature pertaining to root Zn^{2+} absorption is reviewed, and speculative models for the mechanisms of Zn^{2+} uptake are presented. The possibility that phytosiderophores, which are low-molecular weight organic molecules that complex iron and are released by roots, play a significant role in Zn^{2+} absorption in grasses is discussed. For dicots and non-graminaceous monocots, a speculative model is presented whereby Zn^{2+} influx into root cells is mediated by a divalent cation channel. In this model, gating of the channel is influenced by the activity of the plasma membrane reductase involved in ferric reduction, that has recently been shown to be induced by the imposition of micronutrient deficiencies other than Fe (including Zn).

2. Introduction

Zinc (Zn) is a micronutrient that is essential for a number of different aspects of plant physiology and biochemistry, including hormone biosynthesis, structural stability of organelles, cytochrome c synthesis, activation and proper function of a number of enzymes, protein synthesis, and stability and integrity of the root cell plasma membrane (Cakmak and Marschner, 1988; Hewitt, 1984; Marschner, 1986; Welch et al., 1982). Zinc enters the plant primarily via root absorption of Zn^{2+} ions from the soil solution. It is required at quite low levels in the plant; the critical deficiency concentration in plant tissues is 15-30 mg kg^{-1} (Marschner, 1986). Zn^{2+} concentrations in conventional hydroponic solutions that support adequate plant growth range from 0.05 to 0.25 mmol m^{-3}, with Zn toxicities observed above concentrations of 3 to 6 mmol m^{-3} (Carroll and Loneragan, 1968). More recent studies based on "chelator-buffered" nutrient solution techniques (see Chaney, 1988; Norvell, 1991; Parker et al., 1993), that have been used to control free Zn^{2+} activities at appropriately low activities, indicate that the critical Zn^{2+} activity in the rhizosphere is much lower, in the range of 25-160 nmol m^{-3} for tomato and barley (Chaney et al., 1989; Norvell and Welch, 1993). The concentration of soluble inorganic Zn in equilibrium with soil-Zn can also be quite low, particularly at high pH values. In calcareous, alkaline soils, soluble Zn can range from 10^{-2} to 10^{-4} mmol m^{-3}, while next to roots that are absorbing Zn, the concentration of Zn would be even lower due to the generation of diffusional depletion gradients for Zn (Lindsay, 1991).

Solution culture techniques have been widely used to study root mineral ion absorption. In general, the methods employed to quantify and characterize root ion fluxes have focused on either radiotracer flux studies, determination of net ion fluxes via measurement of the rate of depletion of an ion in the uptake solution, or through measurement of changes in root or plant ion content during absorption of that ion. These types of studies have been used primarily to study the absorption of macronutrient mineral ions, such as K^+, and NO_3^-. As I noted in a recent review of micronutrient uptake and translocation, much less research has been conducted on plant micronutrient transport, with the possible exception of iron (Kochian, 1991). This has been certainly true for Zn, for which there are very few recent studies attempting to elucidate mechanisms of Zn uptake into plant roots. In this paper, I will review the available information on root Zn^{2+} absorption, which has focused primarily on characterization of the concentration-dependent kinetics for Zn^{2+} uptake, the role of metabolism in Zn^{2+} uptake, and interactions of other ions with Zn^{2+} absorption. The problems and pitfalls of the techniques that have been employed to study Zn^{2+} uptake will be discussed. Subsequently, based on some recent findings from research on iron (Fe) absorption by plants that have a potential bearing on Zn^{2+} uptake, some speculations concerning the role of ion channels (for dicots) and phytosiderophores (for grasses) in Zn^{2+} transport across the root cell plasma membrane will be presented.

3. Solution culture techniques

In the investigation of root ion transport processes, the utility of solution culture techniques to provide roots as experimental material that are free of mineral coatings and soil particles is obvious. As mentioned above, in the studies on Zn^{2+} absorption, three different approaches have been used. These are: 1) Use of the zinc radioisotope, $^{65}Zn^{2+}$, in short-term uptake experiments to quantify unidirectional Zn^{2+} influx. Additionally, roots can be exposed to $^{65}Zn^{2+}$ for long periods until tracer flux equilibrium is achieved. Then the roots can be transferred to identical solutions lacking $^{65}Zn^{2+}$ for measurement of $^{65}Zn^{2+}$ efflux. Compartmental analysis of these data can provide information concerning Zn^{2+} fluxes and compartmentation in the cell wall, cytoplasm, and vacuole. 2) If the ratio of root/uptake solution is carefully manipulated, conditions can be set up where the roots can rapidly deplete the uptake solution of Zn^{2+}. This change in external $[Zn^{2+}]$ can be monitored, and from these data, rates of net Zn^{2+} uptake at different Zn^{2+} concentrations can be determined. 3) Long term Zn^{2+} absorption rates can be determined by growing plants in a solution containing a constant Zn^{2+} level, and measuring the change in amount of Zn in plants harvested at two successive time points. If the root weights are known at each harvest, the analysis first developed by Williams (1948) can be employed to determine the mean rate of Zn absorption per unit weight of root per unit time.

It is often difficult to compare and analyze Zn^{2+} transport data from studies employing these three experimental approaches because each technique actually measures a different type of flux, and plants employed for each type of study usually are at a different nutritional status, especially for Zn. Also, it is difficult with conventional hydroponic techniques to control Zn^{2+} activities at the low levels that exist in soils, particularly when synthetic chelates such as EDTA or DTPA are used to provide a chelated form of Fe. These compounds will also effectively chelate Zn^{2+}, thus reducing the activity of free Zn^{2+} in solution (Norvell, 1991). Therefore, particularly for the concentration dependent kinetic studies for Zn^{2+} uptake, the actual Zn^{2+} activity at the root surface can be quite a bit lower than the "bulk solution" Zn^{2+} activity used in these studies.

Comparison of the results obtained from two early studies using different approaches to measure Zn^{2+} uptake is illustrative. Schmid et al. (1965) were the first to use short-term $^{65}Zn^{2+}$ uptake experiments to quantify unidirectional Zn^{2+} influx; in their work uptake into excised roots from low salt-grown barley seedlings was studied. They found that Zn^{2+} uptake from either 1 or 5 mmol m^{-3} $ZnCl_2$ was linear over the first 2 hr, with a large component (60%) of this uptake being binding in the cell wall. These results point out the importance for the use of proper desorption techniques when attempting to quantify flux across the plasma membrane, particularly in radiotracer flux studies with a divalent cation that can bind tightly in the wall. At about the same time, Loneragan's group at the University of Western Australia was conducting some detailed and careful studies of Zn^{2+} uptake in a number of crop species. In the study by Carroll and Loneragan (1968), a flowing culture technique was used to grow plants at low Zn^{2+} concentrations that might reflect levels of soluble Zn in the soil (0.01 to 6.25 mmol m^{-3}). They found that even at 0.01 mmol m^{-3} Zn^{2+}, good growth was observed with maximal growth occurring at 0.25 mmol m^{-3} Zn^{2+}. In a subsequent study, they grew plants in the same flowing culture system and measured long-term net Zn^{2+} uptake by the methods of Williams (1948) discussed above. The rates of Zn^{2+} uptake obtained in this study were considerably lower than the Zn^{2+} influx values presented in the Schmid et al. (1965) paper. This is not surprising, for they were measuring net Zn^{2+} uptake into roots equilibrated for days in solutions containing Zn^{2+}, while in the radiotracer studies by Schmid and colleagues, unidirectional Zn^{2+} influx was measured in low salt barley roots that were far from equilibrium for Zn^{2+} in relation to the uptake solution.

Another problem in interpreting the results from many of the investigations of Zn^{2+} uptake is that very high Zn^{2+} concentrations were often employed in uptake solutions; the physiological relevance of studies of Zn^{2+} uptake from solutions containing toxic levels of Zn^{2+} has to be questioned. It should be mentioned that in some of these studies, particularly in some of the more recent studies, high levels (up to 1000 mmol m^{-3}) were used because investigations of heavy metal toxicity were being conducted. But for the scientist interested in gaining a better understanding of the mechanisms plants employ to acquire Zn^{2+} for normal nutritional needs, the early studies from Loneragan's group and other researchers clearly indicate that levels around 0.25 mmol m^{-3} Zn^{2+} are sufficient. In fact, the recent studies using chelate buffer techniques suggest that Zn^{2+} transport needs to be examined in solutions containing much lower Zn^{2+} activities.

4. Kinetics of Zn^{2+} uptake

Many of the investigations of root Zn^{2+} absorption have dealt with either the time-dependent or concentration-dependent kinetics of Zn^{2+} uptake. When Zn^{2+} accumulation in plant roots is studied as a function of time, biphasic uptake characterized by an initial, rapid entry and binding within the root cell wall, followed by a slower linear phase of transport across the plasma membrane is observed (for example, see Santa Maria and Cogliatti, 1988; Schmid et al., 1965; Veltrup, 1978). The initial, rapid phase can be desorbed with other divalent cations or nonradiolabeled Zn^{2+} solutions, is relatively insensitive to metabolic inhibition, and can account for a significant proportion of the total Zn^{2+} accumulated by the tissue. Over longer time periods of accumulation of radiolabeled

Zn^{2+}, evidence for a significant Zn^{2+} efflux from the symplasm has been presented (Santa Maria and Cogliatti, 1988).

A number of studies on the concentration dependence of Zn^{2+} uptake have been conducted, based on the application of an enzyme kinetic analysis approach to root mineral ion absorption that was pioneered by Epstein and coworkers (Epstein and Hagen, 1952). The utility of this approach is that saturating, Michaelis-Menten kinetics can be used to indicate that transport is facilitated by an enzyme-like transport protein with specific transport characteristics (K_m and V_{max}). As mentioned above, many of these studies on Zn^{2+} uptake kinetics have been conducted over a very wide range of Zn^{2+} concentrations, and often at excessively high and nonphysiological Zn^{2+} levels (as high as 1000 mmol m^{-3}). In these studies conducted over a wide range of Zn^{2+} activities. the kinetics of Zn^{2+} uptake often are complex and difficult to interpret. However, there are reports of saturation kinetics for Zn^{2+} absorption over this wide concentration range, with K_m values that are quite high, approx. 50 mmol m^{-3} (Bowen, 1981, 1986; Ramani and Kannan, 1978).

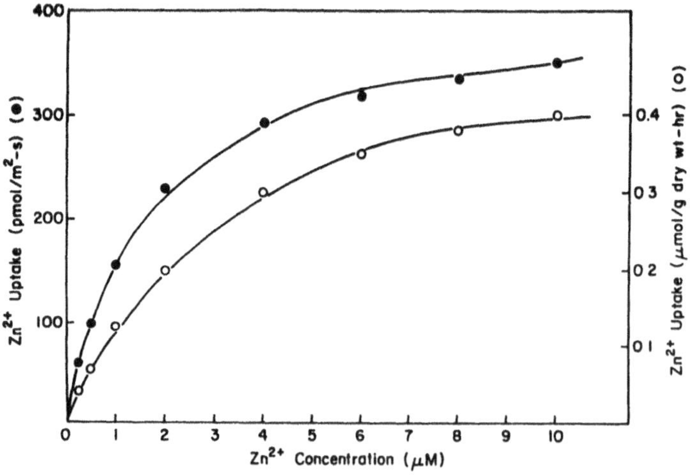

Figure 1. Concentration dependent kinetics for Zn^{2+} uptake into roots. The closed circles are data for net Zn^{2+} uptake into maize roots. Uptake was determined by the solution depletion technique of Claassen and Barber (1974). Uptake curve drawn from data from Mullins and Sommers (1986). The open circles are for Zn^{2+} uptake into barley roots, and curve was drawn from data from Veltrup (1978).

There have been several studies that employed a more realistic range of Zn^{2+} concentrations in the uptake solutions. These studies have yielded Michaelis-Menten kinetics for Zn^{2+} uptake with K_m values in the low mmol m^{-3} range. In their investigation on the effects of alkaline earth cations on Zn^{2+} absorption into wheat roots, Chaudhry and Loneragan (1972) used short-term $^{65}Zn^{2+}$ flux techniques to study the kinetics of uptake. They found that Zn^{2+} uptake followed Michaelis-Menten kinetics with a K_m of about 3 mmol m^{-3}. Veltrup (1978) studied Zn^{2+} uptake in barley roots over a very wide concentration range and obtained complex transport kinetics. However, at the lowest Zn^{2+} concentrations (up to 7.6 mmol m^{-3}), the kinetics followed a single, saturating curve with a K_m of 3 mmol m^{-3} (Figure 1). More recently, kinetic parameters for maize root Zn^{2+}

uptake were investigated using the solution depletion method of Claassen and Barber (1974), and employing a realistic concentration range for Zn^{2+} in the uptake solution (0 to 10 mmol m^{-3}) (Mullins and Sommers, 1986). In this study, they attempted to obtain more precise kinetic data by running their experiments in uptake solution lacking any synthetic chelates (to prevent chelation of Zn^{2+}). Furthermore, the Zn^{2+} activity in the uptake solution was computed using the GEOCHEM speciation computer program (Sposito and Mattigod, 1980). As shown in Figure 1, they obtained saturation kinetics with a K_m of approx. 1.5 mmol m^{-3}. These studies indicate that at Zn^{2+} levels that approximate concentrations of soluble Zn in soils, uptake is mediated by a transport protein with a fairly high affinity for Zn^{2+}.

Compartmental analysis of $^{65}Zn^{2+}$ transport in wheat seedling roots was employed by Santa Maria and Cogliatti (1988), in order to determine the unidirectional Zn^{2+} fluxes and half-times for exchange of Zn^{2+} in subcellular compartments. Under steady-state conditions, a significant efflux of Zn^{2+} across the plasma membrane was found, that was 65 to 85% of the Zn^{2+} influx. The compartmental analysis yielded results consistent with three compartments in series. The compartment assigned to the apoplasm had a $t_{1/2}$ of 0.08 h and occupied from 8 to 14% of the root volume. Based on this assessment, a Zn concentration in the apoplasm of around 0.5 mol m^{-3} was estimated. This high concentration suggests that significant Zn adsorption is occurring in the Donnan free space. The next fastest exchanging compartment had a $t_{1/2}$ of 0.55 h, accounted for about 8% of the total root content, and is suggested to be the cytoplasm. The slowest exchanging compartment had a $t_{1/2}$ of 134 h and contained 76% of the total root Zn. Some anomalous kinetic aspects of this slowest component suggest that it could be more complex than simply vacuolar Zn, and may be partially due to the binding of Zn in organic complexes in the vacuole or cytoplasm.

4.1. Metabolic requirements and transport interactions with other ions

If Zn is transported across the plasma membrane as the Zn^{2+} ion, then a large inwardly directed electrochemical gradient for uptake exists, due to the large, interior negative electrical potential gradient across the plant cell plasma membrane, and the assumption that the free Zn^{2+} activity in the cytosol would remain low as absorbed Zn^{2+} is chelated by low molecular weight organic compounds. Thus, Zn^{2+} uptake would be a thermodynamically passive process, and one would not need to invoke an ion pump (ATPase) or secondarily active transport system (H^+ symport or antiport) to mediate Zn^{2+} uptake. As discussed later, Zn^{2+} uptake could occur via a divalent cation channel, with the driving force due primarily to the negative membrane potential. These passive transport systems are still coupled to metabolism, and the use of metabolic inhibitors should elicit an inhibition of uptake via a reduction or depolarization of the membrane potential.

There appear to be some contradictory findings in the literature concerning the coupling of Zn^{2+} uptake to metabolism. There have been a number of reports in algae, barley roots, and bean plants, suggesting that Zn^{2+} uptake is not metabolically-dependent, based on a lack of response to metabolic inhibitors (Broda et al., 1964; Gutknecht, 1961,1963; Rathore et al., 1970). Others, however, have presented strong evidence in support of a metabolically-controlled transport of Zn^{2+}. In the previously mentioned study by Schmid et al. (1965), Zn^{2+} transport across the barley root cell plasma membrane was quite sensitive to low temperature, anaerobiosis, DNP, and azide. Bowen (1969) demonstrated that Zn uptake in sugarcane leaf tissue was dependent on metabolic energy, based on observations of significant inhibitions by low temperature and a variety of metabolic inhibitors. Giordano et al. (1974), also found that DNP severely depressed Zn uptake from solutions containing either 0.005 or 5 mmol m^{-3} Zn^{2+}, in intact rice seedlings. They suggested that the contradictory results concerning the role of metabolism in Zn^{2+} uptake may be due to differences in experimental conditions (Zn concentration, inhibitor concentration, influence of other ions), and particularly, the use of unreasonably high Zn^{2+} concentrations in some of the earlier studies.

In terms of nutrient interactions, Chaudhry and Loneragan (1972) showed that all of the alkaline earth cations (Ca^{2+}, Mg^{2+}, Sr^{2+}, and Ba^{2+}) inhibited Zn^{2+} uptake in wheat roots. Kinetic analysis indicated that these ions noncompetitively inhibited Zn^{2+} uptake, and thus did not share the same transport mechanism. The main competitive interaction for Zn^{2+} is with Cu^{2+}. Schmid et al. (1965) showed that Cu strongly inhibited Zn^{2+} uptake, while Mn had no effect. Bowen (1969) showed that Cu competitively inhibited Zn^{2+} uptake, and suggested that Zn^{2+} and Cu^{2+} shared the same transport system. Although transport physiologists like to think that there are separate transport proteins for each important nutrient, it is possible that for some of the divalent micronutrient cations with similar hydrated radii, such as Cu^{2+} and Zn^{2+} (hydrated radii = 0.42 and 0.43 nm, respectively), a single cation channel could mediate the transport of both ions into the cell. Obviously, this speculation awaits the test of future research efforts.

4.2. Speculations concerning the mechanism of Zn^{2+} absorption across the root cell plasma membrane

Based on the kinetic properties, inhibitor sensitivity, interactions with other ions, and thermodynamics of root Zn^{2+} absorption, it is possible to speculate about the nature of the transport protein that facilitates the uptake of Zn^{2+} into cells of the plant root. If Zn is absorbed as the Zn^{2+} ion, then Zn^{2+} uptake is a thermodynamically passive process as Zn^{2+} would be transported into the cell down its electrochemical potential gradient. This thermodynamic constraint eliminates the need for involving complex, active transport systems (ATPase or H^+-coupled transport system) in Zn^{2+} uptake. Thus, Zn^{2+} could be transported via a cation channel, and based on the studies on Zn^{2+} and Cu^{2+} transport, this channel could be permeable to Cu^{2+} and possibly some other divalent cations. The other factor to consider would be the role of Zn^{2+} chelation in uptake. In soil solutions, up to 50% of the soluble Zn is in chelated form (Tinker, 1981), and as discussed below, chelation of Zn^{2+} by low molecular weight organic compounds might play an important role in Zn^{2+} uptake in grasses. If Zn^{2+} is absorbed into the root cell as a complex with an organic ligand, then the transport system presumably would have sites for recognition of the Zn^{2+} complex at the plasma membrane surface, as a prerequisite for absorption of the complex. Both possible types of Zn^{2+} transport system are discussed below.

4.3. Zn^{2+} uptake in grasses

In grasses, Fe deficiency triggers the synthesis and release of non-protein amino acids, called phytosiderophores, into the rhizosphere (Takagi, 1976). These compounds, which include mugeneic and avenic acids, have a relatively high affinity for Fe^{3+} and thus can effectively chelate Fe^{3+} ion and transport it to the outer face of the root-cell plasma membrane. The ferric phytosiderophore complex is transported into the cell via a transport system which has been suggested to be specific for Fe (Marschner et al., 1989; Römheld and Marschner, 1986), although this point is clearly controversial (Crowley et al., 1991).

Recent findings indicate that phytosiderophore release is not specific for Fe deficiency and also occurs during Zn deficiency. Zhang et al. (1989, 1991a,b) found that in barley and wheat, Zn deficiency triggered the enhanced release of the same phytosiderophores that were released during imposition of Fe deficiency. It was shown that these compounds were effective in solubilizing Zn as well as Fe from the soil. In wheat, the authors showed that under Fe deficiency, enhanced release of phytosiderophores resulted in an enhanced uptake of Zn^{2+} (using $^{65}Zn^{2+}$). Although these compounds are called phytosiderophores, they have the ability to chelate Zn^{2+}, Cu^{2+} and Mn^{2+}. According to Nomoto et al. (1987), the log stability constants of mugeneic acid with Zn(II), Cu(II) and Fe(III) are 10.7, 18.3, and 18.1, respectively. Treeby et al. (1989) showed that exudates from Fe deficient barley roots mobilized Zn, Mn, Cu and Fe from a calcareous soil. The lack of specificity of these compounds to chelate and mobilize only Fe have caused some to suggest a more general name for these compounds, such as phytochelate (Crowley et al., 1987) or phytometallophore (Welch, 1993).

Based on these findings, a model for Zn^{2+} uptake in grasses is presented in Figure 2. In response to either Zn or Fe deficiency, the micronutrient metal chelators (labeled phytosiderophore or phytometallophore) are synthesized and released from the root into the rhizosphere. These compounds will function to chelate and increase the concentration of soluble Zn and Fe at the outer surface of the plasma membrane. In this model, both the

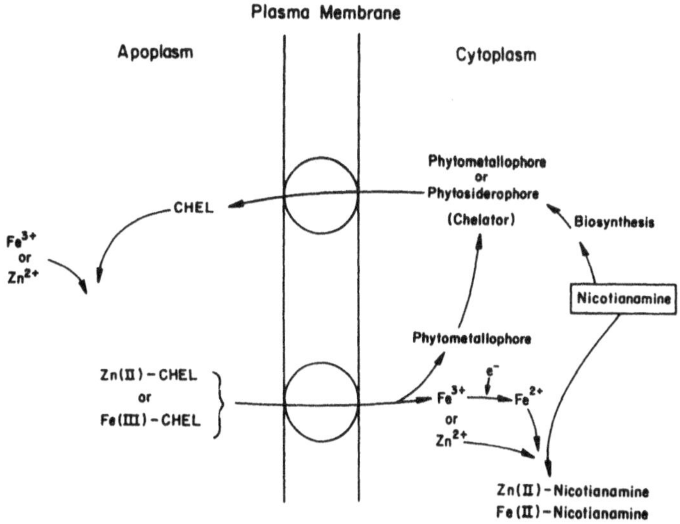

Figure 2. Speculative model for phytosiderophore (or phytometallophore) based uptake of Zn^{2+} into root cells. In response to either Fe or Zn deficiency, phytosiderophore synthesis (from nicotianamine) and release into the rhizosphere is stimulated. In the rhizosphere, these chelates can complex and mobilize either Zn(II) or Fe(III). Transport of the Zn(II) or Fe(III) chelate into the root cell is depicted as occurring via the same transport protein, which is a point of controversy in the literature.

Zn-chelate and Fe-chelate are transported into the root cell via the same transport protein, which would indicate that the transporter has recognition sites for either complex; following binding of the complex to the recognition site, the transport system would facilitate the transport of either complex into the cell. The mechanism by which these metal-chelate complexes are transported into the cell is poorly understood and a certain amount of controversy surrounds this topic. Marschner and coworkers (Marschner et al., 1989; Zhang et al., 1989) suggest that the transporter is highly specific for the ferric chelate complex. This is based on the observation that when Fe is supplied to roots of Fe deficient barley plants as ferric 3-hydroxymugeneic acid, Fe uptake is 100 to 1000 times larger than when Fe is supplied as Fe(III)EDTA or ferric hydroxide. When Zn is supplied as either $ZnSO_4$, Zn(II)EDTA, or as the complex with mugeneic acid to the same roots, there was no difference in the rate of Zn absorption. However, as Crowley et al. (1991) points out, in these experiments, both the Zn-mugeneic acid and Fe-mugeneic acid complexes were absorbed at approximately the same rate. This observation indicates that the transporter can as readily transport Zn as well as Fe, and does not have a specific recognition site for the Fe-mugeneic acid complex. This is an area of research that clearly awaits further investigation.

4.4. The involvement of ion channels in Zn^{2+} uptake

One of the major areas of progress in plant ion transport in recent years has been in our understanding of the functioning of ion channels. This progress is the result of improvements in techniques for studying transport systems, including the patch clamp technique that allows one to study the functioning of single ion channels in a patch of membrane. Most of the research on plant ion channels has focused on K^+ channels, with some research on anion channels and more recently, Ca^{2+} channels. Ion channels are characterized primarily by their ability to transport many ions per second. Ion channels have turnover numbers of around 10^6 ions/sec, which is about 1000 times faster than carrier-like transporters such as ion pumps. They are also characterized by their ability to open and close rapidly. A number of factors control channel opening, or gating, and include voltage gradients, ions, phosphorylation, G proteins and various ligands (Tester, 1990). It has been mentioned previously in this review that Zn^{2+} might be transported into root cells via an ion channel. Because ion channels have the ability to transport a large number of ions in a short amount of time, some questions arise when considering the possibility of a channel mediating Zn^{2+} influx. Because the plant requirement for Zn^{2+} is low, and Zn^{2+} fluxes are small, one would expect that if a specific Zn^{2+} channel exists, it would be in low abundance in the membrane. Of course, because the free Zn^{2+} activity in the rhizosphere is quite low, Zn^{2+} flux through these channels could be small. Also, the possibility exists that a relatively nonspecific divalent cation channel functions, for example, to transport Mg^{2+} into the cell; because ion channels do not generally exhibit perfect selectivity for one ion, possibly Zn^{2+} (as well as Cu^{2+} and even Fe^{2+} and Mn^{2+}) moves into the cell on a channel that functions to transport a macronutrient divalent cation.

Based on recent findings from research on iron reduction and absorption in peas, we have suggested, quite speculatively, that cation channels might be involved in the absorption of a number of micronutrient cations, including Zn^{2+}, Fe^{2+}, Cu^{2+}, and Mn^{2+} (Kochian et al., 1991; Welch et al., 1993). Our research indicated that, as was discussed above for grasses, the response of peas, a dicot, to Fe deficiency also influences the absorption of micronutrient cations other than Fe. Iron absorption in these plant species is a two-step process. First, Fe(III) is reduced to Fe(II) at the surface of the root cell plasma membrane by the activity of a reductase that is induced by Fe deficiency, and whose functioning is a critical prerequisite for Fe absorption. Subsequently, Fe^{2+} is transported into the cell by a transport protein that is presumably separate from the reductase (Kochian, 1991). Our findings indicate that the plasma membrane reductase might play a more general role in micronutrient nutrition than just reducing Fe. It appears that the activity of this reductase might be important for the gating of ion channels that mediate the uptake of certain divalent cations, including Zn^{2+}.

Table 1. Influence of Fe Deficiency on Root Divalent Cation Concentrations.

Treatment	**Root Cation Concentration ($\mu g\ g^{-1}$ dry wt.)**				
	Ca^{2+}	Mg^{2+}	Zn^{2+}	Mn^{2+}	Cu^{2+}
Sparkle +Fe	3235	5307	176	298	8
Sparkle -Fe	3986	9793	528	350	104

Cation concentrations were determined for roots harvested from 15 day-old pea (*Pisum sativum* L., cv. Sparkle) seedlings grown without and with 2 mmol m^{-3} Fe (as Fe[III]-EDDHA; EDDHA: N,N'-ethylenebis[2-(2-hydroxyphenyl)glycine) in 1/4 strength modified Johnson's solution. It should be noted that roots were washed for 5 min in distilled water but were not desorbed for cations binding to the cell wall. Cation concentration values are the means for 4 replicate plants. Standard errors (not presented) were not greater than 5% for any mean value. There was no statistically significant difference in root biomass (measured either as fresh or dry weight) for root systems from +Fe and -Fe grown seedlings. Data from Kochian et al. (1991).

It was found that when pea seedlings were made Fe deficient, thus inducing or stimulating the ferric chelate reductase, the accumulation of a number of divalent mineral cations was increased. As shown in Table 1, significant increases in the concentrations of Zn^{2+}, Cu^{2+}, Mn^{2+}, and Mg^{2+} were found in Fe deficient pea roots, when compared with roots from Fe sufficient plants. Certainly for Zn^{2+} and Mg^{2+}, which do not change their oxidation state in biological systems, reduction should have no direct effect on their absorption. Also, using chelate buffer techniques to control Fe^{2+} activity in solution, we have studied ^{59}Fe-labeled Fe^{2+} influx in roots from Fe deficient and sufficient pea plants (Kochian et al., 1991). It was found that there was a 2-fold increase in Fe^{2+} influx in the Fe deficient roots, again indicating that the activity of the putative ferric reductase stimulates the uptake of ions that are not being reduced. Additionally, Welch et al. (1993) found that induction of the ferric reductase was not specific for Fe deficiency. It was observed that imposition of Cu deficiency also induced this reductase, and the reductase could reduce Cu(II) in addition to Fe(III). Jolley and Brown (1991) also presented evidence in support of the concept that the ferric reductase is not regulated only by plant Fe status. They found that Zn deficiency in some genotypes of *Phaseolus vulgaris* also stimulated ferric reductase activity by the roots. Recent preliminary studies from our labs suggests that the pea root ferric reductase might also be induced by Zn^{2+} deficiency.

What is the connection between induced ferric reductase activity, increased absorption of Zn^{2+} (and other divalent cations), and ion channels? There is recent evidence in the literature that regulation of some cation channels involves reduction of critical sulfhydryl groups in the channel protein. Bertl and Slayman (1992) recently described a nonspecific cation channel in the yeast vacuolar membrane that did not function properly unless the cytoplasmic face of the membrane was exposed to sulfhydryl reducing agents (mercaptoethanol or dithiothreitol). When these agents were applied, the channel was then properly gated, suggesting that the membrane redox state might be involved in ion channel gating, possibly through reduction of specific sulfhydryl groups.

All of these different pieces of information have been used to present a very speculative model for Zn^{2+} (and other divalent cation) absorption via a cation channel that is influenced by the activity of the plasma membrane ferric reductase. As depicted in Figure 3, in response to deficiency of certain micronutrient metals (Zn, Fe, Cu), the reductase is induced and can transfer electrons to exogenous Fe(III), and possibly other,

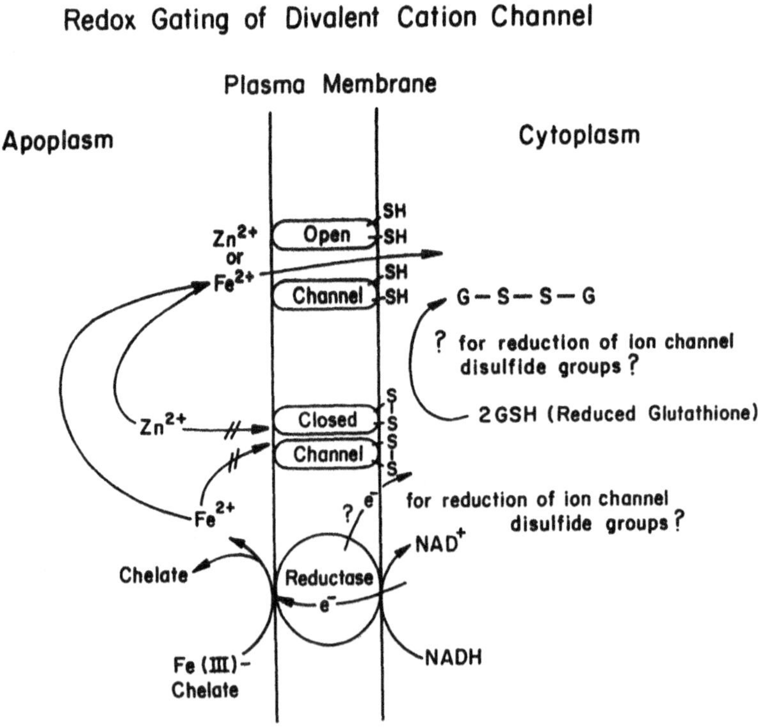

Figure 3. Speculative model for Zn^{2+} (and possibly other divalent cation) uptake into root cells of dicots and nongraminaceous monocots. In response to deficiency of Fe, Cu, or Zn, the plasma membrane reductase that functions in ferric reduction and Fe uptake is induced. Stimulation of this reductase results in the reduction of sulfhydryl groups involved in gating a divalent cation channel that mediates Zn^{2+} influx. It is not clear how sulfhydryl reduction is linked to the plasma membrane reductase. In this model, the reduction either occurs directly, with electrons flowing down an unknown pathway, or indirectly, via an increase in the level of reduced glutathione in the cytoplasm.

unknown electron acceptors. Also functioning in the root cell plasma membrane is a divalent cation channel that mediates the absorption of Zn^{2+} and possibly other divalent cations. When the reductase is induced by micronutrient deficiency, critical sulfhydryl groups in the cation channel are reduced, thus opening the channel and allowing Zn^{2+} to enter the cell. It is not clear how the reductase facilitates the reduction of the channel sulfhydryl groups. In this model, it is suggested that this could be done directly, by shunting some of the electrons through the membrane (via unknown carriers) to the channel protein. Alternatively, the sulfhydryl reduction could proceed via a cytoplasmic sulfhydryl modifier, such as reduced glutathione, that might increase in concentration when the reductase is turned on. Lesuisse and Labbe (1992) have shown that when yeast cells are made Fe deficient and the ferric reductase is turned on, levels of reduced glutathione increase significantly in the cell.

Quite obviously, this model for regulation of Zn^{2+} absorption via a cation channel is very speculative, and it contains many gaps. However, it does explain some of the recent observations pertaining to micronutrient nutrition, and is a useful starting point for further experimentation. Certainly the hypothesis that Zn^{2+} transport involves a divalent cation channel is testable, using the patch clamp technique and possibly research with isolated membrane vesicles. The challenge for the future will be to use recently advanced technologies, such as chelate buffer techniques, in conjunction with tracer flux studies and patch clamp and other microelectrode approaches, to begin to understand how plants absorb and transport this essential mineral element.

References

Bertl A and Slayman C L 1992 Cation-selective channels in the vacuolar membrane of *Saccharomyces*: Dependence on calcium, redox state, and voltage. Proc. Natl. Acad. Sci. USA 87, 7824-7828.

Bowen J E 1969 Absorption of copper, zinc and manganese by sugarcane leaf tissue. Plant Physiol. 44, 255-261.

Bowen J E 1981 Kinetics of active uptake of boron, zinc, copper, and manganese in barley and sugarcane. J. Plant Nut. 3, 215-233.

Bowen J E 1986 Kinetics of zinc uptake by two rice cultivars. Plant Soil 94, 99-107.

Broda E, Dresser E and Findenegg G 1964 Wirkung von dinitrophenol, azid, und anaerobiose auf die zinkaufnahme durch algen. Naturwissenschaften 51, 361-362.

Cakmak I and Marschner H 1988 Increase in membrane permeability and exudation in roots of zinc-deficient plants. J. Plant Physiol. 132, 356-361

Carroll M D and Loneragan JF 1968 Response of plant species to concentrations of zinc in solution. I. Growth and zinc content of plants. Aust. J. Agric. Res. 19, 859-868.

Chaney R L 1988 Metal speciation and interactions among elements affect trace element transfer in agricultural and environmental food chains. *In* Metal Speciation: Theory, Analysis, and Application. Eds. J R Kramer and H E Allen. pp 219-260. Lewis publisher, Inc, Chelsea, MI

Chaney R L, Bell P F, and Coulombe B A 1989 Screening strategies for improved nutrient uptake and utilization by plants. HortScience 24, 565-572.

Chaudhry F M and Loneragan J F 1972 Zinc absorption by wheat seedlings and the nature of its inhibition by alkaline earth cations. J. Expt. Bot. 23, 552-560.

Claassen N and Barber S A 1974 A method for characterizing the relation between nutrient concentration and flux into roots of intact plants. Plant Physiol. 54, 564-568.

Crowley D E, Reid C P P and Szaniszlo P J 1987 Microbial siderophores as iron sources for plants. *In* Iron Transport in Animals, Plants, and Microorganisms. Eds. G Winkelmann, D Van der Helm and J B Neilands. pp 370-286. VCH Chemie, Weinheim, Germany.

Crowley D E, Wang Y C, Reid, C P P and Szaniszlo P J 1991 Mechanisms of iron acquisition from siderophores by microorganisms and plants. *In* Iron Nutrition and Interactions in Plants. Eds. Y Chen and Y Hadar. pp. 213-232. Kluwer Academic Publishers, Netherlands.

Epstein E and Hagen C E 1952 A kinetic study of the absorption of alkali cations by barley roots. Plant Physiol.

27, 457-474.

Giordano P M, Noggle J C and Mortvedt J J 1974 Zinc uptake by rice as affected by metabolic inhibitors and competing cations. Plant Soil 41, 637-646.

Gutknecht J 1961 Mechanism of radioactive zinc uptake by *Ulva lactuca*. Limnol. Oceanogr. 6, 426-431.

Gutknecht J 1963 ^{65}Zn uptake by benthic marine algae. Limnol. Oceanogr. 8, 31-38.

Hewitt E J 1984 The essential and functional mineral elements. *In* Diagnosis of Mineral Disorders in Plants, Vol 1. Eds. C Bould, E J Hewitt and P Needham. pp 7-53. Chemical Publishing, NY.

Jolley, V D and Brown J C 1991 Factors of iron-stress response mechanism enhanced by Zn-deficiency stress in Sanilac, but not Saginaw navy bean. J. Plant Nutr. 14, 257-265.

Kochian L V 1991 Mechanisms of micronutrient uptake and translocation in plants. *In* Micronutrients in Agriculture, Second Edition. Eds. J J Mortvedt, F R Cox, LM Shuman and R M Welch. pp 229-296. Soil Science Society of America, Madision, WI.

Kochian L V, Norvell W A, Shaff J E and Chaney R L 1991 Characterizing root iron uptake using a ferrous chelate to buffer free Fe^{2+} ion activity in solution. Plant Physiol 96, S142.

Lesuisse E and Labbe P 1992 Iron reduction and trans-plasma membrane electron transfer in the yeast *Saccharomyces cerevisiae*. Plant Physiol. 100, 769-777.

Lindsay W L 1991 Inorganic equilibria affecting micronutrients in soils. *In* Micronutrients in Agriculture, Second Edition. Eds. J J Mortvedt, F R Cox, LM Shuman and R M Welch. pp 89-112. Soil Science Society of America, Madision, WI.

Marschner H 1986 Mineral Nutrition of Higher Plants. Academic Press, NY. 674 p.

Marschner H, Treeby M and Römheld V 1989 Role of root-induced changes in the rhizosphere for iron acquisition in higher plants. Z. Pflanzen. Bodenk. 152, 197-204

Mullins G L and Sommers L E 1986 Cadmium and zinc influx characteristics by intact corn (*Zea mays* L.) seedlings. Plant Soil 96, 153-164.

Nomoto K, Sugiura Y and Takagi S 1987 Mugeneic acids, studies on phytosiderophores *In* Iron Transport in Animals, Plants, and Microorganisms. Eds. G Winkelmann, D Van der Helm and J B Neilands. pp. 401-424. VCH Chemie, Weinheim, FRG.

Norvell W A 1991 Reactions of metal chelates in soils and nutrient solutions. *In* Micronutrients in Agriculture, Second Edition. Eds. J J Mortvedt, F R Cox, LM Shuman and R M Welch. pp 187-227. Soil Science Society of America, Madision, WI.

Norvell W A and Welch R M 1993 Growth and nutrient uptake by barley (*Hordeum vulgare* L. cv. Herta): Studies using an N-(2-Hydroxyethyl)ethylenedinitrilotriacetic acid-buffered nutrient solution technique. I. Zinc ion requirements. Plant Physiol. 101, 619-625.

Parker D R, Aguilera J J and Thomason D N 1993 Zinc-phosphorus interactions in two cultivars of tomato (*Lycopersicon esculentum* L.) grown in chelator-buffered nutrient solutions. Plant Soil 143, 163-177.

Ramani S and Kannan S 1978 Zinc absorption and transport in young peanut seedlings. Commun. Soil Sci. and Plant Anal. 9, 311-316.

Rathore, V S, Wittwer S H, Jyung W H, Bajaj Y P S and Adams M W 1970 Mechanisms of zinc uptake in bean (*Phaseolus vulgaris*) tissues. Physiol. Plant. 23, 908-919.

Römheld V and Marschner H 1986 Evidence for a specific uptake system for iron phytosiderophores in roots of grasses. Plant Physiol. 80, 175-180.

Santa Maria G E and Cogliatti D H 1988 Bidirectional Zn-fluxes and compartmentation in wheat seedling roots. J. Plant Physiol. 132, 312-315.

Schmid W E, Haag H P and Epstein E 1965. Absorption of zinc by excised barley roots. Physiol. Plant. 18, 860-869.

Sposito G and Mattigod S V 1980 GEOCHEM: A computer program for the calculation of chemical equilibria in soil solutions and other natural water systems. Kearney Foundation of Soil Science, Univ. of California, Riverside.

Takagi S 1976 Naturally occurring iron-chelating compounds in oat and rice root washings. I. Activity measurement and preliminary characterization. Soil Sci. Plant Nutr. 22, 423-433.

Tester M 1990 Plant ion channels: whole-cell and single-channel studies. New Phytol. 114, 305-340.

Tinker R B 1981 Levels, distribution and chemical forms of trace elements in food plants. Phil. Trans. Roy Soc., London B294, 41-55.

Treeby M, Marschner H and Römheld V 1989 Mobilization of iron and other micronutrient cations from a calcareous soil by plant-borne, microbial, and synthetic chelators. Plant Soil 114, 217-226.

Veltrup W 1978 Characteristics of zinc uptake by barley roots. Physiol. Plant. 42, 190-194.

Welch R M 1993 Micronutrient nutrition of plants. *In* Critical Reviews in Plant Sciences. Ed. B V Conger. CRC

Press, Boca Raton, FL, (*In Press*).

Welch R M, Webb M J and Loneragan J F 1982 Zinc in membrane function and its role in phosphorous toxicity. *In* Plant Nutrition 1982, Proceedings of the Ninth International Plant Nutrition Colloquium. Ed. A Scaife. pp 710-715. Commonwealth Agricultural Bureau, UK.

Welch R M, Norvell W A, Schaefer S C, Shaff J E and Kochian L V 1993 Induction of iron(III) and copper(II) reduction in pea (*Pisum sativum* L.) by Fe and Cu status: Does the root-cell plasmalemma Fe(III)-chelate reductase perform a general role in regulating cation uptake? Planta (*In Press*).

Williams R F 1948 The effects of phosphorus supply on the rates of intake of phosphorus and nitrogen and upon certain aspects of phosphorus metabolism in graminaceous plants. Aust. J. Sci. Res. 131, 333-361.

Zhang F, Römheld V and Marschner H 1989 Effect of zinc deficiency in wheat on the release of zinc and iron mobilizing root exudates. Z. Pflanzen. Bodenk. 152, 205-210.

Zhang F, Römheld V and Marschner H 1991a Release of zinc mobilizing root exudates in different plant species as affected by zinc nutritional status. J. Plant Nutr. 14, 675-686.

Zhang F, Römheld V and Marschner H 1991b Diurnal rhythm of release of phytosiderophores and uptake rate of zinc in iron-deficient wheat. Soil Sci. Plant Nutr. 37, 671-678.

Chapter 5.

Zinc Uptake from Soils

H. MARSCHNER

1. Abstract

Characterizing zinc availability by soil testing provides important information on the pool size of zinc potentially available for uptake. Concentrations of zinc in soil solution, particularly at high soil pH, however, are very low and mobility and transport to the root surface are usually rate limiting factors of soil supply. Utilization of potentially available zinc is thus mainly or exclusively confined to rhizosphere soil. Root-induced changes in the rhizosphere are of particular imporatance for zinc uptake from soils. In soils of high pH, rhizosphere acidification by supply of ammonium nitrogen, or for legumes by N_2 fixation, are effective mechanisms in enhancing zinc mobilization. The same holds true for rhizosphere acidification or enhanced excretion of organic acids and chelators as root responses to deficiency of phosphorus or iron. Root colonization by VA mycorrhizae increases spatial availability of zinc similarly to that of phosphorus. Mycorrhizal plants usually have higher zinc contents in the shoot dry matter and are less sensitive to zinc deficiency than non-mycorrhizal plants. As a rule, all factors which impair root colonization by VA mycorrhizae, including high levels of soil or fertilizer phosphorus, tend to decrease zinc contents in plants and increase the risk of zinc deficiency in plants grown on soils low in extractable zinc. Marked genotypical differences in zinc efficiency are usually caused by differences in zinc acquisition from soils. In lowland rice zinc deficiency is widespread in neutral and alkaline soils. Elevated bicarbonate concentrations are the major factor responsible for low zinc contents in rice plants grown on high pH soils, high in organic matter. In such soils the high bicarbonate concentrations impair zinc uptake by direct inhibition of root growth and activtiy.

2. Zinc availability in soils

2.1. General

For characterization of the available fraction of zinc in soils DTPA is frequently used as extractant (see Chapter 12). Such extraction methods merely characterize the pool size (capacity, quantity, potential availability) of a "labile" fraction in soils comprising water soluble, exchange- able, adsorbed, chelated, or occluded zinc (Fig. 1). Most of the zinc in soils occurs on surfaces of clays, hydrous oxides, and organic matter, rather than in soil solution (Armour et al., 1990). However, zinc is primarily taken up by roots from the soil solution. The concentration in the soil solution (intensity) and the rates of replenishment are therefore of key importance in determining plant availability (Fig. 1). The DTPA extractable zinc can therefore only estimate the probability of whether soils

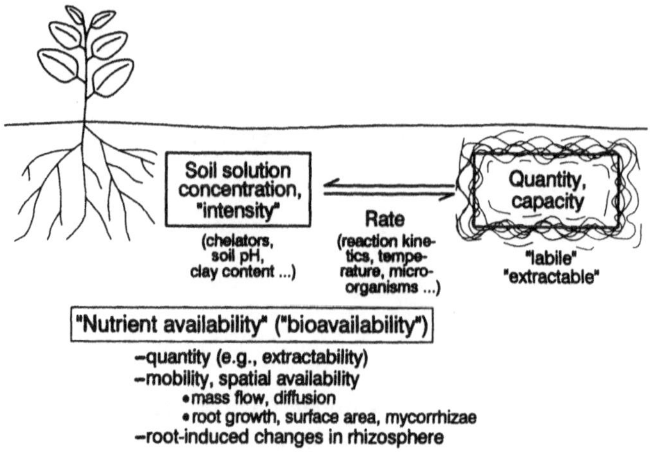

Figure 1. Quantity/intensity ratio of nutrient availability, and factors determining the "bioavailability" of mineral nutrients such as zinc.

can provide sufficient zinc to the roots to meet plant demand. As a working average value 1 mg extractable Zn kg^{-1} would represent a pool size (quantity) of about 3 kg Zn ha^{-1} in the top soil (0-20 cm depth).

2.2. *Plant demand for zinc*

In the shoots, critical deficiency contents are in the range of 10-15 mg Zn kg^{-1} dry wt. in most graminaceous species, and 20-30 mg Zn kg^{-1} dry wt. in most dicotyledenous species. For an annual biomass production of 10 tons ha^{-1} the demand of zinc is therefore in the range of 100-300 g ha^{-1}, which is quite low as compared with the pool size of the critical level of DTPA extractable zinc (~0.3-1.4 mg kg^{-1} soil; Sims and Johnson, 1991).

There are characteristic differences between plant species to zinc deficiency when grown on soils at a given low level of extractable zinc. For example, the likelihood increases that maize, sorghum or soybean become zinc deficient (Adriano, 1986), or that the responses to zinc fertilizer increase in the order maize < soybean < flax < bean, whereas wheat is not responsive (Moraghan, 1984). Many fruit trees such as apples, peaches and citrus are particularly sensitive to zinc deficiency when grown on soils low in zinc. In all these instances the differences between plant species in sensitivity to zinc deficiency are not related to differences in demand in the shoot tissue. Thus, for a given pool size in the soil the accessibility to the plants, the "bioavailability" (Barber, 1984) varies among plant species as well as genotypes within a species (see Chapter 8).

3. Bioavailability of zinc

3.1. *Supply by mass flow and diffusion*

For zinc as for any other mineral nutrient, in addition to the quantity, the concentration in the soil solution and transport to the root surface are of key importance for uptake. Root growth and surface area are also major factors since they determine distances between "source" (quantity, e.g. exchangeable fraction) and "sink", the root surface (Fig. 1). As most zinc is bound to the soil matrix, the concentration in soil solution is usually very low and thus also the supply by mass flow to the roots. Data on zinc concentrations in soil solutions are scarce. In non-polluted soils the concentrations are in the range of 3×10^{-8} - $5\times10^{-7}M$ (Barber, 1984). Zinc concentration in soil solution is highly dependent on soil pH, decreasing to very low levels at high soil pH (Jeffery and Uren, 1983; Brümmer et al., 1986). For a given soil the zinc concentration in the soil solution of the rooting zone may also follow a distinct seasonal pattern increasing from ~5×10^{-8} M in spring to ~2×10^{-7} in summer (McGrath et al., 1988; Sinclair et al., 1990). In soil solutions most of the zinc is present as free metal ion and as labile complex (Jeffery and Uren, 1983). The proportion of free metal ion (15-30%) seems to be rather independent of soil organic matter and dissolved organic carbon (McGrath et al., 1988).

Because of the low concentration of zinc in the soil solution, supply by mass flow to the roots accounts only for a minor fraction of plant demand. Assuming a transpiration coefficient of 300 L kg^{-1} dry matter and a zinc concentration in the bulk soil solution of $10^{-7}M$, about 2 mg zinc can be supplied by mass flow as compared with a demand of 10-30 mg Zn kg^{-1} dry wt. As in most soils, particularly calcareous soils, the zinc concentration in the soil solution is probably an order of magnitude lower (i.e. in the range of $10^{-8}M$), supply by mass flow is clearly of minor importance.

Supply of zinc to the roots is mainly confined to diffusion and thus, to a zone around the root which does not extend beyond the root hair cylinder. In this respect the situation is similar to that for phosphorus (Barber, 1984; Jungk and Claassen, 1989; Jungk, 1991). Depletion of zinc around roots, typical for supply by diffusion, has been demonstrated for wheat using ^{65}Zn and autoradiographic methods (Wilkinson et al., 1968). In agreement with this, and irrespective of soil pH, a depletion of zinc in the water extractable fraction has been found in the rhizosphere soil from oat plants as compared with the bulk soil (Table 1). This depletion was confined to a "labile" fraction, however, not the total water extractable fraction.

Table 1. Concentrations (Mol x 10^{-9}) of total and labile zinc in water extracts of bulk and rhizosphere soil of oat plants grown in soils with different pH*

	Soil pH (soil-water mixture, 1:1 ratio)		
	4.37	5.78	6.44
Bulk soil			
total zinc	1652	965	637
labile zinc	471	372	127
Rhizosphere soil			
total zinc	1516	1092	773
labile zinc	185	111	40

*From Sarong et al., (1989)

Based on studies in flowing nutrient solutions using different plant species, the adequate range of zinc concentrations in the rhizosphere (soil) solution have been shown to be in the range between ~$6\times10^{-8}M$ and ~$8\times10^{-6}M$ (Asher, 1987). These are concentrations much above those to be expected in the bulk soil solutions in most soils, and high pH soils in particular. By using chelate-buffered solutions, free Zn^{2+} activities in the solution in the range between 10^{10} and $10^{11}M$ were found to be adequate for growth of barley (Laurie et al., 1991) and tomato (Parker et al., 1992). However, such extremely low adequate concentrations of Zn^{2+} required a simultaneous excess of about 100 μM zinc chelate as buffer at the plasma membrane of the root cells, i.e. an unlimited pool size for replenishment of Zn^{2+} at the plasma membrane. In soil-grown plants a buffer of such size does not exist, and total zinc concentrations (free Zn^{2+} and chelated zinc) are at least three orders of magnitude lower. Critical deficiency, or sufficiency concentrations obtained in chelator buffered nutrient solutions cannot therefore be applied to soil-grown plants.

3.2. Role of root growth

For zinc, bioavailability is limited mainly by low mobility in the soil solution and thus low spatial availability. Accordingly, root growth and surface area are important parameter for bioavailability of zinc. For a given amount of DTPA extractable zinc, for example 0.5 mg kg^{1} soil, the spatial accessibility of this fraction increases about twofold by doubling the root surface area. In top soils with average root length densities of 2-4 cm cm^{3}, about 10% of the soil might be expected to supply zinc to the roots by diffusion, which brings the 0.5- 1.0 mg DTPA zinc as (=~ 3 kg Zn ha^{1}) critical level closer to the amount of zinc supplied to roots of soil-grown plants. In the long list of environmental factors in general and soil factors in particular which are known to influence the correlations between extractable soil zinc and plant uptake (Lindsay and Cox, 1985), root growth and surface area should also therefore be considered.

Figure 2. Schematic presentation of main soil and plant factors decreasing or increasing zinc availability and uptake by soil-grown plants.

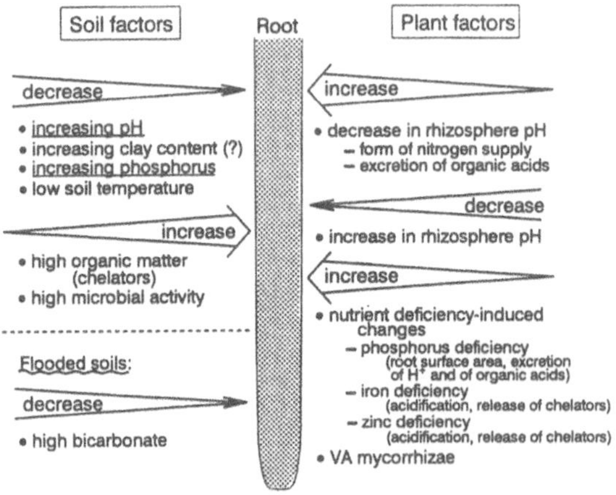

3.3. Soil factors affecting zinc supply to, and uptake by roots

Of the soil factors, pH has the most marked influence on plant availability of zinc (Fig. 2). In the pH range of 5.5-7.0 the equilibrium concentration of zinc may decrease 30 to 45 times for each unit increase in soil pH (Barber, 1984; Moraghan and Mascagni, 1991). Diffusion coefficients for zinc in calcareous soils may therefore be about 50 times lower than in acid soils, and liming of acid soils leads to a decrease in diffusion coefficient of zinc similar to values found in calcareous soils (Moraghan and Mascagni, 1991). Liming acid soils sharply decreases zinc contents in plants (Grove and Sumner, 1985; Parker and Walker, 1986) with the risk of inducing zinc deficiency in plants grown in highly weathered tropical soils low in zinc (Duguma et al., 1988). Despite the sharp decrease in zinc content in plants, the DTPA extractable zinc in soils is often not, or only slightly decreased (Lins and Cox, 1988), suggesting that the "rate" and "intensity" of zinc (Fig. 1) limit the supply to the root surface. On the other hand, liming is an effective procedure to decrease zinc uptake and avoid zinc toxicity in plants grown on zinc polluted soils (Lee and Craddock, 1969; El-Kherbawy et al., 1989).

Soil extraction methods respond less sensitively to changes in soil pH than the concentrations of zinc in soil solution and thus also the diffusion coefficient and plant uptake. Correlations between extractable zinc and plant uptake can thus be very much improved by inclusion of soil pH (Lins and Cox, 1988; Brennan and Gartrell, 1990) into the calculations. Consideration of soil organic matter content is also a factor which improves correlations (Brennan, 1992a). Soil organic matter may increase the diffusion rate of zinc in soils (Sharma and Deb, 1988), for example, by desorption of zinc and formation of soluble complexes, thereby increasing release, concentration and thus supply rate to the root surface (Fig. 1). The same is certainly true for microbial activity in the bulk soil and the rhizosphere in particular. The rise in zinc concentration in the soil solution that occurs with increased soil temperature is suggested to be an expression of enhanced mobilization of zinc from insoluble forms by biologically produced chelators (Linehan et al., 1989).

The interactions between phosphorus content in soils and zinc availability to plants are highly complex (see Chapter 9). Generally, with increase in soil content or supply of fertilizer phosphorus, plant uptake of zinc decreases more or less sharply and often beyond a level which can be attributed to dilution effects due to growth enhancement. In contrast, extractable zinc in soils is either not or only slightly decreased by high phosphorus supply. In acid tropical soils the risk of phosphorus-induced zinc deficiency increases when phosphorus fertilizer application is combined with liming (Friesen et al., 1980). Although liming overcomes aluminum toxicity and thereby increases root growth, the sharp decrease in zinc concentrations in the soil solution combined with enhanced shoot growth and thus zinc demand, requires addition supply of zinc fertilizer to prevent a reduction in growth with high phosphorus and lime supply (Friesen et al., 1980).

Table 2. Influence of phosphorus supply and soil temperature on shoot growth and phosphorus and zinc contents in the shoots of flax (*Linum usitatissimum* L.)*

Treatment** ($mg\ kg^{-1}$ soil)		Dry weight ($g\ plant^{-1}$)			Contents in the shoot dry matter P (%)			Zn ($mg\ kg^{-1}$)		
P	Zn	7°	15°	24°	7°	15°	24°	7°	15°	24°
0	2	0.36	2.71	7.13	0.05	0.19	0.21	14.9	23.0	29.3
40	2	0.92	5.40	9.73	0.09	0.43	0.48	11.7	18.3	14.7
80	2	1.28	6.25	10.25	0.20	0.46	0.54	9.7	15.7	16.3

*Based on Moraghan (1980)
**Soil pH 8.1; 0.6 mg Zn kg^{-1} (DTPA extr.).

As the soil water content falls, diffusion coefficients of ions in soils steeply decline. These relationships have not been investigated for zinc but extensively studied with phosphorus (Jungk, 1991) where the same principles should apply (Barber, 1984) and deductions may be made for zinc even though experimental data are lacking. Contrary to expectation, uptake of significant amounts of zinc has been shown by roots growing through a layer of soil drier than the wilting point, provided that the roots had access to water elsewhere (e.g. in the subsoil; Nambiar, 1976a). In the dry soil more mucilage is released by the roots in response to mechanical impedance, and this probably facilitates zinc transport at the soil-root interface (Nambiar, 1976b).

For soils relatively low in extractable zinc, low soil temperatures often enhance incidence and severity of zinc deficiency symptoms, for example, in maize, bean and potato (Moraghan and Mascagni, 1991). Increasing phosphorus supply increases the likelihood of low temperature-induced zinc deficiency as shown in Table 2 for flax. At low temperatures zinc uptake from soils is not specifically impaired and the same holds true for phosphorus. Impaired root activity and growth are most likely the main factors involved, but presumably a lower root colonization with VA mycorrhizae (see below) also plays a part. Inhibited zinc transport from roots to shoots at low temperatures may be a contributing factor (Schwartz et al., 1987).

When phosphorus fertilizer is applied phosphorus uptake and shoot growth increase leading to further decrease in the zinc contents in the shoots mainly by dilution (Table 2). At higher temperatures, however, the decrease in zinc content exceeds the dilution effect and this can probably be attributed to a decrease in VAM colonization of the roots and root surface area. Decrease in root growth, and in root-shoot dry weight ratio in particular are typical plant responses to improved phosphorus nutritional status (Bell et al., 1989). Thus, in the responsive range zinc demand of plants increases but not the spatial availability of zinc in soils which may even decrease. In addition, mobilization of zinc in the rhizosphere often declines with improvement of the phosphorus nutritional status of plants.

4. Concentration and dynamics of zinc in the rhizosphere

4.1. General

Conditions in the rhizosphere differ in many respects from those in the bulk soil, as for example, in pH, the presence of root exudates, and microbial activity. In many instances root-induced changes in the rhizosphere are plant responses to mineral nutrient deficiency, and those caused by phosphorus or iron deficiency are also of particular relevance to the dynamics and mobilization of zinc in the rhizosphere. The importance of root-induced changes in the rhizosphere for availability of micronutrients has been summarized elsewhere (Marschner, 1991), and the changes leading to increase in availability and plant uptake of zinc are shown schematically in Fig. 2.

4.2. Root induced changes in pH

Rhizosphere pH may differ from that of the bulk soil by up to two units, depending on plant and soil factors. The most important factors for root-induced changes in rhizosphere pH are imbalance in cation/anion uptake ratio and the corresponding differences in net excretion of H^+ and HCO_3^- (or OH^-), excretion of organic acids, and indirectly, microbial acid production from root release of organic carbon, and enhanced CO_2 production.

The form of nitrogen supply has the most prominent influence on cation/anion uptake ratio, and thus on rhizosphere pH (Marschner and Römheld, 1983). In neutral and alkaline soils rhizosphere acidification in plants fed with ammonium can enhance not only mobilization and uptake of phosphorus from sparingly soluble calcium phosphates but also of micronutrients such as zinc. For example, in a Luvisol (pH 6.8) rhizosphere pH of bean plants was increased to 7.3 in nitrate fed plants and decreased to 5.4 in ammonium fed plants, and the zinc contents increased from 34 to 49 mg kg^{-1} shoot dry matter in the ammonium fed plants (Marschner et al., 1989; Marschner, 1991). When in ammonium fed bean plants nitrification is inhibited by addition of N-Serve, rhizosphere pH of this soil is even further decreased to 4.5 (Thomson et al., 1993). The corresponding uptake rates of zinc per unit root length increased from 7 $\mu g\ m^{-1}$ in nitrate fed plants to 13 $\mu g\ m^{-1}$ in ammonium fed plants, and to 19 $\mu g\ m^{-1}$ in ammonium fed plants + N-Serve (Thomson et al., 1993).

Increase in rhizosphere pH in nitrate fed plants decreases both the concentration and mobility of zinc in the soil solution. In nitrate fed plants this increase in rhizosphere pH is particular prominent in plant species which reduce nitrate predominantly or exclusively in the roots, as the case for many perennial species. Both the high susceptibility of many fruit trees to zinc deficiency, and their high requirement for extractable zinc in soils, are presumably causally related to increa sed rhizosphere pH.

The effect of NH_4 nutrition in bringing about rhizosphere acidification enhances zinc availability for plants growing in soil at high bulk soil pH and in soils low in extractable zinc. In soils with high contents of extractable zinc, however, ammonium supply may induce zinc uptake in excess of zinc demand and result in growth depression by zinc toxicity. This is true in particular for plants which have an inherently high efficiency in zinc acquisition, as for example buckwheat (Rothmeier and Amberger, 1985; Marschner, 1991).

Rhizosphere acidification is an inherent property of certain plant species such as chickpea (Marschner and Römheld, 1983) even when supplied with nitrate-nitrogen (Fig. 3). This acidification is related to some extent to high cation/anion uptake ratio even with nitrate supply, but enhanced excretion of organic acids accounts for most of this rhizosphere acidification (see below).

When relying on symbiotic N_2 fixation rather than nitrate nutrition, the cation/anion uptake ratio increases, although somewhat less than with ammonium nutrition (Raven et al., 1991). Accordingly, rhizosphere pH of symbiotic legumes lies between that for ammonium fed and nitrate fed plants (Fig. 3). It can be predicted therefore that, compared with nitrate fed legumes the uptake of zinc by N_2 fixing legumes should be higher as is known for the uptake of iron and manganese, for example, in soybean (Wallace, 1982).

4.3. Root exudation of organic acids and chelators

At least in dicots root-induced rhizosphere acidification is also a widespread phenomena of plant responses to phosphorus deficiency, either due to increase in cation/anion uptake ratio, for example in sunflower (Hoffland et al., 1989a) or enhanced excretion of organic acids, for example in rape (Hoffland et al., 1989b) and in various

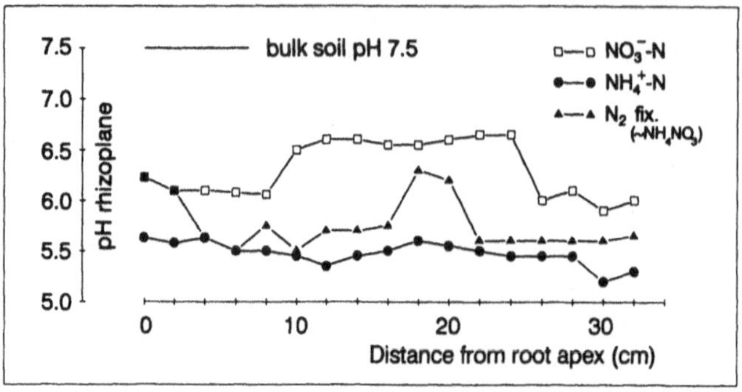

Figure 3. Effect of form of nitrogen supply on rhizoplane pH of chickpea (*Cicer arietinum*) grown in a calcareous soil of pH 7.5 (Dinkelaker and Marschner, unpublished).

legume species (Lipton et al., 1987; Ohwaki and Hirata, 1992). A particular root response to phosphorus deficiency exists in cluster-rooted plants such as members of the Proteaceae (e.g. *Banksia* spp; Grierson and Attiwill, 1989) and in annual species such as *Lupinus albus* (Gardner et al., 1983). In the zones of root clusters, the so-called proteoid roots, large amounts of citric acid are excreted by *Lupinus albus*, leading to sharp decline in rhizosphere pH, even in highly calcareous soils (Dinkelaker et al., 1989). Besides mobilization of sparingly soluble calcium phosphates, as a side-effect large amounts of iron, manganese and zinc are also mobilized in the rhizosphere soil. Supply of soluble phosphorus drastically depresses proteoid root formation as well as the uptake of iron (Marschner et al., 1987) and of manganese and zinc. When grown on calcareous soil a supply of soluble phosphorus decreases manganese contents in the shoots of *Lupinus albus* from 4970 to 833 $\mu g\ g^{-1}$ dry wt and of zinc from 30 to 16 $\mu g\ g^{-1}$ dry wt. Enhanced mobilization of zinc by citric acid is also to be expected in acid soils by desorption of zinc and the formation of negatively charged complexes (Chairidchai and Ritchie, 1993).

Enhanced rhizosphere acidification by an increase in net proton excretion, together with stimulation of reducing capacity of the roots, often associated with enhanced release of phenolics from the roots, are typical root responses to iron deficiency in dicots and monocots, with the exception of the Graminaceae (Römheld, 1987). This root response not only enhances mobilization of iron in the rhizosphere and uptake by plants but also of manganese (Marschner et al., 1986). In calcareous soils low in both available iron and manganese this response to iron deficiency may thus prevent manganese deficiency (Moraghan, 1985), whereas in calcareous soils with low iron and high manganese availability, excessive amounts of manganese may be mobilized in the rhizosphere and manganese toxicity may occur (Moraghan, 1979). Reports on a corresponding iron deficiency-induced inc toxicity are rare (Gupta and Chipman, 1976), but iron deficiency-induced enhancement of zinc mobilization in the rhizosphere and uptake of zinc are probably more widespread on calcareous soils than considered at present.

Graminaceous species respond to iron deficiency by the release of non-proteinogenic amino acids, so-called phytosiderophores (Römheld, 1987; Takagi et al., 1988). These compounds form stable chelates not only with iron (FeIII) but also with zinc and copper and are as effective as DTPA in mobilizing zinc from calcareous soils (Table 3). The response of graminaceous species grown in calcareous soils low in extractable iron is thus one which might mobilize not only iron but also other micronutrients thereby

Table 3 Mobilization of micronutrients from a calcareous soil by DTPA ($10^{-5}M$) and phytosiderophores ($10^{-5}M$, from iron deficient barley plants)*

Chelator	Mobilization (nmoles g^{-1} soil) Fe	Cu	Zn
DTPA	292	173	656
Phytosiderophores	577	73	699

*Treeby et al.(1989)

increasing availability and, for example, enhances the zinc nutritional status of plants. A range of root-induced changes in the rhizosphere also occur in response to zinc deficiency. In zinc deficient dicots rhizosphere acidification occurs even when nitrate is the form of N supply. This acidification is brought about by an increase in the cation/anion uptake ratio (Cakmak and Marschner, 1990). Simultaneously, membrane permeability of the root cells is increased with the efflux of low-molecular-weight organic solutes such as sugars, amino acids and phenolics (Cakmak and Marschner, 1988). However, only the root exudates of graminaceous species are effective in enhancing the mobilization of zinc either from a zinc-loaded resin (Table 4) or from calcareous soils. The effective compounds in root exudates of zinc deficient graminaceous species which bring about this mobilization are the same phytosiderophores released under iron deficiency (Zhang et al., 1989; 1991). Whilst this zinc deficiency-induced increase in the release of phytosiderophores and other organic solutes might be an effective mechanism of mobilization of zinc and other micronutrients in the rhizosphere, it seems also to increase the risk of infection of the roots by soil-borne pathogens such as *Gaeumannomyces graminis* (take-all) in wheat (Brennan, 1992b).

Table 4 Effect of zinc nutritional status of cotton and wheat plants on root exudation and zinc mobilization from a zinc-loaded resin by root exudates*

Zinc nutritional status (Zn supply)	Amino acids	Root exudates Sugars ($\mu g\ g^{-1}$ root dry wt h^{-1})	Phenolics	Zn mobilization by root exudates ($\mu mol\ g^{-1}$ root dry wt 4 h^{-1})
Cotton: + Zn	32	250	78	0.6
- Zn	110	500	107	0.6
Wheat: + Zn	21	315	34	0.4
- Zn	48	615	80	4.9

*Marschner et al. (1990)

Besides the root-induced changes which can directly increase solubility and mobility of zinc in the rhizosphere, the associated presence of a much higher activity of non-infecting rhizosphere microorganisms may indirectly affect solubility and mobility of zinc, for example by the production of chelators. A reflection of such a role of rhizosphere microorganisms might be the raised zinc concentration in the soil solution in the rooting zone of barley during rapid vegetative growth (Linehan et al., 1989) or the increase in complexing properties of soil-extracts for zinc from 9.9% from fallow soil to 16% in soil extracts after six weeks growth of maize (Merckx et al., 1986).

4.4 Zinc efficiency

When plants are grown on soils low in extractable zinc, marked genotypical differences can be observed in the incidence of zinc deficiency symptoms and the depression of growth. The mechanisms responsible for differences in zinc efficiency are not fully understood. In most instances, efficient species and genotypes are characterized by a greater zinc acquisition from soils (see Chapter 8). Differences in efficiency between plant species are probably related to inherent differences in rhizosphere pH, root exudation or root colonization by VA mycorrhizae (see below). Within a plant species interesting relationships occur between zinc efficiency and iron deficiency-induced root responses. Under zinc deficiency in *Phaseolus vulgaris* the roots of the zinc-inefficient cultivar Salinac have a higher reducing capacity and take up more iron than the zinc-efficient cultivar Saginaw (Brown, 1979; Jolley and Brown, 1991). In graminaceous species zinc and iron efficiency may either be associated for a particular cultivar as has been found in oat (Brown and McDaniel, 1978) or not associated as has been shown in wheat for the cultivars Aroona and Durati (Cakmak et al., 1993). The zinc inefficient Durati grows poorly on soils low in zinc, in contrast to the zinc efficient Aroona (Graham et al., 1992). In agreement with this, under zinc deficiency, release of phytosiderophores is much higher in Aroona than in Durati, whereas, under iron deficiency the release of phytosiderophores is similar in both cultivars (Cakmak et al., 1993).

4.5 Role of vesicular-arbuscular (VA) mycorrhizae in zinc uptake

The most distinct effects of VA mycorrhizae on growth enhancement occur by improving the supply of mineral nutrients of low mobility in the soil, in particular phosphorus. The same mycorrhizal effect may also be achieved with the micronutrients zinc and copper. A large number of examples occur in the literature demonstrating this beneficial effect of VA mycorrhizae on zinc uptake by the host plant. The effects are usually reported as being more distinct in soils low in extractable zinc, or low zinc mobility (e.g. high soil pH), and in plants with coarse root systems, for example fruit trees such as peaches (Gilmore, 1971), citrus (Menge et al., 1982) and apple (Runjin, 1989), or in tree legumes such as *Leucaena leucephala* (Manjunath and Habte, 1988). In graminaceous species too such as maize and wheat grown in soils with low zinc contents, VA mycorrhizae increase zinc uptake and shoot contents in the dry matter, despite an increase in shoot biomass (Swamvinathan and Verma, 1979; Faber et al., 1990). This enhancement of zinc uptake is not confined to pot experiments but also occurs under field conditions, for example in barley (Jakobsen, 1983). Liming acid soils depresses uptake of zinc in both non-mycorrhizal and VA mycorrhizal plants (Table 5). However, in mycorrhizal plants the zinc contents are always higher and thus, there is less risk in mycorrhizal plants of lime-induced zinc deficiency. As a rule VA mycorrhizae also increase copper contents in plants but depress manganese contents (Pacovsky, 1986; Kothari et al., 1991a) usually much more than the results shown in Table 5.

The external hyphae of VA mycorrhizae can absorb and translocate zinc, similar to phosphorus, to the host from outside the rhizocylinder. The capacity of the external

Table 5 Effect of soil pH (liming) and VAM inoculation (Glomus macrocarpum) on shoot growth and contents of micronutrients in teff (*Eragrostis* tef) plants*

Soil pH	VAM inoc.	Shoot dry wt (g pot^{-1})	Contents in the shoot dry matter (mg kg^{-1}) Zn	Cu	Fe	Mn
5.49	-VAM	1.15	47.0	12.6	82	133
	+VAM	1.35	57.7	15.3	95	114
6.91	-VAM	1.25	14.7	13.6	87	86
	+VAM	1.36	22.9	16.8	88	80

*Tekalign Mamo and Killham (1987)

hyphae for delivery of zinc and copper is high and may account for 50-60% of the total uptake in white clover, and 25% in maize (Fig. 4).

This capacity of the external hyphae of VA mycorrhizae to supply a relatively large proportion of the plant demand of zinc has implications for the zinc nutrition of plants when phosphorus fertilizers are applied. In low phosphorus soils with application of phosphorus fertilizer in non-mycorrhizal plants growth is usually distinctly enhanced as well as the phosphorus content in the shoot dry matter, whereas zinc contents decrease, for example in soybean (Lambert and Weidensaul, 1991) or maize (Table 6). Growth enhancement and corresponding dilution effects can often account for the decrease in zinc content. However, in VA mycorrhizal plants, despite of a further growth enhancement both phosphorus and zinc contents increase in the shoot dry matter, reflecting the importance of VA mycorrhizae in providing zinc and phosphorus to the host plant (Table 6).

As a rule at high soil contents of phosphorus, or levels of fertilizer phosphorus, root colonization by VA mycorrhizae decreases (Table 7). For phosphorus the decrease in VA mycorrhizal root colonization and delivery by the external hyphae is compensated by higher root uptake due to elevated soil solution concentrations and diffusion rates. However, for zinc such a compensation does not occur unless zinc is simultaneously supplied. On soils with low zinc contents a depression in mycorrhizal activity in roots is therefore often associated with decrease in zinc contents in the shoots which exceed values that can be explained by dilution (Singh et al., 1986). The decrease in zinc contents

Table 6 Growth and shoot contents of phosphorus and zinc in non-mycorrhizal and VA mycorrhizal (*Glomus etunicatum*) maize grown in an Oxisol*

Treatment P-fertilizer	VAM inoc.	Shoot dry wt dry matter (g plant^{-1})	Contents in the shoot dry matter P (g kg^{-1})	Zn (mg kg^{-1})	Zn (μg plant^{-1})
-P	-VAM	1.75	1.0	107	187
+P	-VAM	5.05	2.3	69	348
+P	+VAM	6.02	3.2	108	650

*Eiwazi and Weir (1989)

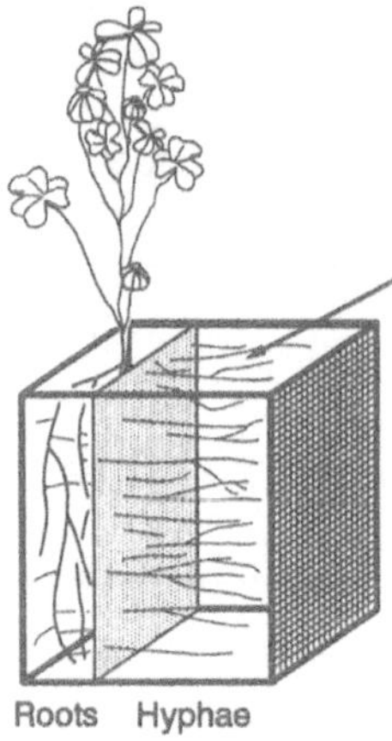

Estimated contribution of extraradical hyphae (% of shoot content)

	Clover	Maize
P	79	20
Zn	50	25
Cu	60	25

Figure 4. Contribution of extraradical mycorrhizal hyphae (*Glomus mosseae*) to the uptake of phosphorus, zinc, and copper in white clover and maize plants grown in compartmented boxes (Compiled data of Kothari et al., 1991b; Li et al., 1991).

may be less distinct or absent when sparingly soluble phosphate fertilizers are supplied, and phosphate fertilizers which contain substantial amounts of zinc as byproduct (Pairunan et al., 1980; Brennan and Gartrell, 1990).

Many examples from pot and field experiments in the last decade suggest that in VA mycorrhizal host plants low root colonization by VA mycorrhizae is often a major factor responsible for low zinc contents in plants. There are, however, a few reports where such a role of VA mycorrhizae was not observed (Lu and Miller, 1989; Weber et al., 1993). As a rule, suppression of VA mycorrhizal root colonization by high levels of soil or fertilizer phosphorus, and also by certain crop rotations increase the likelihood of zinc deficiency on soils low in extractable zinc (Thompson, 1990). In contrast to maize, a previous crop of sugar beet (i.e. a non-host plant) is a known factor increasing the likelihood of zinc deficiency in subsequent crop such as bean (Table 8). When compared with maize as previous crop, sugar beet behaves similarly to fallow, both treatments are associated with lower zinc contents in the shoots and total uptake of zinc in bean. For this experiment soil DTPA extractable zinc was in the range of 0.42-0.65 mg kg^{-1} and not significantly different following sugar beet, fallow or maize (Leggett and Westermann, 1986). Absence of zinc deficiency symptoms in wheat growing on soils with low contents of extractable zinc (0.29 mg kg^{-1}) can be attributed to a large extent to zinc delivery by

Table 7 Residual effect of phosphorus application in 1979 on shoot contents of phosphorus and zinc, and on root infection with VAM in wheat grown in 1984*

Phosphorus application (kg P ha^{-1})	Contents in the shoot dry matter		VAM root infection (%)
	P (g kg^{-1})	Zn (mg kg^{-1})	
0	1.82	34.3	~38
80	2.18	18.4	~20
160	2.69	16.2	~15

*Based on Singh et al. (1986)

Table 8 Growth, zinc content and zinc uptake in bean (*Phaseolus vulgaris* L. cv Sanilac) as affected by previous crops and zinc fertilization*

Previous crops	Zn applied ($kg\ ha^{-1}$)	Dry wt** ($kg\ ha^{-1}$)	Zn content** ($mg\ kg^{-1}$ dry wt)	Zn uptake** ($g\ ha^{-1}$)
Fallow	0	61	13.0	0.83
	11.2	125	13.2	1.67
Sugar beet	0	106	15.0	1.59
	11.2	144	16.2	2.32
Maize	0	163	20.7	3.38
	11.2	194	22.7	4.37

*Based on Leggett and Westermann (1986)
**Developmental growth stage V4

VAM mycorrhizae to the host plant (Thompson, 1990). In principle the same holds true for citrus, only at a generally higher level of extractable zinc (Menge et al., 1982).

Increasing the proportion of non-host plants and length of periods that the soil lies fallow brings about a decline in the VA mycorrhizal infection potential in the soil which may favour root infection with soil-borne pathogens (take-all) in wheat (Thompson and Wildermuth, 1988). Such root infection by pathogens further impairs zinc acquisition from soils.

Various lists on "mycorrhizal dependency" for crop and pasture species have been drawn up depending on how these species benefit by enhanced phosphorus acquisition from low phosphorus soils (Howeler et al., 1987; Khasa et al., 1992; Schweiger et al., 1993). As a first approximation, on soils with low zinc contents and high pH soils in particular, these lists on mycorrhizal dependency may also provide useful information about potential beneficial effects of VA mycorrhiza on zinc nutrition of the host plants.

The effectiveness of VA mycorrhizae in acquisition and delivery of zinc seems to be retained also when plants are grown at high zinc supply (Dueck et al., 1986; Schuepp et al., 1987), at least when the soil pH is also high (El-Kherbawy et al., 1989). It is therefore to be expected that in absence of distinct stimulatory effects on growth in plants grown on zinc-polluted soils VA mycorrhizae may further increase the zinc content of plants and the risk of zinc toxicity.

For ectomycorrhizae precise information is lacking on their effects on zinc uptake and shoot contents of the host plants grown on soils with low zinc contents. However, at high zinc content in the substrate certain ectomycorrhizal fungi can effectively bind heavy metals such as zinc in their extramatrical hyphae (Denny and Wilkins, 1987) and thereby markedly decrease the zinc content in the shoot of the host plant (Table 9). However, this capacity for selective binding of zinc in the hyphae differs much between fungal species, despite a similar fungal biomass.

5. Zinc availability in flooded soils

Zinc deficiency is widespread in flooded rice, and the zinc contents in the plants are

Table 9 Shoot and root contents of zinc in non-mycorrhizal and ectomycorrhizal seedlings of *Pinus sylvestris* supplied with high zinc concentrations*

Treatment (fungus)	Shoot dry wt (g plant^{-1})	Shoot zinc content (mg plant^{-1})	Shoot zinc content (μg g^{-1} dry wt)	Fungal biomass (% of short roots)	Short root Zn content (μg g^{-1} dry wt)
Non-mycorrhizal	16.2	3.14	197	-	273
Paxillus involutus	14.3	1.52	106	54	708
Thelephora terrestris	16.5	3.89	240	66	309

*Based on Colpaert and Van Assche (1992)

often poorly correlated with the contents of extractable zinc in soils. Adsorption of zinc on surfaces of hydrated iron and manganese oxides in the oxygenated rhizosphere of rice roots is probably an important factor in decreasing the concentration and mobility of zinc in the rhizosphere of flooded rice (Singh and Ballu, 1983). Zinc deficiency is usually more distinct in soils with high pH and soils with high content of organic matter (Forno et al., 1975a; Moraghan and Mascagni, 1991) or when organic substances are applied (Mandal et al., 1988).

In neutral and alkaline soils there is strong negative correlation between soil pH and rice yield when no zinc fertilizer is applied (Table 10). As in the pH range of 6.5 to 8.0 the content of DTPA extractable zinc in paddy soils only slightly declines (Sedberry et al., 1980), the drastic decrease in zinc uptake in plants without zinc fertilizer supply (Table 10) has to be attributed to other factors, mainly elevated levels of bicarbonate (HCO_3^-). High bicarbonate concentrations (15-40m*M*) strongly inhibit zinc uptake by rice roots and particularly transport to the shoots (Forno et al., 1975a; Dogar and Hai, 1980).

Rice cultivars differ much in their sensitivity to zinc deficiency, especially when growing in soils of high pH. It had been suggested that genotypical differences in susceptibility to bicarbonate are the responsible factor (Forno et al., 1975b). In zinc inefficient cultivars grown in high bicarbonate concentrations, as well as low root zone

Table 10 Influence of soil pH and zinc supply on grain yield and zinc content in leaves (mid tillering) in flooded rice*

Treatment Soil pH	kg Zn ha^{-1}	Grain yield (kg ha^{-1})	Zinc content in leaves (mg kg^{-1} dry wt)
6.8	0	5934	9
	1.9	7212	17
7.3	0	5265	9
	1.9	6171	18
7.7	0	2788	10
	1.9	6637	14

*Based on Sedberry et al. (1988)

temperature, the contents in the shoot dry matter not only of zinc, but also iron and manganese are lower as compared with the efficient cultivars, or high root zone temperatures (Yang et al., 1993a). Thus, zinc efficiency in flooded rice is causally related to high tolerance to elevated bicarbonate concentrations. Even concentration of 5-10 m*M* bicarbonate inhibit root growth of the zinc inefficient cultivar, but slightly stimulate root growth of the efficient cultivar. Close positive correlations exist between accumulation of organic acids in the roots and inhibition of root growth by high bicarbonate concentrations (Yang et al., 1993b).

Acknowledgement

The author is grateful to E. A. Kirkby for critical reading of the manuscript and correcting the English text.

References

Adriano, D.C.: Trace Elements in the Terrestrial Environment. Springer-Verlag, New York (1986)

Armour, J.D.; G.S.P. Ritchie and A.D. Robson: Extractable zinc in particle size fractions of soils from Western Australia and Queensland. Aust. J. Soil Sci. 28, 387-397 (1990)

Asher, C.J.: Effects of nutrient concentration in the rhizosphere on plant growth. Proc. XIIIth Congress Int. Soc. Soil Sci. (ISSS) Hamburg, August 1986, Symposium Vol. 5, pp. 209-216 (1987)

Barber, S.A.: Soil Nutrient Bioavailability. A Mechanistic Approach. John Wiley and Sons, New York (1984)

Bell, M.J.; K.J. Middleton and J.P. Thompson: Effects of vesicular-arbuscular mycorrhizae on growth and phosphorus and zinc nutrition of peanut (*Arachis hypogaea* L.) in an Oxisol from subtropical Australia. Plant and Soil 117, 49-57 (1989)

Brennan,R.F.: The relationship between critical concentrations of DTPA-extractable zinc from the soil for wheat production and properties of south-western Australian soils responsive to applied zinc. Commun. Soil Sci. Plant Anal. 23, 747-759 (1992a)

Brennan, R.F.: The effect of zinc fertilizer on take-all and grain yield of wheat grown on zinc-deficient soils of the Esperance region, Western Australia. Fert. Res. 31, 215-219 (1992b)

Brennan, R.F. and J.W. Gartrell: Reaction of zinc with soil affecting its availability to subterranean clover. I. The relationship between critical concentrations of extractable zinc and properties of Australian soils responsive to applied zinc. Aust. J. Soil Res. 28, 293-302 (1990)

Brown, J.C.: Effect of zinc stress or factors affecting iron uptake in navy bean. J. Plant Nutr. 1, 171-183 (1979)

Brown, J.C. and M.E. McDaniel: Factors associated with differential response of two oat cultivars to zinc and copper stress. Crop Science 18, 817-820 (1978)

Bruemmer, G.W.; J. Gerth and U. Herms: Heavy metal species, mobility and availability in soils. Z. Pflanzenernähr. Bodenk. 149, 382-398 (1986)

Cakmak, I. and H. Marschner: Increase in membrane permeability and exudation of roots of zinc deficient plants. J. Plant Physiol. 132, 356-361 (1988c)

Cakmak, I. and H. Marschner: Decrease in nitrate uptake and increase in proton release in zinc deficient cotton and sunflower plants. Plant and Soil 129, 261-268 (1990)

Cakmak, I.; K.Y. Gülüt; H. Marschner and R.D. Graham: Effect of zinc and iron deficiency on phytosiderophore release in wheat genotypes differing in zinc efficiency. J. Plant Nutrition (*submitted*) (1993)

Chairidchai, P. and G.S.P. Ritchie: Zinc adsorption by sterilized and non-sterilized soil in the presence of citrate and catechol. Commun.Soil Sci. Plant Anal. 24, 261-275 (1993)

Colpaert, J.V. and J.A. Van Assche: Zinc toxicity in ectomycorrhizal *Pinus sylvestris*. Plant and Soil 143, 201-211 (1992)

Denny, H.J. and D.A. Wilkins: Zinc tolerance in *Betula spp.* VI. The mechanism of ectomycorrhizal amelioration of zinc toxicity. New Phytol. 106, 545-553 (1987)

Dinkelaker, B.; V. Römheld and H. Marschner: Citric acid excretion and precipitation of calcium citrate in the rhizosphere of white lupin (*Lupinus albus* L.). Plant, Cell Environ. 12, 285-292 (1989)

Dogar, M.A. and T. van Hai: Effect of P, N and HCO_3^- levels in the nutrient solution on rate of Zn absorption by rice roots and Zn content in plants. Z. Pflanzenphysiol. 98, 203-212 (1980)

Dueck, T A , P Visser, W H O Ernst and H Schat Vesicular-arbuscular mycorrhizae decrease zinc-toxicity in grasses in zinc polluted soil Soil Biol Biochem 18, 331-333 (1986)

Duguma, B , B T Kang and D U U Okeli Effect of liming and phosphorus application on performance of *Leucaena leucoce- phala* in acid soils Plant and Soil 110, 57-61 (1988)

Eivazi, F and C C Weir Phosphorus and mycorrhizal interaction on uptake of P and trace elements by maize Fert Res 21, 19-22 (1989)

El-Kherbawy, M , J S Angle, A Heggo and R L Chaney Soil pH, rhizobia, and vesicular-arbuscular mycorrhizae inoculation effects on growth and heavy metal uptake of alfalfa (*Medicago sativa* L) Biol Fertil Soils 8, 61-65 (1989)

Faber, B A , R J Zasoski, R G Burau and K Uriu Zinc uptake by corn as affected by vesicular-arbuscular mycorrhizae Plant and Soil 129, 121-130 (1990)

Forno, D A , S Yoshida and C J Asher Zinc deficiency in rice 1 Soil factors associated with the deficiency Plant and Soil 42, 537-550 (1975a)

Forno, D A , S Yoshida and C J Asher Zinc deficiency in rice II Studies on two varieties differing in susceptibility to zinc deficiency Plant and Soil 42, 551-563 (1975b)

Friesen, D K , M H Miller and A S R Juo Liming and lime-phosphorus-zinc interactions in two Nigerian Ultisols II Effects on maize root and shoot growth Soil Sci Soc Am J 44,, 1227-1232 (1980)

Gardner, W K , D A Barber and D G Parbery The acquisition of phosphorus by *Lupinus albus* III The probable mechanism by which phosphorus movement in the soil/root interface is enhanced Plant and Soil 70, 107-124 (1983)

Gilmore, A E The influence of endotrophic mycorrhizae on the growth of peach seedlings J Amer Soc Hort Sci 96, 35-38 (1971)

Graham, R D , J S Ascher and S C Hynes Selecting zinc-efficient cereal genotypes for soils low in zinc status Plant and Soil 146, 241-250 (1992)

Grierson, P F and P M Attiwill Chemical characteristics of the proteoid root mat of *Banksia integrifolia* L Aust J Bot 37, 137-143 (1989)

Grove, J H and M E Sumner Lime induced magnesium stress in corn impact of magnesium and phosphorus availability Soil Sci Soc Am J 49, 1192-1196 (1985)

Gupta, U C and E W Chipman Influence of iron and pH on the yield and iron, manganese, zinc, and sulfur concentrations of carrots grown on sphagnum plat soil Plant and Soil 44, 559-566 (1976)

Hoffland, E , G R Findenegg and J A Nelemans Solubilization of rock phosphate by rape I Evaluation of the role of the nutrient uptake pattern Plant and Soil 113, 155 160 (1989a)

Hoffland, E , G R Findenegg and J A Nelemans Solubilization of rock phosphate by rape II Local root exudation of organic acids as a response to P-starvation Plant and Soil 113, 161-165 (1989b)

Howeler, R H , E Sieverding and S Saif Practical aspects of mycorrhizal technology in some tropical crops and pastures Plant and Soil 100, 249-283 (1987)

Jakobsen, I Vesicular-arbuscular mycorrhiza in field-grown crops II Effect of inoculation on growth and nutrient uptake in barley at two phosphorus levels in fumigated soil New Phytol 94, 595-604 (1983)

Jeffery, J J and N E Uren Copper and zinc species in the soil solution and effects of soil pH Aust J Soil Res 21, 479-488 (1983)

Jolley, V D and J C Brown Factors in iron-stress response mechanism enhanced by Zn deficiency stress in Sanilac, but not Saginaw navy bean J Plant Nutr 14, 257-265 (1991)

Jungk, A Dynamics of nutrient movement at the soil-root interface *In* Plant Roots, The Hidden Half (eds J Waisel, A Eshel and U Kafkafi), pp 455-481 Marcel Dekker, Inc New York (1991)

Jungk, A and N Claassen Availability in soil and acquisition by plants as the basis for phosphorus and potassium supply to plants Z Pflanzenernahr Bodenk 152, 151-157 (1989)

Khasa, P , V Furlan and J A Fortin Response of some tropical plant species to endomycorrhizal fungi under field conditions Trop Agric (Trinidad) 69, 279-283 (1992)

Kothari, S K , H Marschner and V Romheld Effect of a vesicular-arbuscular mycorrhizal fungus and rhizosphere microorganisms on manganese reduction in the rhizosphere and manganese concentration in maize (*Zea mays* L) New Phytol 117, 649-655 (1991a)

Kothari, S K , H Marschner and V Romheld Contribution of VA mycorrhizal hyphae in acquisition of phosphorus and zinc by maize grown in a calcareous soil Plant and Soil 131, 177-185 (1991b)

Lambert, D H and T C Weidensaul Element uptake by mycorrhizal soybean from sewage-sludge-treated soil Soil Sci Soc Am J 55, 393-398 (1991)

Laurie, S H , N P Tancock, S P McGrath and J R Sanders Influence of complexation on the uptake by plants of iron, manganese, copper and zinc I Effect of EDTA in a multi-metal and computer simulation study J

Exp. Bot. 42, 509-513 (1991)
Lee, C.R. and G.R. Craddock: Factors affecting plant growth in high-zinc medium. II. Influence of soil treatments on growth of soybeans on strongly acid soil containing zinc from peach sprays. Agron. J. 61, 565-567 (1969)
Leggett, G.E. and D.T. Westermann: Effect of corn, sugarbeets, and fallow on zinc availability to subsequent crops. Soil Sci. Soc. Am. J. 50, 963-968 (1986)
Li, X.-L.; H. Marschner and E. George: Acquisition of phosphorus and copper by VA-mycorrhizal hyphae and root-to-shoot transport in white clover. Plant and Soil 136, 49-57 (1991)
Lindsay, W.L. and F.R. Cox: Micronutrient soil testing for the tropics. Fert. Res. 7, 169-200 (1985)
Linehan, D.J; A.H. Sinclair and M.C. Mitchell: Seasonal changes in Cu, Mn, Zn and Co concentrations in soil in the root-zone of barley (*Hordeum vulgare* L.). J. Soil Sci. 40, 103-115 (1989)
Lins, I.D.G. and F.R. Cox: Effect of soil pH and clay content on the zinc soil test interpretation for corn. Soil Sci. Soc. Am. J. 52, 1681-1685 (1988)
Lipton, D.S.; R.W. Blanchar and D.G. Blevins: Citrate, malate, and succinate concentration in exudates from P-sufficient and P-stressed *Medicago sativa* L. seedlings. Plant Physiol. 85, 315-317 (1987)
Lu, S. and M.H. Miller: The role of VA mycorrhizae in the absorption of P and Zn by maize in field and growth chamber experiments. Can. J. Soil Sci. 69, 97-109 (1989)
Mandal, B., G.C. Hazra and A.K. Pal: Transformation of zinc in soils under submerged conditions and its relation with zinc nutrition of rice. Plant and Soil 106, 121-126 (1988)
Manjunath, A. and M. Habte: Development of vesicular-arbuscular mycorrhizal infection and the uptake of immobile nutrients in *Leucaena leucocephala*. Plant and Soil 106, 97-103 (1988)
Marschner, H.: Root-induced changes in the availability of micronutrients in the rhizosphere. *In* Plant Roots. The Hidden Half (eds. Y. Waisel; A. Eshel and K. Kafkafi), pp. 503-528, Marcel Dekker Inc., New York (1991)
Marschner, H. and V. Römheld: In-vivo measurement of root-induced pH changes at the soil-root interface: Effect of plant species and nitrogen source. Z. Pflanzenphysiol. 111, 241-251 (1983)
Marschner, H.; V. Römheld and I. Cakmak: Root-induced chages of nutrient availability in the rhizosphere. J. Plant Nutr. 10, 1175-1184 (1987a)
Marschner, H.; V. Römheld and F.S. Zhang: Mobilization of mineral nutrients in the rhizosphere by root exudates. Transact. 14th Int. Congr. Soil Science, Kyoto 1990, Vol. II, pp. 158-163 (1990)
Marschner, H.; M. Treeby and V. Römheld: Role of root-induced changes in the rhizosphere for iron acquisition in higher plants. Z. Pflanzenernähr.Bodenk. 152, 197-204 (1989)
Marschner, H.; V. Römheld, W. J. Horst and P. Martin: Root-induced changes in the rhizosphere: Importance for the mineral nutrition of plants. Z. Pflanzenernähr. Bodenk. 149, 441-456 (1986)
McGrath, S.P.; J.R. Sanders and M.H. Shalaby: The effect of soil organic matter levels on soil solution concentrations and extractabilities of manganese, zinc and copper. Geoderma 42, 177-188 (1988)
Menge, J.A.; W.M. Jarrell; C.K. Labanauskas; J.C. Ojala; C. Huszar; E.L.V. Johnson and D. Sibert: Predicting mycorrhizal dependency of troyer citrange on *Glomus fasciculatus* in California citrus soils and nursery mixes. Soil Sci. Soc. Am. J. 46, 762-768 (1982)
Merckx, R.; J. H. van Ginkel; J. Sinnaeve and A. Cramers: Plant-induced changes in the rhizosphere of maize and wheat. II. Complexation of cobalt, zinc and manganese in the rhizosphere of maize and wheat. Plant and Soil 96, 95-107 (1986)
Moraghan, J.T.: Manganese toxicity in flax growing on certain calcareous soils low in available iron. Soil Sci. Soc. Am. J. 43, 1177-1180 (1979)
Moraghan, J.T.: Effect of soil temperature on response of flux to P and Zn fertilizers. Soil Sci. 129, 290-296 (1980)
Moraghan, J.T.: Differential response of five species to phosphorus and zinc fertilizers. Commun. Soil Sci. Plant Anal. 15, 437-447 (1984)
Moraghan, J.T.: Manganese deficiency in soybeans as affected by FeEDDHA and low soil temperature. Soil Sci. Soc. Am. J. 49, 1584-1586 (1985)
Moraghan, J.T. and H.J. Mascagni, jr.: Environmental and soil factors affecting micronutrient deficiencies and toxities. *In* Micronutrients in Agriculture (eds. J.J. Mortvedt, F.R. Cox, L.M. Shuman, R.M. Welch) pp. 371-425, SSSA Book Series No.4, Madison, WI (1991)
Nambiar, E.K.S.: Uptake of Zn^{65} from dry soil by plants. Plant and Soil 44, 267-271 (1976a)
Nambiar, E.K.S.: The uptake of zinc-65 by oats in relation to soil water content and root growth. Aust. J. Soil Res. 14, 67-74 (1976b)
Ohwaki, Y. and H. Hirata: Differences in carboxylic acid exudation among P-starved leguminous crops in relation to carboxylic acid contents in plant tissues and phospholipid level in roots. Soil Sci. Plant Nutr. 38,

235-343 (1992)

Pacovsky, R.S.: Micronutrient uptake and distribution in mycorrhizal or phosphorus-fertilized soybeans. Plant and Soil 95, 379-388 (1986)

Pairunan, A.K.; A.D. Robson and L.K. Abbott: The effectiveness of vesicular-arbuscular mycorrhizas in increasing growth and phosphorus uptake of subterranean clover from phosphorus sources of different solubilities. New Phytol. 84, 327-338 (1980)

Parker, M.B. and M.E. Walker: Soil pH and manganese effects on manganese nutrition of peanut. Agron. J. 78, 614-620 (1986)

Parker, D.R.; J.J. Aguilera and D.N. Thomason: Zinc-phosphorus interactions in two cultivars of tomato (*Lycopersicon esculentum* L.) grown in chelator-buffered nutrient solutions. Plant and Soil 143, 163-177 (1992)

Raven, J.A.; C. Rothemund and B Wollenweber: Acid-base regulation by *Azolla* spp. with N_2 as sole N source and with supplementation by NH_4^+ or NO_3^-. Acta Botanica 104, 132-138 (1991)

Römheld,V.: Different strategies for iron acquisition in higher plants. Physiol. Plant. 70, 231-234 (1987)

Rothmeier, H. und A. Amberger: Zink-Aufnahme verschiedener Pflanzen unter Einfluß von NH_4^- Ernährung und DCD. *In* VDLUFA-Schriftenreihe, Heft 11, 126-133, Symposium Nitrifikationshemmstoffe (1983)

Runjin,L.: Effects of vesicular-arbuscular mycorrhizas and phosphorus on water status and growth of apple. J. Plant Nutr. 12, 997-1017 (1989)

Sarong, L.Q.; D.R. Bouldin and W.S.Reid: Total and labile zinc concentrations in water extracts of rhizosphere and bulk soils of oats and rice. Commun. Soil Sci. Plant Anal. 20, 271-289 (1989)

Schüepp, H.; B. Dehn und H. Sticher: Interaktionen zwischen VA-Mykorrhizen und Schwermetallbelastungen. Angew. Bot. 61, 85-96 (1987)

Schwartz, S.M.; R.M. Welch; D.L. Grunes; E.E.Cary; W.A. Norvell; M.D. Gilbert; M.P.Meridith and C.A. Sauchirico: Effect of zinc, phosphorus and root-zone temperature on nutrient uptake by barley. Soil Sci. Soc. Am. J. 51, 371-375 (1987)

Schweiger, P.; A. Robson; J. Barrow and L. Abbott: Root hair length determines beneficial effect of a *Glomus sp.* on shoot growth of some pasture species. *In* Management of Mycorrhizas in Agriculture, Horticulture and Forestry. In press (1993)

Sedberry, J.E. jr.; D.P. Bligh; F.J. Peterson and M.C. Amacher: Influence of soil pH and application of zinc on the yield and uptake of selected nutrient elements by rice. Commun. Soil Sci. Plant Anal. 19, 597-615 (1988)

Sedberry, J.E. jr.; F.J. Peterson; F.E. Wilson; D.B. Mengel; P.E.Schilling and R.H. Brupbacher: Influence of soil reaction on application of zinc on yields and zinc contents of rice plants. Commun. Soil Sci. Plant Anal. 11, 283-295 (1980)

Sharma, K.N. and D.L. Deb: Effect of organic manuring on zinc diffusion in soils of varying texture. J. Indian Soc. Soil Sci. 36, 219-224 (1988)

Sims, J.T. and G.V. Johnson: Micronutrients soil tests. *In* Micronutrients in Agriculture, 2nd Edition (eds. J.J. Mortvedt; F.R. Cox; L.M Shuman; R.M. Welch) pp. 427-476, SSSA Book Series No. 4, Madison, WI (1991)

Sinclair, A.H.; L.A. Mackie-Dawson and D.J.Linehan: Micronutrient inflow rates and mobilization into soil solution in the root zone of winter wheat (*Triticum aestivum* L.). Plant and Soil 122, 143-146 (1990)

Singh, B. and R.P. Bollu: Effect of EDTA and sulphate carried zinc and manganese on uptake and content of zinc in paddy. Indian J. Agric. Chem. 16, 191-198 (1983)

Singh, J.P.; R.E. Karamanous and J.W.B. Stewart: Phosphorus-induced zinc deficiency in wheat on residual phosphorus plants. Agron. J. 78, 668-675 (1986)

Swamvinathan, K. and B.C. Verma: Response of three crop species to vesicular arbuscular mycorrhizal infections in zinc deficient Indian soils. New Phytol. 82, 481-487 (1979)

Takagi, S.; S. Kamei and Ming-Ho Yu: Efficiency of iron extraction from soil by mugineic acid family phytosidero- phores. J. Plant Nutrition 11, 643-651 (1988)

Tekalign Mamo and K.S. Killham: Effect of soil liming and vesicular-arbuscular-mycorrhizal inoculation on the growth and micronutrient content of the teff plant. Plant and Soil 102, 257-259 (1987)

Thompson, J.P.: Soil sterilization methods to show VA-mycorrhizae aid P and Zn nutrition of wheat in Vertisols. Soil Biol. Biochem. 22, 229-240 (1990)

Thompson, J.P. and G.B. Wildermuth: Colonization of crop and pasture species with vesicular-arbuscular mycorrhizal fungi and a negative correlation with root infection by *Bipolaris sorokiniana*. Can. J. Bot. 69, 687-693 (1988)

Thomson, C.J.; H. Marschner and V. Römheld: Effect of nitrogen fertilizer form on pH of the bulk soil and rhizosphere, and on the growth, phosphorus, and micronutrient uptake of bean. J. Plant Nutr. 16, 493-506 (1993)

Treeby, M.; H. Marschner and V. Römheld: Mobilization of iron and other micronutrients from a calcareous soil by plant-borne, microbial, and synthetic metal chelators. Plant and Soil 114, 217-226 (1989)

Wallace, A.: Effect of nitrogen fertilizer and nodulation on lime-induced chlorosis in soybean. J. Plant Nutr. 5, 363-368 (1982)

Weber, E.; M.C. Saxena; E. George and H.Marschner: Effect of vesicular-arbuscular mycorrhiza on vegetative growth and harvest index of chickpea grown in northern Syria. Field Crops Res. 32, 115-128 (1993)

Wilkinson, H.F.; J.F. Loneragan and J.P. Quirk: The movement of zinc to plant roots. Soil Sci. Soc. Amer. Proc. 32, 831-833 (1968a)

Yang, X.; V. Römheld and H. Marschner: Effect of bicarbonate and root zone temperature on uptake of Zn, Fe, Mn and Cu in different rice varieties (*Oryza sativa* L.) growing in calcareous soil. J. Plant Nutr. (*submitted*). (1993a)

Yang, X.; V. Römheld and H. Marschner: Effect of bicarbonate on root growth and accumulation of organic acid in Zn-inefficient and Zn-efficient rice varieties (*Oryza sativa* L.) J. Plant Nutrit. (*submitted*). (1993b)

Zhang, F.; V. Römheld and H. Marschner: Effect of zinc deficiency in wheat on the release of zinc and iron mobilizing root exudates. Z. Pflanzenernähr. Bodenk. 152, 205-210 (1989)

Zhang, F.S.; V. Römheld and H. Marschner: Diurnal rhythm of release of phytosiderophores and uptake rate of zinc in iron-deficient wheat. Soil Sci. Plant Nutr. 37, 671-678 (1991)

Chapter 6.

Distribution and Transport of Zinc in Plants

NANCY E. LONGNECKER and ALAN D. ROBSON

1. Abstract

Zinc (Zn) distribution and transport in plants is affected by the level of Zn supply and plant species. When plants have low to adequate Zn supply, Zn concentrations are usually higher in growing tissue than in mature tissue; this is true for roots, vegetative shoots and reproductive tissues. In plants tolerant of toxic levels of Zn, accumulation has been observed in the root cortex and in leaves. In these tissues, Zn accumulates in cell walls or is sequestered in vacuoles.

Zinc transport in the xylem does not necessarily coincide with that of water. Zinc is a nutrient with variable mobility that is retranslocated to a greater extent when in adequate supply. Zinc movement out of old leaves coincides with their senescence; both can be delayed by Zn deficiency. Species differ widely in their ability to load Zn into seeds; some native plants adapted to nutrient-poor soils have 10 times greater Zn concentrations in their seeds than most cultivated species, but have similar or lower Zn concentrations in their leaves.

2. Introduction

We will describe the different pools of Zn in plants, how much Zn is in the pools, the pattern of accumulation in them and the movement of Zn between them. Then we will describe pathways of transport between the pools and control of Zn movement.

3. Pools of zinc

The first pool of Zn in plants is seed Zn which is mobilized to the growing seedling. Zinc taken up by roots moves through the stems to the leaves. It can be stored in stems and mobilized later to growing tissue. Zinc can move from senescing tissue to growing vegetative tissue and reproductive organs and seeds. The movement of Zn between these pools is affected by the amount of Zn in them.

3.1. Seeds

Loading of Zn into seeds is discussed in the section on Zn in reproductive tissues. *Location of seed pools:* Zinc content in mature seeds is localized in the embryo. For example, 47% of the total Zn content in maize kernels was present in the "germ" even though the germ was only 8% of the kernel dry weight (Massey and Loeffel, 1967). In that study, Zn concentration was also highest in the embryo, with 138 mg Zn kg^{-1} in the germ, 46 mg Zn kg^{-1} in the pericarp and 10 mg Zn kg^{-1} in the endosperm.

Mobilization of zinc from seed reserves: Zinc is readily mobilized from seed reserves. Using soybean seeds labelled with ^{65}Zn, Sudia and Green (1972) found that 24% of the ^{65}Zn in the first generation seed ended up in seed of the subsequent generation. Eighty to 90% of the Zn in lupin cotyledons was mobilized to seedlings (Hocking and Pate, 1978; Hocking, 1980). By the time the cotyledons abscised, this represented 15% of the total seedling Zn (Hocking and Pate, 1978). For plants grown with deficient Zn supply, seed reserves could represent a larger proportion of the total seedling Zn content. The mobilization of Zn from cotyledons was similar in Zn-deficient and Zn-adequate *Pinus radiata* (McGrath and Robson, 1984a).

3.2. *Distribution between roots and shoots*

When wheat plants were supplied with ^{65}Zn-labelled nutrient solution, ^{65}Zn accumulated in the roots for the first 2 hours (Santa Maria and Cogliatti, 1988). After 4 hours, the rate of ^{65}Zn accumulation in the roots decreased and transport to the shoot accounted for a greater proportion of the net Zn uptake (54% over the 27 hour period). Similarly, when Zn was resupplied to Zn-deficient cotton plants, there was a lag phase between Zn uptake and transport to the shoots (Cakmak and Marschner, 1990). After 22 hours of Zn supply, the Zn concentration of roots increased from 15 to 128 mg Zn kg^{-1} while that of the total shoots only increased from 13 to 28 mg Zn kg^{-1}. In another study, there was a localized increase in Zn concentration of the shoot tip and young leaves from 13 to 151 mg kg^{-1} within 24 hours of supply of Zn to Zn-deficient *Phaseolus vulgaris* (Cakmak et al., 1989).

With Zn toxicity, there is evidence that excess Zn accumulates in root cortical cell walls or vacuoles (see next section). However, tolerant genotypes do not exclude Zn from their shoots and can accumulate Zn in their leaves.

3.3. *Roots*

The Zn content and concentration of plant roots is affected by the availability of Zn in the root zone (see Chapter 5) and by the interactions between Zn and other nutrients (see Chapter 9). Zinc accumulation in roots of subclover grown with high Zn supply was predictably greater than that of plants grown with no added Zn (Reuter, 1980). With toxic levels of Zn supply, a higher proportion of the total plant Zn may accumulate in the roots. For example, with adequate Zn supply, soybeans had 92 μg Zn in their trifoliate leaves and only 35 μg Zn in their roots (White et al., 1979). With toxic Zn supply, the soybeans had 133 μg Zn in trifoliate leaves and 1335 μg Zn in their roots.

A higher Zn concentration in the roots does not necessarily reflect Zn accumulation. When Ruano et al. (1987) grew bean plants with toxic levels of Zn supply, the Zn concentration of roots was higher but their Zn content was lower. The higher root Zn concentration correlated with decreased root growth, not with an accumulation of Zn.

Location of zinc in roots: Autodradiographs of subclover plants supplied with adequate levels of ^{65}Zn 20 and 43 days after germination showed a high level of radioactivity at 77 days after germination in root tips and points thought to be lateral root initials (Riceman and Jones, 1958c). In roots of *Silene vulgaris* which is tolerant to high levels of Zn supply, the Zn concentration was higher in the cortex (500 mg kg^{-1}) than in the stele (100 mg kg^{-1}) (Sieghardt, 1990).

Most of the Zn in the subcellular fractions of homogenized faba bean roots was present in the "soluble" fraction (Polar, 1976). This is probably due to the association of Zn with enzymes and low molecular weight organic compounds (see Chapter 7). The next largest fraction of root Zn was in the cell-wall debris. There are numerous reports of high Zn concentrations in cell walls from studies of plants grown with adequate to toxic levels of Zn. Santa Maria and Cogliatti (1988) calculated that the Zn concentration in apparent free space of wheat roots grown with adequate Zn supply could be as high as 0.5 m*M*. The greatest Zn content in roots of Zn-tolerant *Agrostis* was in the cell wall fraction (Turner, 1970). Zinc accumulation by isolated cell walls was rapid and not affected by temperature (Turner and Marshall, 1971). Zinc accumulation in root cell walls was high in the Zn-tolerant clone of *Agrostis*, but not in the Cu-tolerant clone or the Zn-intolerant clone (Turner, 1970), indicating that Zn accumulation was due to specific Zn adsorption sites in the Zn-tolerant clone. In studies with Zn-tolerant plants, Zn was detoxified by localization in globules in small vacuoles, probably complexed with phytate (Van Steveninck et al., 1987, 1990).

Patterns of zinc accumulation in roots: More recently absorbed Zn was retained in roots of Zn-deficient *Lupinus angustifolius* and soybean than in roots of Zn-adequate plants (S.M. Shah and A.D. Robson, unpublished results). When Zn supply is increased, there is usually a transient increase in root Zn concentration. After Zn addition, the Zn concentration of the total roots of subclover increased from 43 to 65 mg kg^{-1} within the first week, then decreased to 30 mg kg^{-1} in the subsequent 3 weeks (Riceman and Jones, 1958b). The ^{65}Zn concentration of faba bean roots was 2.7 times higher in young root tissue than in old root tissue (Polar, 1976). The decrease in ^{65}Zn in old root tissue was greatest in the "soluble" and ribosomal fractions.

Mobilization of zinc in roots: Split root experiments have shown that there is transport of Zn from root zones with Zn supplied to root zones with no external Zn supply (Loneragan et al., 1987). However even when Zn supplied in one root compartment was adequate for maximum shoot growth, the movement of Zn to the second root compartment did not compensate for the lack of external Zn supply in the second compartment (Loneragan et al., 1987; Robson and Snowball, 1989; Nable and Webb, 1993).

3.4 Stems

Stems are an important, often overlooked component of the transport pathway in plants. There are numerous reports of the temporary accumulation of Zn in stems. In bush beans grown with increasing levels of Zn supply (ranging from 0.13 to 0.75 mg Zn L^{-1}), the Zn concentration of the stems increased from 105 to 347 mg kg^{-1} (Ruano et al., 1987). This corresponded to an accumulation of Zn content in the stems from 14 to 66 μg Zn. At lower levels of Zn supply, sheaths of maize and barley had higher ^{65}Zn content than roots or leaves (Singh and Steenberg, 1974). However, even when stems accumulate Zn, the Zn concentration is not necessarily high because of their large size. Of all the plant parts in one study, lupin stems had the lowest Zn concentration: 12 mg kg^{-1} in *L. albus* stems compared to 28 mg kg^{-1} in leaves and 37 mg kg^{-1} in young shoots (Reay and Waugh, 1981).

Location of zinc in stems: The concentration of ^{65}Zn was highest in the nodes of subclover stems (Riceman and Jones, 1958c) and rice (Obata and Kitagishi, 1980b). The higher ^{65}Zn concentration in the subclover nodes was due to an accumulation in the vascular tissue and was highest where there were vascular traces to petioles or laterals (Riceman and Jones, 1958c).

Pattern of zinc accumulation in stems: Zinc concentration of apple stems was greater with increasing Zn supply (from deficient to adequate levels) and decreased with time over the first year of growth at all levels of supply (Watkins, 1982). When solution-grown *P.radiata* seedlings were given a pulse of ^{65}Zn, there was an accumulation of ^{65}Zn in stems of plants receiving low or adequate Zn supply (McGrath and Robson, 1984b). Within 3 days, the label in the stem of Zn-adequate plants had decreased from 52% of the total plant ^{65}Zn to 19%. In contrast, the movement of the pulse was slower from stems of Zn-deficient plants. When the Zn supply was changed from nil to adequate at the time of the pulse, the rate of movement of the ^{65}Zn through the stems was increased to that of the controls.

Mobilization of zinc from stems: In studies using excised stems of *P. radiata*, solution containing 100μM $CaSO_4$ and either 3 or 30 μM $ZnSO_4$ labelled with ^{65}Zn was drawn through 10 cm lengths of stems (McGrath and Robson, 1984c). Ninety-nine % of the ^{65}Zn retained by the stems was in the first 2 cm. Decreasing the temperature from 22°C to 1°C had no effect on the distribution of ^{65}Zn, indicating that the retention of Zn by the stems was a non-metabolic process. Zinc was more evenly distributed in different stem segments when the calcium concentration of the inflow solution was increased to 3000 μM. Addition of EDTA to the inflow solution greatly decreased the retention of Zn in the stems while addition of citrate had a smaller effect.

Application of Zn as a spray during the dormant period has been recommended in horticulture (Harley et al., 1956). Evidence from work with ^{32}P suggested that nutrients can be absorbed by bark, particularly through pruning wounds or ruptures in bark due to growth bursts in spring (Harley et al., 1956). However other work concluded that foliar sprays were more effective than dormant-season sprays for relieving Zn deficiency (Orphanos, 1975).

3.5 Leaves

Zinc concentration in healthy leaves generally ranges from 15 to 100 mg Zn kg^{-1} (Table 1). In Zn-deficient leaves, the Zn concentrations of many cultivated plants were 10- 15 mg Zn kg^{-1} (Table 1). However, Zn concentration was very low in leaves of some indigenous Proteaceae which are adapted to nutrient-poor soils (*e.g.* 2.5 mg Zn kg^{-1} in *Hakea victoriae* ; Kuo et al., 1982). It is unknown if these native species have a lower internal requirement for Zn.

In a Zn toxicity study, soybeans with leaf Zn concentrations of 400 to 730 mg Zn kg^{-1} in plants suffered yield reductions of 10-33% (White et al., 1979). The genotypes most tolerant of high Zn supply were not necessarily those with the lowest leaf Zn concentrations. The Zn concentration of wild plants growing on soils contaminated from mining can be much higher (*e.g.* 2900 mg kg^{-1} for *Minuartia*; Sieghardt, 1990).

Location of zinc in leaves: In Zn-deficient plants, more of the total Zn in leaves is present in the leaf blade than in the petiole (Riceman and Jones, 1958b). With increasing Zn supply, the Zn content is more equally apportioned between the leaf parts. This does not mean the Zn concentrations are equal in the blade and the petiole. The petiole Zn concentration stayed about 7 mg Zn kg^{-1} in soybean plants which showed Zn deficiency symptoms, even though Zn concentration of the leaf blades increased in response to increasing Zn supply (Table 2).

Table 1. Zinc content in seeds and concentrations in leaves and seeds of various species.

Species	Zn supply	Leaf Zn concentration (mg kg^{-1})	Zn Zn content (µg per seed)	Seed Zn concentration (mg kg^{-1})	Seed weight (mg per seed)	Ref.
Trifolium	ø added	10	0.1	17	6	a
subterraneum	0.1 mg Zn/L^{-1}	24	0.1	12	5	
Lupinus angustifolius	*na*		*na*	*30*	*na*	*b*
Lupinus albus	*na*		*na*	*29*	*na*	
Triticum aestivum	*na*		*na*	*22*	*na*	
Glycine max cv York	ø added	100				c
	262 mg Zn/kg^{-1} soil	500				
	524 mg Zn/kg^{-1} soil	1000				
Zea mays L.			na	18 - 10.	na	d
Zea mays L.	ø added	14				e
	1.3 mg Zn/l	40				
Banksia coccinea	cs	11	1.5	110	14	f
Banksia hookeriana	cs	11	4.7	123	38	
Hakea victoriae	cs	2	4.2	150	28	
Minuartia verna	Soil collected from mine dumps	2900				g

na = not available
cs = collected specimens
a = Riceman and Jones, 1958a,b
b = White et al., 1981a
c = White et al., 1979
d = Mozafar, 1990
e = Gibson and Leece, 1981
f = Kuo et al., 1982
g = Sieghardt, 1990

In a study comparing *Lupinus albus* and *L. angustifolius*., there were genotypic differences in partitioning between leaf blade and petiole (the Zn concentrations of leaflets and petioles were 28 and 24 mg Zn kg^{-1} for *L. albus* compared to 63 and 22 mg Zn kg^{-1} for *L. angustifolius;* Reay and Waugh, 1981). The study used only one level of Zn and it is unclear whether the differences were due to genotypic differences in partitioning or to differences in Zn status because of different uptake at the given Zn supply.

In some Zn-tolerant plants, there can be an accumulation of Zn in the leaves (Mathys, 1977; Sieghardt, 1990). Krotz et al. (1989) demonstrated the likelihood that Zn

Table 2. Zinc concentrations (mg kg^{-1}) of the fifth leaf blade and petiole of soybean plants receiving different Zn supply (data from Ohki, 1977)

	Zn supply (μg Zn L^{-1})							
	ø[ab]	20[ab]	50[ab]	100[a]	200	400	700	1000
Blade	12	19	17	23	32	55	75	115
Petiole	7	6	7	7	12	28	41	76

[a]deficiency symptoms noted on other leaves on the plant.
[b]deficiency symptoms noted on this leaf and dry weight of plant decreased by Zn deficiency

is sequestered in the vacuole as complexes with endogenous organic acids.

Pattern of accumulation: The pattern of accumulation of Zn in leaves depends on the level of supply. In Zn-deficient plants, young, vegetative meristems generally have higher Zn concentration than older leaves (*e.g.* Table 3). Plants appear able to supply Zn to their meristems and do not continue to accumulate Zn in their older leaves under Zn-deficient conditions. With high Zn supply, Zn accumulates in older leaves of plants and the Zn concentration of older leaves is much higher than that of the new growth.

In plants with adequate Zn supply, Zn concentration of leaf blades of different ages is similar, with little indication of Zn accumulation or remobilization (Table 3). Riceman and Jones (1958b) showed that the Zn concentration of an individual subclover leaf with adequate Zn supply reached a maximum before it was fully expanded. Thus, Zn content reached its maximum before dry weight did. Zinc concentration subsequently decreased as the leaf grew and diluted its Zn. For example, the Zn concentration of the second leaf went from 50 mg Zn kg^{-1} at 32 days after sowing (DAS) to 30 mg Zn kg^{-1} at 41 DAS. At the same time, there was a decrease in Zn content of leaves once the maximum value had been reached; the total leaf Zn content went from 0.45 to 0.25 μg Zn and the petioles went from 0.23 to 0.10 μg Zn (Riceman and Jones, 1958b).

Mobilization of zinc from leaves: Zinc fertilizer can be applied as a foliar spray. Foliar sprays of Zn-deficient apple trees increased the Zn concentration of leaves which expanded after the spray (Orphanos, 1975). However the increased Zn concentration was

Table 3. Zinc concentration (mg kg^{-1}; 42 DAS) of new growth (NG), youngest folded leaf blade (YFL), youngest open leaf blade (YOL) and older leaf blade (YOL +3) of *Trifolium subterraneum* receiving different Zn supply (Reuter et al., 1982)

	Zn supply (μg Zn per 3 kg soil)				
	0	400	800*	1600	6400
NG	23	25	30	41	60
YFL	13	12	23	41	62
YOL	8	9	19	44	121
YOL + 3	-	8	24	98	487

*Maximum dry matter achieved at the 800 μg Zn supply.

due to direct spray on the apices, not to translocation from the sprayed leaves. It is unknown if there is greater Zn translocation from leaves of plants which are less severely Zn-deficient at the time of spray application. This would have implications for scheduling foliar applications of Zn for increased effectiveness.

Zinc mobilization out of old leaves depends on the Zn status of the plant. Riceman and Jones (1958a) showed that the total Zn in leaves and petioles decreased as Zn accumulated in inflorescences, burrs and seeds of subclover. The decrease of Zn in old leaves was greater in plants with adequate Zn supply than with deficient supply. Mobilization of Zn from old leaves to new growth was also greater with adequate than deficient Zn supply to *P.radiata* (McGrath and Robson, 1984a), soybean and *L. angustifolius* (S.M. Shah and A.D. Robson, unpublished data). At toxic levels of Zn supply, there was a higher ratio of Zn in mature leaves to developing leaves, indicating a greater retention of Zn (Ruano et al., 1987). In wheat with adequate Zn supply, Hill et al. (1979) showed that when old leaves senesced, there was a sharp decline in leaf N, Zn and Cu. When senescence was delayed by copper deficiency, there was no decrease in content of Cu or Zn of old leaves and little of N.

3.6. Reproductive organs

With low to adequate Zn supply, the Zn concentration of reproductive organs tends to be higher than that of vegetative organs (Riceman and Jones, 1958a; Hocking and Pate, 1978). Reported values of Zn concentrations in flowers and fruits depend on the stage of development of the tissue. Young, developing tissue tends to have higher Zn concentration than mature tissue (except for seeds).

Seed loading of Zn is affected by plant species and availability of Zn to the parent plant. Different species load different amounts of Zn into their seeds when grown under the same conditions. Examples of the quantities of Zn in seeds are given in Table 1. In a study of eight species of Proteaceae which are well adapted to nutrient-deficient soils, Kuo et al. (1982) found that leaves of these species had similar or lower Zn concentration than did crop species. However, the concentration in seeds was much higher, up to 100 times greater than that of leaves. White et al. (1981a) conducted a survey of wheat and two lupin species grown at a wide range of sites in Western Australia and found that there was a small but significant difference between the species in the Zn concentration of their seeds.

Site of growth can be a more important determinant of seed Zn concentration than species (White et al., 1981a). Unfortunately, grain weights were not given in the results of that study. It is possible that lower concentrations at some sites were due to better yield and grain filling and thus dilution of Zn. However, Riceman and Jones (1958c) showed that the Zn concentration of subclover seeds increased in response to increased Zn supply (from 17 mg Zn kg^{-1} at the lowest Zn supply to 51 mg Zn kg^{-1} at the highest). Thus, while nutrient content of seeds is relatively constant (compared to the variation seen in other plant tissues; Welch, 1986), the Zn contents and concentrations in seeds do vary significantly.

Differential seed loading of Zn has ramifications at both the deficient and toxic ends of the spectrum. With Zn deficiency, low Zn in seeds may affect seedling growth. This has been shown with other nutrients (Welch, 1986), but not with Zn to our knowledge. For plants growing in soil with high Zn availability, there is concern that Zn could accumulate in the seeds of some plants in levels unacceptable for either human or livestock consumption. However, Spitzer et al. (1981) found that while the Zn concentration of wheat seeds grown on sludge-amended soils in Canada doubled, the highest values were only 78 mg Zn kg^{-1} which would not represent a health risk (see Chapter 10). Spitzer et al. (1981) did not observe an accumulation of Zn in the globoid crystals of wheat grains which store other nutrients, even in seeds grown on the sludge-amended soil. Data from Kuo et al. (1982) suggested that Zn only accumulated in globoid-rich regions of cotyledons of some species.

Location of zinc in reproductive tissues: Autoradiographs showed high concentrations of ^{65}Zn in developing florets and seeds of subclover (Riceman and Jones, 1958c). High concentrations were evident in the stamens, in the base of the floret (where the ovaries are located) and the calyx. The ^{65}Zn concentration in the corollas was low. These results corresponded to those of chemical analysis which showed most of the Zn in the reproductive parts of subclover was loaded into the seed at all levels of Zn supply. The Zn concentration of young burrs increased with increasing Zn supply (Riceman and Jones, 1958a). The Zn concentration of the mature burr was high, but most of this was present in the seed (12 mg kg^{-1} at low Zn supply and 51 mg kg^{-1} at adequate Zn supply).

In apple fruit, the Zn concentration decreased from the skin (6 mg kg^{-1}) to the core (2 mg kg^{-1}) (Faust et al., 1967). Zinc concentrations in ears of maize were higher in the apical section (424- 456 mg kg^{-1}) than in the basal section (240- 330 mg kg^{-1}) (Mozafar, 1990).

Pattern of zinc accumulation in reproductive tissues: The flower parts differed in their Zn concentrations in a detailed study of subclover inflorescences at different stages of development (Riceman and Jones, 1958c). The high ^{65}Zn concentration of the stamens persisted even after they were shed. This contrasted with the calyx which was high initially, but decreased during senescence. The ^{65}Zn of the corolla was always low.

In young subclover burrs, peduncles contained high ^{65}Zn concentrations which decreased as the burrs matured (Riceman and Jones, 1958c). Zinc concentrations of inflorescences and immature burrs ranged from 6 mg kg^{-1} at low Zn supply to 35 mg kg^{-1} at adequate Zn supply. These dropped to 4 and 14 mg kg^{-1} in the mature burr "chaff" while the Zn concentration of the seeds was always higher. The Zn content of pods of *L.albus* reached its maximum more quickly than dry weight or most other minerals studied (including N, K, P, Mn, Cu or Mg; Hocking and Pate, 1977). The pod Zn content then decreased.

The ^{65}Zn concentration was high at all times in the developing seed (Riceman and Jones, 1958c). The testa had an initially high ^{65}Zn concentration which decreased as the seed matured. The ^{65}Zn concentration appeared high in all parts of the developing embryo.

In studies of developing legume and cereal seeds, the pattern of Zn accumulation was similar. Duffus and Rosie (1976) observed that Zn accumulated steadily in the whole grain of *Hordeum distichum* L. In contrast, Zn accumulated rapidly in the embryo, reached a maximum of 200 ng Zn per embryo 55 days after anthesis and then decreased to 80 ng Zn per embryo at maturity. Zinc may accumulate early in the embryo because of the role of Zn in protein synthesis (see Chapter 7) and then be transported to storage tissue at seed maturity. The accumulation of Zn in the endosperm was more sigmoidal in nature and reached a maximum of ca. 700 ng per endosperm. Zinc content also increased in developing lupin embryos more rapidly than the dry weight or other nutrients examined (Hocking and Pate, 1977).

4. Transport of zinc

Zinc deficiency causes decreased internode elongation and stunting of young leaves. These observations have frequently led to the incorrect conclusion that Zn is immobile in the phloem. In fact, a much more complex and still incomplete picture of Zn transport is emerging.

Xylem

Zinc absorbed by the roots is rapidly transported to the shoots (Riceman and Jones, 1958c). However, the transport of Zn to shoots is not simply via passive transport in the transpiration stream. Reay (1987) used the ratio of different elements to silicon as a measure of passive transport in *L. albus* shoots and found that the flux of Zn was 5 times greater than that of silicon. Obata and Kitagishi (1980a) used ^{65}Zn autoradiography to show that ^{65}Zn accumulated in expanding leaves of rice, not in fully expanded leaves which had the highest rates of transpiration.

The form of Zn which is actually transported in xylem sap remains unclear. Zinc concentration in xylem sap was 2- 6 μM in stem exudate of soybeans and tomatoes receiving adequate (0.5 μM Zn) Zn supply and 15- 85 μM in plants receiving high (3- 8 μM Zn) Zn supply (White et al., 1981b). From electrophoresis of ^{65}Zn in xylem sap, Tiffin (1967) concluded that Zn is transported as a cation in xylem sap. However, as noted by McGrath (1982), complexes with low stability could have been dissociated because of changes in pH or redox potential during electrophoresis. At either 3 or 30 μM Zn, Zn moved less readily through excised stems of *P. radiata* when it was supplied as an inorganic salt ($ZnSO_4$) than as an anionic or uncharged complex (ZnEDTA or Zn-citrate; McGrath and Robson, 1984c).

Using the metal, organic acid and amino acid concentrations in xylem sap (White et al., 1981b), White et al. (1981c) predicted that Zn is mainly transported in xylem of soybean as Zn-citrate complex and in tomato as Zn-citrate or malate complexes. Using the model, CHELATE, White et al. (1981c) predicted that more Zn was present as uncomplexed Zn^{2+} in tomato xylem sap (21%) than in soybean (4%). Mullins et al. (1986) compared predictions of Zn complexation in xylem and phloem sap using two computer programs, CHELATE and GEOCHEM. Even though both programs have a similar theoretical basis, their predictions on the form of Zn in xylem sap differ significantly. For example, CHELATE predicted 80-90% of the Zn in soybean sap to be complexed with citrate while the GEOCHEM prediction was 33-50%. The difference in these results is too large to predict proportions with confidence, but the evidence suggests that future studies should determine the proportion of free Zn in the sap and should focus on citrate and malate as probable complexing agents.

Zinc accumulates in stems of plants receiving marginal or deficient Zn supply (McGrath and Robson, 1984b). Autoradiography has shown the accumulation to be in the nodes of stems (Riceman and Jones, 1958c; Singh and Steenberg, 1974; Obata and Kitagishi, 1980b). Obata and Kitagishi (1980b) provided evidence that ^{65}Zn was absorbed in the vascular bundles of rice nodes and suggested a role for transfer cells. This coincided with greater accumulation of ^{65}Zn in expanding leaves than in mature leaves and was interpreted to be the mechanism whereby plants can maintain a high supply of Zn to meristematic tissues which apparently have a higher requirement than mature tissues. Less ^{65}Zn was transported to the meristems of Zn-deficient than Zn-adequate rice, but the meristems still had a higher content than the mature leaves of Zn-deficient plants (Obata and Kitagishi, 1980a). The observation that fewer transfer cells formed in nodes of Zn-deficient than control *L.angustifolius* and *Phaseolus vulgaris* (C. Revell, J. Kuo and A.D. Robson, unpublished data) is consistent with the role of transfer cells in nodes in modifying the content of the xylem stream. The concept of transfer cells playing a role in modifying the content of xylem sap destined for different plant tissues is also supported by work on nitrogen transport (Layzell et al., 1981).

4.1. Phloem

The retranslocation of Zn in plants is variable and depends on the Zn supply to the plant. Zinc accumulation in reproductive structures of subclover coincided with a decrease in the total Zn in leaves and petioles (Riceman and Jones, 1958c). Zinc decreases more in senescing leaves of plants with adequate Zn supply than with deficient supply. When Zn is applied by foliar spray, there is conflicting evidence about whether it is retranslocated to other tissues. When developing apices were protected from the spray, Zn did not move from the sprayed leaves to the apices (Orphanos, 1975). However, when Wallihan and Heymann-Herschberg (1956) applied drops of ^{65}Zn labelled $ZnCl_2$ to citrus leaves or coated whole leaves, they observed movement of ^{65}Zn out of leaves which was greater when the ^{65}Zn was applied on or near the midrib or lateral veins. They also observed some ^{65}Zn in roots. There was more ^{65}Zn in the roots 5 days after application when young leaves were coated than old leaves. Because no mention is made of precautions to prevent ^{65}Zn dropping from the leaves to the soil, this evidence is not conclusive.

Reports of Zn concentration in phloem sap range from 3 to 170 μM Zn (Robson and Pitman, 1983). Using paper electrophoresis of phloem exudate from *Ricinus communis* grown with $^{65}ZnCl_2$ in the nutrient solution, Van Goor and Wiersma, 1976 showed that most of the ^{65}Zn was present in a negative form and was present in two peaks distinct from the reference $^{65}Zn^{2+}$ peak. GEOCHEM was used with data on Zn and organic compound concentrations in phloem sap of *Yucca*, to predict that 53% of the Zn in phloem would be complexed with asparagine (or aspartate; there is an unfortunate confusion in the symbols used) and about 15% each with glutamine and tyrosine (Mullins et al., 1986).

Zn accumulates in meristematic tissues and seeds, even under Zn-deficient conditions (see previous sections on distribution). It is generally assumed that tissues with low transpiration rates must be supplied via the phloem. However, it is also possible that xylem sap destined for such structures arrives enriched in Zn which is absorbed and transported to the tissue symplastically.

The role of the fruit in receiving Zn from the xylem and transporting it to the seed via the phloem varies with species. Hocking and Pate (1977) estimated that *L.angustifolius* pods contributed approximately 20% of the embryos' Zn accumulation compared to 8% in field pea. Zinc transported to seeds at different nodes apparently comes from different pools of Zn within the plant. Sudia and Green (1972) grew soybeans from ^{65}Zn-labelled seed and examined the ^{65}Zn loading in the seed of the subsequent generation. They found that seeds at lower nodes accumulated greater proportions of the ^{65}Zn. Ten per cent of the original activity was recovered in seed produced at the lowest 4 nodes, dropping to 0.2% in seed produced at the uppermost nodes.

5. Conclusions

The forms of Zn in both xylem and phloem are still uncertain. However, evidence from examination of sap contents (Van Goor and Wiersma, 1976; White et al., 1981b,c), from theoretical predictions (White et al., 1981c; Mullins et al., 1986) and from transport studies (McGrath and Robson, 1984c) all support the importance of complexation in the transport streams.

The control of long-distance transport of Zn is far from understood. Does the Zn status of the plant control phloem loading of Zn? Is Zn transport to the meristem mainly via xylem or phloem? Does apoplastic transport play a major role in the transport of Zn, especially between different xylem pathways? Are there specific adsorption sites which pull Zn from the transpiration stream in the largely expanded cell wall of transfer cells?

References

Cakmak I and Marschner H 1990 Decrease in nitrate uptake and increase in proton release in zinc-deficient cotton, sunflower and buckwheat plants. Plant Soil. 129, 261-268.

Cakmak I, Marschner H , Bangerth F 1989 Effect of zinc nutritional status on growth, protein metabolism and levels of indole-3-acetic acid and other phytohormones in bean (*Phaseolus vulgaris* L.) J. Exp. Bot. 40, 405-412.

Duffus C M and Rosie R 1976 Changes in trace element composition of developing barley grain. J. Agric. Sci. 87, 75-79.

Faust M, Shear C B and Smith C B 1967 Investigations of corking disorders of apples. I. Mineral element gradients in 'York Imperial' apples. Am. Soc. Hort. Sci. 91, 69-72.

Gibson T S, Leece D R 1981 Estimation of physiologically active zinc in maize by biochemical assay. Plant Soil. 63, 395-406.

Harley C P, Regeimbal L O and Moon H H 1956 Absorption of nutrient salts by bark and woody tissues of apple and subsequent translocation. Am. Soc. Hort. Sci. 67, 47-57.

Hill J, Robson A D and Loneragan J F 1979 The effect of copper supply on the senescence and the retranslocation of nutrients of the oldest leaf of wheat. Ann. Bot. 44, 279-287.

Hocking P J 1980 Redistribution of nutrient elements from cotyledons of two species of annual legumes during germination and seedling growth. Ann. Bot. 45, 383-396.

Hocking P J and Pate J S 1977 Mobilization of minerals to developing seeds of legumes. Ann. Bot. 41, 1259-1278.

Hocking P J and Pate J S 1978 Accumulation and distribution of mineral elements in the annual lupins *Lupinus albus* L and *Lupinus angustifolius* L. Aust. J. Agric. Res. 29, 267-280.

Jolley V D, Brown J C 1991 Factors in iron-stress response mechanism enhanced by Zn-deficiency stress in Sanilac, but not Saginaw navy bean. J. Plant Nutr. 14, 257-265.

Krotz R M, Evangelou B P and Wagner G J 1989 Relationships between cadmium, zinc, Cd-peptide and organic acid in tobacco suspension cells. Plant Physiol. 91, 780-787.

Kuo J, Hocking P J and Pate J S 1982 Nutrient reserves in seeds of selected Proteaceous species from south-western Australia. Aust. J. Bot. 30, 231-249.

Layzell D B, Pate J S, Atkins C A and Canvin D T 1981 Partitioning of carbon and nitrogen and the nutrition of root and shoot apex in a nodulated legume. Plant Physiol. 67, 30-36.

Loneragan J F, Kirk G J and Webb M J 1987 Translocation and function of zinc in roots. J Plant Nutr 10, 1247-1254.

Massey H G and Loeffel F A 1967 Factors in interstrain variation in zinc content of maize (*Zea mays* L.) kernels. Agron. J. 59, 214-217.

Mathys W 1977 The role of malate, oxalate and mustard oil glucosides in the evolution of zinc resistance in herbage plants. Physiol. Plant. 40, 130-136.

McGrath J F 1982 Some aspects of the zinc nutrition of *Pinus radiata* D. Don. Ph. D. thesis, The University of Western Australia, Nedlands, Western Australia.

McGrath J F and Robson A D 1984a The distribution of zinc and the diagnosis of zinc deficiency in seedlings of *Pinus radiata*, D. Don. Aust. For. Res. 14, 175-186.

McGrath J F and Robson A D 1984b The influence of zinc supply to seedlings of *Pinus radiata* D Don on the internal transport of recently absorbed zinc Aust J Plant Physiol 11, 165-178

McGrath J F and Robson A D 1984c The movement of zinc through excised stems of seedlings of *Pinus radiata* D Don Ann Bot 54, 231-242

Mozafar A 1990 Kernel abortion and distribution of mineral elements along the maize ear Agron J 82, 511-514

Mullins G L, Sommers L E and Housley T L 1986 Metal speciation in xylem and phloem exudates Plant Soil 96, 377-391

Nable R O and Webb M J 1993 Further evidence that zinc is required throughout the root zone for optimal plant growth and development Plant Soil (*in press*)

Obata H and Kitagishi K 1980a Time course of zinc or manganese accumulation within individual leaves J Sci Soil Manur, Jpn 51, 292-296 (*in Japanese*)

Obata H and Kitagishi K 1980b Investigation on pathway of zinc transport in vegetative node of rice plant by autoradiography J Sci Soil Manur, Jpn 51, 297-301 (*in Japanese*)

Ohki K 1977 Critical zinc levels related to early growth and development of determinate soybeans Agron J 69, 969-974

Orphanos P I 1975 Spray application of zinc to young apple trees Hort Res 15, 23-30

Polar E 1976 Variations in zinc content of subcellular fractions from young and old roots, stems and leaves of broad beans (*Vicia faba*) Physiol Plant 38, 159-165

Reay P F 1987 The distribution of nine elements in shoots of *Lupinus albus* L and *Lupinus angustifolius* L compared with that of silicon as a measure of passive transport Ann Bot 59, 219-225

Reay P F and Waugh C 1981 Mineral-element composition of *Lupinus albus* and *Lupinus angustifolius* in relation to manganese accumulation Plant Soil 60, 435-444

Reuter D J 1980 Distribution of copper and zinc in subterranean clover in relation to deficiency diagnosis Ph D Thesis, Murdoch University, Western Australia

Reuter D J, Loneragan J F, Robson A D and Plaskett D 1982 Zinc in subterranean clover (*Trifolium subterraneum* L cv Seaton Park) I Effects of zinc supply on distribution of zinc and dry weight among plant parts Aust J Agric Res 33, 989-999

Riceman D S and Jones G B 1958a Distribution of zinc and copper in subterranean clover (*Trifolium subterraneum* L) grown in culture solutions supplied with graduated amounts of zinc Aust J Agric Res 9, 73-122

Riceman D S and Jones G B 1958b Distribution of dry weight and of zinc and copper among the individual leaves of seedlings of subterranean clover (*Trifolium subterraneum* L) grown in complete culture solution and in a culture solution deficient in zinc Aust J Agric Res 9, 446-463

Riceman D S and Jones G B 1958c Distribution of zinc in subterranean clover (*Trifolium subterraneum* L) grown to maturity in a culture solution containing zinc labelled with the radioactive isotope ^{65}Zn Aust J Agric Res 9, 730-744

Robson A D and Pitman M G 1983 Interactions between nutrients in higher plants *In* Enc Pl Physiol, New Series, Vol 15b Eds A Lauchli and R L Bieleski pp 147-180

Robson A D and Snowball K 1989 The effect of 2-(4-2',4-dichlorophenoxy-phenoxy)-methyl propanoate on the uptake and utilization of zinc by wheat Aust J Agric Res 40, 981-990

Ruano A, Barcelo J and Poshcenrieder C 1987 Zinc toxicity-induced variation of mineral element composition in hydroponically grown bush bean plants J Plant Nutr 10, 373-384

Santa Maria G E and Cogliatti D H 1988 Bidirectional Zn-fluxes and compartmentation in wheat seedling roots J Plant Physiol 132, 312-325

Sieghardt H 1990 Heavy-metal uptake and distribution in *Silene vulgaris* and *Minuartia verna* growing on mining-dump containing lead and zinc Plant Soil 123, 107-111

Singh B R and Steenberg K 1974 Plant response to micronutrients I Uptake, distribution and translocation of zinc in maize and barley plants Plant Soil 40, 637-646

Spitzer E W, Webber M and Lott J N A 1981 Elemental composition of globoid crystals in protein bodies of wheat grain grown on soil treated with sewage sludge Can J Bot 59, 403-409

Sudia T W and Green D G 1972 The translocation of Zn-65 and Cs-134 between seed generations in soybean (*Glycine max* (L) Merr) Plant Soil 37, 695-697

Tiffin L O 1967 Translocation of manganese, iron, cobalt, and zinc in tomato Plant Physiol 42, 1427-1432

Turner R G 1970 The subcellular distribution of zinc and copper within the roots of metal tolerant clones of *Agrostis tenuis* Sibth New Phytol 69, 725-731

Turner R G and Marshall C 1971 The accumulation of Zn-65 by root homogenates of zinc-tolerant clones of

Agrostis tenuis Sibth. New Phytol. 70, 539-545.
Van Goor B J and Wiersma D 1976 Chemical forms of manganese and zinc in phloem exudates. Physiol. Plant. 36, 213-216.
Van Steveninck R F M, Van Steveninck M E, Fernando D R, Godbold D L, Horst W J and Marschner H 1987 Identification of zinc-containing globules in roots of a zinc-tolerant ecotype os Deschampsia caespitosa. J. Plant Nutr. 10, 1239-1246.
Van Steveninck R F M, Van Steveninck M E, Wells A J and Fernando D R 1990 Zinc tolerance and the binding of zinc as zinc phytate in Lemna minor, x-ray microanalytical evidence J. Plant Physiol. 137, 140-146.
Wallihan E F and Heymann-Herschberg L 1956 Some factors affecting absorption and translocation of zinc in citrus plants. Plant Physiol. 31, 294-299.
Watkins P A 1982 Copper and zinc nutrition of Granny Smith apple plants. M.S. Thesis. The University of Western Australia, Nedlands, Western Australia.
Welch, R M 1986 Effects of nutrient deficiencies on seed production and quality", in B. Tinker and A. Läuchli (eds) Advances in Plant Nutrition, Praeger, Connecticut, pp. 205-247.
White C L, Robson A D and Fisher H M 1981a Variation in nitrogen, sulphur, selenium, cobalt, manganese, copper and zinc contents of grain from wheat and two lupin species grown in a range of Mediterranean environments. Aust. J. Agric. Res. 32, 47-59.
White M C, Decker A M and Chaney R L 1979 Differential cultivar tolerance in soybean to phytotoxic levels of soil Zn. I. Range of cultivar response. Agron. J. 71, 121-126.
White M, Decker A and Chaney R 1981b Metal complexation in xylem fluid. I. Chemical composition of tomato and soybean stem exudate. Plant Physiol. 67, 292-300.
White M, Baker F, Chaney R and Decker A 1981c Metal complexation in xylem fluid. II. Theoretical equilibrium model and computational computer program. Plant Physiol. 67, 301-310.

Chapter 7.

Form and Function of Zinc Plants

PATRICK H. BROWN, ISMAEL CAKMAK and QINGLONG ZHANG

1. Abstract

The essential micronutrient zinc occurs in plants either as a free ion, or as a complex with a variety of low molecular weight compounds. Zinc may also be incorporated as a component of proteins and other macromolecules. As a component of proteins, zinc acts as a functional, structural, or regulatory cofactor of a large number of enzymes. Many of the physiological perturbations resulting from zinc deficiency are associated with the disruption of normal enzyme activity, thus zinc-deficiency induced inhibition of photosynthesis is coincident with a decrease in activity of key photosynthetic enzymes. Zinc deficiency also increases membrane leakiness by inhibiting the activity of enzymes involved in the detoxification of membrane damaging oxygen radicles. Recent evidence suggests that zinc plays a key role in stabilizing RNA and DNA structure, in maintaining the activity of DNA synthesizing enzymes and controlling the activity of RNA degrading enzymes. Thus, zinc may play a role in controlling gene expression. Though our understanding of the function of zinc has increased greatly in the last thirty years, there are still many aspects of zinc metabolism that remain controversial. In the following review we summarize the current knowledge of the physiology of zinc and illustrate areas in which our knowledge remains incomplete.

2. Introduction

The biological role of Zn was first identified by Raulin in 1869, who observed that common bread mold (*Aspergillus niger*) did not grow in the absence of Zn. Soon thereafter, Zn was identified as an ubiquitous component of both animal and plant tissue. This observation stimulated Zn research in crops, and in 1914 the first demonstration of Zn deficiency in plants was made (Mazé, 1914). Evidence that Zn was generally essential for plants followed in 1926 (Sommer and Lipman), while the first identification of Zn deficiency in field conditions was reported in 1937 in the deciduous orchards of California (Chandler, 1937). Zinc deficiency is now recognized as one of the most common micronutrient deficiencies and is becoming increasingly significant in crop production. Whereas the agricultural significance of Zn was recognized in the early part of the 20th century, a specific role for Zn in plants was not identified until the late 1960's. Since then a series of Zn-containing enzymes have been identified and considerable progress has been made in identifying the chemical form and physiological effects of Zn deficiency in plants.

In the last 20 years several reviews have dealt with aspects of Zn in soils and plants. The two most comprehensive of these provide a good account of the reactions of Zn in soils (Lindsay, 1972) and forest ecosystems (Boardman and McGuire, 1990). In this chapter we discuss the chemical forms of Zn in the plant and the role of Zn in plant metabolism and physiology. Particular attention will be paid to the physiological and morphological effects of Zn deficiency in higher plants.

3. Forms of zinc in plants

In plants, Zn does not undergo valency changes, and its chemistry differs from Mg^{2+}, Ca^{2+}, or Mn^{2+} in that it forms more stable complexes with a particular affinity for tetrahedral complexes. The majority of Zn in the leaf is associated with low molecular weight complexes, storage metalloproteins, free ions, and insoluble forms associated with the cell wall. Zinc can become inactivated within the cell either by ligand formation (Leece, 1978) or by complexation with phosphorous (Olsen, 1972). Depending on plant species, anywhere from 58% to 91% of plant Zn may be soluble (Peterson, 1969; Welch et al., 1976). This water-soluble Zn fraction is usually considered to be the physiologically active fraction and is regarded as a better indicator of Zn status than is total Zn content (Cakmak and Marschner, 1987). Among these soluble Zn forms, low molecular weight complexes are frequently in the greatest abundance and are probably the most significant form of active Zn. Though more than 70 Zn metalloenzymes have been identified, functional metalloproteins represent only a small proportion of the total Zn present in the plant.

3.1. Low molecular weight complexes and free Zn

Identification of the dominant low molecular weight Zn complexes has been performed with a range of electrophoretic and chromatographic techniques and more recently has been estimated by using chemical equilibrium computer programs (CHELATE, GEOCHEM). Most of the soluble Zn in the cell is associated with low molecular weight anionic complexes, with only a small percent present as free ions.

In plant leaves, soluble Zn exists largely as an anionic compound possibly associated with amino acids. In lettuce (*Lactuca sativa* L.) the major Zn fraction had a molecular weight of 1250 and contained S, reducing sugars, and amino acids (Walker and Welch, 1987). The isolated low molecular weight fractions represented 73% of total soluble Zn, which is 58% of the total Zn in lettuce leaf. Similar results have been observed in seed, where 62% to 70% of seed Zn was soluble (Welch et al., 1974; Khan and Weaver, 1989). Free Zn ion present in leaf tissues is generally low and ranges from 5.8% of total Zn in tomato (Bowen et al., 1962) to 6.5% in alfalfa (Johnson and Schrenk, 1964).

The cell wall may also play a role in controlling free Zn activity. A relationship between the affinity of cell wall extracts for free Zn, and the relative tolerance of diverse species to excess Zn, suggests a role for the cell wall in controlling free Zn in the cell (Turner, 1970; Turner and Marshall, 1971). As reviewed by Torre et al., (1991) various cell wall compounds (i.e. lignin, cellulose, hemicellulose, etc.) possess a high binding affinity to Zn. Accordingly, it is assumed that 90% or more of the total Zn in roots is adsorbed in the apoplast of rhizodermal and cortical cells (Schmid et al. 1965). The significance of cell wall binding of Zn, however, remains controversial, particularly as the total binding capacity of the cell wall may be too small to be physiologically significant (Wainright and Woolhouse, 1978).

The presence of significant levels of low molecular weight Zn complexes may play an important role in supplying Zn to physiologically active macromolecules. The pool of low molecular weight Zn complexes would be easily degradable and as such may be regarded as 'physiologically active' in the same way that Zn incorporated in enzymes is 'physiologically active' (Olsen, 1972). Low molecular weight Zn-ligands may also have catalytic activity themselves - e.g., amide hydrolysis by Cu and Zn (Groves and Dias, 1979) - though it is unlikely that these compounds have sufficient specificity or activity to play a significant catalytic role in higher plants (Walker and Welch, 1987).

Finally, low molecular weight species may play a role in Zn detoxification. Various ligands in the cell may act as a buffer system to absorb excessive metal concentrations; an example of this is the 'phytochelatins" isolated by Grill et al. (1985). Phytochelatins are low molecular weight metal-binding peptides with the structure of (γ-glutamylcysteine)n-glycine, which have been identified in a wide range of species and are synthesized in response to exposure to excess levels of heavy metals such as Cd, Zn, and Hg. These compounds are believed to play a role in the homeostasis and detoxification of Zn through metal thiolate formation.

Most of the soluble Zn in plants seems to be bound to low molecular weight ligands. However, isolation of a Zn-binding compound in plant extracts does not necessarily imply a Zn binding role *in vivo*, since redistribution and dissociation of Zn among ligands during isolation is possible (Walker and Welch, 1987). The presence of these low molecular weight ligands in Zn binding *in vivo* has not been demonstrated and remains to be elucidated further.

3.2. Zinc in proteins

In plants, Zn acts as a functional, structural, or regulatory cofactor of a large number of enzymes. The Zn atom is usually tightly bound to the apoenzyme and can be removed only with severe chemical treatment. It forms strong complexes with radicals of polar groups containing O_2, N, and S (Figure 1).

The main Zn-containing enzymes, and enzymes that require Zn for activity, are remarkably diverse in their properties, though most of the enzymes bind Zn through imidazole and cysteine. X-ray analysis shows that catalytic Zn is bound with three protein ligands and a water molecule, whereas in proteins where Zn plays a structural and regulatory role, Zn is fully coordinated by four protein ligands (Vallee, 1983). The presence of a water molecule bound to catalytic Zn, signifying an open coordination site, is considered essential for the function of Zn in catalysis. The coordinate geometry of catalytic Zn is highly distorted and can fluctuate between tetracoordinate and pentacoordinate - i.e., it is *entatic*, while noncatalytic Zn is more regular in its coordinate geometry.

Since Zn has a highly localized charge and electron affinity, Zn enzymes have a very effective attacking group. The property of *entasis* enables the Zn metalloenzymes to have a lower energy barrier for the transition state and hence accelerates the conversion of substrate to products (Vallee, 1983). Enzymes with less aggressive attacking centers than Zn can only attack large substrates by using many points of attachment to the substrate. Zn enzymes are mainly functional with (a) low molecular weight substrates which cannot be handled by organic groups in a multi-point attachment and for (b) less specific but very rapid attack involving relatively weak binding of the substrate (Williams, 1989).

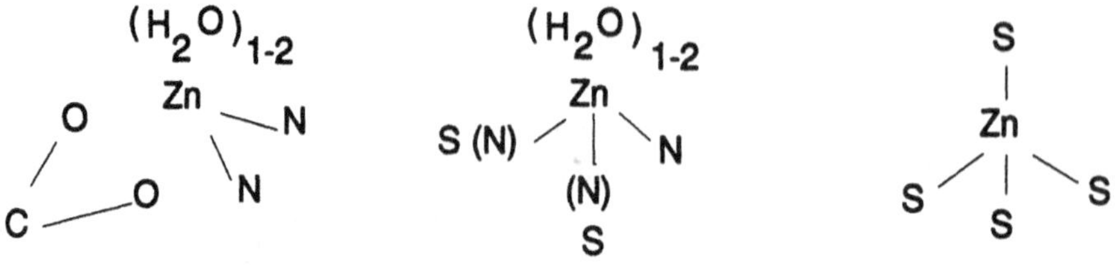

Figure 1. The geometries at the active sites of some enzymes. The geometries are usually irregular, to the left carboxypeptidases, centre carbonic anhydrase and alcohol dehydrogenase, in contrast with zinc sites which are just structural, shown on right. (Reprinted with permission from Williams, 1989)

4. The functions of zinc in physiological processes

Though the precise mode of action of Zn in enzymes is still not fully understood, two types of mechanism have been hypothesized to account for the manner in which the metal affects catalysis (Vallee, 1983).

1. Zinc-carbonyl hypothesis: The substrates bind directly to Zn to form an 'enzyme-substrate" complex and displace metal-bound H_2O molecules in the process; i.e., Zn functions to activate the electrophile (e.g., aldolase, peptidase).

2. Zinc-hydroxide mechanism: The substrate does not bind directly to the metal but mediates its function through the metal-bound water molecule. The resultant metal-bound hydroxide ion can then attack the substrate; i.e., Zn acts to activate the nucleophile (e.g., CA).

In addition to the above two mechanisms, it is possible that a third mechanism that integrates the Zn-carbonyl and Zn-hydroxide hypothesis may exist. In this mechanism Zn would act both polarizing the substrate and activating the water molecule. This explanation is favored by some investigators (King and Fife, 1983).

Whatever the potential mechanisms, the involvement of Zn in the activity of various enzymes would indicate that it has profound effects on normal plant metabolism. In general, metabolism of carbohydrates, proteins, auxin, and reproductive process are most severely interrupted under Zn deficiency. The influence of Zn on these and other processes is discussed in the following text.

4.1. Carbohydrate metabolism:

The involvement of Zn in carbohydrate metabolism can be demonstrated through its effect on photosynthesis and sugar transformations. In general, respiration in plants is not affected by Zn status.

1. Photosynthesis: Zn deficiency can cause a reduction of net photosynthesis by 50% to 70%, depending on plant species and degree of deficiency. Several mechanisms can contribute to this reduction.

(a) Carbonic Anhydrase (CA) activity: Zinc is a component of plant CA (Tobin, 1970). The enzyme from dicotyledons is composed of six subunits with a molecular weight of about 180,000 and 6 Zn atoms per molecule (Tobin, 1970), while CA in

monocotyledon is much smaller and contains less Zn. Some species may also contain Zn free CA (Fellner, 1963). CA is localized in the chloroplast in C_3 plants and cytosol of mesophyll in C4 plants. Some CAM plants may have CA in chloroplast while others in cytosol.

When plants encounter increasing Zn stress, a sharp decline in CA activity occurs (Ohki, 1976). This is the most immediate effect among any of the various enzymes of the CO_2 assimilation pathway. Though the function of CA in C_3 plant remains uncertain, the presence of CA in chloroplasts of C_3 plants was considered evidence for its involvement in maintaining the internal bicarbonate pool of the chlorophyll or, possibly, its association with ribose bisphosphate carboxylase in CO_2 fixation (Everson and Slack, 1968). In C_4 plants, the CA present in cytosol of mesophyll cells specifically catalyzes the conversion of CO_2 to HCO_3^-, which is then assimilated by PEP carboxylase (Hatch and Burnell, 1990).

Evidence that CA is not directly involved in C_3 photosynthesis was presented by Randall and Bouma (1973). They found that CA activity in Zn-deficient spinach dropped to about 10% of those in normal plants, while phosphorylation rates were only slightly affected. They suggested that CA has no close relationship with photosynthesis and that the effect of Zn is on the photosynthetic process itself rather than via CA. This conclusion was supported by Boardman (1975). In contrast, Seethambaram and Das (1985) observed that photosynthesis in Zn deficient rice (*Oryza sativa*), could be enhanced by the addition of CA in combination with ATP and NADPH. In the C_4 plant pearl millet (*Pennisetum americanum*), however, the addition of ATP and NADPH alone was sufficient to overcome the reduction in photosynthesis under Zn deficiency.

It is significant that Zn deficiency results in impaired photosynthesis in all plant species whereas the relative importance of CA to the photosynthetic process varies greatly between C_3 and C_4 plants. Further, in some species in which Zn-deficiency results in reduced photosynthesis, the CA appears to be Zn-free (Fellner, 1963). Given the diversity in function and form of CA in plants it seems doubtful that the primary role of Zn in photosynthesis is through its function in CA.

(b) Other photosynthetic enzymes: The effect of Zn on photosynthesis is not only via carbonic anhydrase. In Zn-deficient navy bean, Jyung et al. (1972) observed a reduction in the activity of ribulose 1,5-bisphosphate carboxylase (RuBPC), which catalyzes the initial step of photosynthetic CO_2 fixation. This reduction was confirmed in barley (*Hordeum vulgare*) (Stiborova et al., 1987) and in rice and pearl millet (Seethambaram and Das, 1985). Zinc may stabilize carbamate formation in a manner similar to Mg^{2+} in the enzyme CO_2-Mg^{2+} complex.

(c) Chlorophyll content and chloroplast structure: Reduction in photosynthesis under Zn deficiency can also be a result of a drastic decrease in chlorophyll content, and abnormal structure of both the mesophyll and bundle sheath chloroplasts. Using electromicroscopy, Shrotri et al. (1978) observed that mesophyll chloroplasts in maize became swollen and the number and size of grana thylakoids were reduced, whereas the number and size of osmiophilic globuli were increased under Zn deficiency. The bundle sheath chloroplasts also showed similar phenomena. Such effects of Zn deficiency on chloroplasts have also been reported for beans (*Phaseolus vulgaris*) by Thomson and Weiser (1962) and Jyung et al. (1975).

The exact role of Zn in photosynthesis remains obscure and is complicated by the multiple effects of Zn deficiency on processes closely related to photosynthesis and carbon fixation. More research in this area is needed.

2. *Sucrose and starch formation:* Zn deficiency greatly depresses the activity of aldolase in plant tissue, which then impairs the conversion of fructose 1-6-diphosphate to its subsequent compounds. Singh and Gangwar (1974) reported that Zn deficiency caused a decline in the level of sucrose in sugarbeet (*Beta vulgaris* L.). Such effect of Zn deficiency on sucrose synthesis was confirmed in maize (*Zea mays*) by Shrotri et al. (1980). The lower sucrose synthetase activity was attributed to this reduction in sucrose level.

Zn may play a role in the metabolism of starch. Jyung et al. (1975) found that the starch content, activity of starch synthetase, and the number of starch grains were all depressed under Zn deficient in beans. Whether this effect on starch and sucrose formation is a primary result of Zn deficiency still remains an open question. An increase in reduced sugar and reduction of starch formation under Zn deficiency has been confirmed in other crops (Reed, 1940; Sukhija et al., 1987).

In contrast to these reports, in cabbage (*Brassica olereacea* L.) (Sharma et al., 1982) and bean (Marschner and Cakmak, 1989) leaves, especially in mature leaves, Zn deficiency increased the concentrations of sugars and starches, particularly sucrose, whereas in bean, root concentrations of carbohydrates were strongly decreased. This altered distribution of carbohydrates between leaves and roots indicates impaired sucrose transport from the source leaves of Zn-deficient plants. Correspondingly, it was found that resupply of Zn to deficient plants for 48 h clearly caused restoration of phloem loading of sucrose (Cakmak, 1988; Marschner and Cakmak, 1989). The reason for the impaired sucrose transport from Zn-deficient source leaves is not well understood, though it might be related to the role of Zn in the structural integrity of biomembranes (Welch et al., 1982; Cakmak and Marschner, 1988a).

4.2. *Protein metabolism:*

In general, the amount of protein in Zn-deficient plant is dramatically reduced but the composition remains almost unchanged. Compared with control plants, in Zn-deficient bean leaves, the concentration of free amino acids measured by HPLC is increased by a factor of 6.5, which is decreased to 5.1, 2.7, and 1.4 after a resupply of Zn to deficient plants for 24 h, 48 h, and 72 h, respectively (Cakmak et al., 1989). These decreases are associated with a simultaneous increase in protein concentration, indicating a major role of Zn in protein synthesis. The mechanism of adverse effect of Zn deficiency on protein synthesis is attributed to a sharp reduction in RNA and deformation and reduction of ribosomes(Prask and Plocke, 1971; Kitagishi and Obata, 1986). This has been confirmed by Kitagishi et al. (1987), who reported that RNA level and free 80S ribosomes, in the meristem tissue of rice seedlings, were depressed dramatically under Zn deficiency.

The reduction of RNA content under Zn deficiency can contribute to its effect on RNA polymerase and RNase. Zn is required for the activity of RNA polymerase (Falchuk et al., 1978; Jendrisak and Burgess, 1975) and protects ribosomal RNA from attack by ribonuclease. RNase activity is known to be strongly depressed by the presence of Zn, and high rates of RNase activity are a typical feature of Zn deficiency in higher plants (Dwivedi and Takkar, 1974). Thus the earliest observed causal event of Zn deficiency is a sharp decrease in the level of RNA. This reduction in RNA can occur even before the enhancement of RNase in rice and pearl millet seedlings (Seethambaram and Das, 1984). This indicates that the effect of Zn deficiency is more pronounced on the biosynthesis of RNA and less on the enhancement of RNase activity. The importance of Zn in protein synthesis would suggest that remarkably high concentrations of Zn are required by meristematic tissue, where cell division as well as synthesis of nucleic acid and protein is actively taking place. This has been found to be the case in the meristem tissue in rice (Kitagishi and Obata, 1986).

The most fundamental effect of Zn on protein metabolism is via its involvement in the stability and function of genetic material. In addition to its effect on stabilization of the structure of RNA and DNA, Zn is believed to be essential for at least two chromatic proteins. The chromatin TFIIIA protein is essential for transcription (Hanas et al., 1983) while $g^{32}P$ protein is involved in replication (Giedroc et al., 1986), by substitution of Zn for Mg in the ternary enzyme-CO_2-metal complex or by reaction with enzyme SH groups (Stiborova et al., 1987). These Zn-binding domains of proteins, usually termed "zinc fingers", have been found in higher plants (Vodkin and Vodkin, 1989). Zn is believed to be involved in maintaining the protein conformation of the DNA-binding domain required for function (Hanas et al., 1983). Czupryn et al. (1987) reported that Zn deficiency induces changes in the amount and types of histone and non-histone proteins as well as in their interaction with DNA in *Euglena gracilis*. This further confirms that Zn regulates the activity of genetic material by modulating the binding properties of regulatory proteins to their particular genes. Zinc may act directly as a component of the protein or indirectly as an activator affecting the function of Zn-activated kinase capable of chemically modifying histone and non-histone proteins (Kang et al., 1974). A specific effect of Zn deficiency on gene expression in animals may occur as a result of the failure to induce a group of enzymes required for DNA synthesis (Chesters, 1992). Thus zinc may play a role in facilitating changes in gene expression either as a component of the zinc finger protein required to enhance transcription of specific genes, or through alterations in chromatin structure, or by disrupting the synthesis of DNA. Further work on the role of Zn in gene expression is required.

4.3. Membrane integrity:

In animals, Zn is believed to play a critical physiological role in the structure and function of the biomembrane (Bettger and O'Dell, 1981; Chvapil, 1973). The evidence of Zn involvement in membranes of higher plants has been demonstrated indirectly. Using root exudates as an indicator of root plasma membrane permeability, Welch et al. (1982) found greater leakage of ^{32}P from roots of Zn-deficient wheat than from Zn-adequate roots. Cakmak and Marschner (1988a) also observed that Zn deficiency increased root exudation (net flux) of K^+, amino acids, sugars, and phenolics by a factor of at least 2.5. Resupply of Zn to deficient plants for 12 hours substantially decreased this leakage. From these results, it was concluded that Zn must have a role in maintaining the integrity of cellular membranes. This function may involve the structural orientation of macromolecules within the membrane or the maintenance of ion transport mechanisms.

Zn is known to be required for stabilization of biomembranes by interaction with phospholipids (Sunamota et al., 1980) and sulfhydryl groups of membrane protein (Chvapil, 1973). According to Bettger and O'Dell (1981), loss of membrane integrity is the earliest biochemical change caused by Zn deficiency. Besides its role as a structural component in biomembranes similar to Ca, Zn plays a key role in controlling both generation as well as detoxification of free oxygen radicals, which are potentially damaging to membrane lipids and sulfhydryl groups. Zinc especially exerts an inhibitory action on membrane damage catalyzed by $O_2^{\cdot -}$ - generating NADPH oxidase, as shown in microsomal membranes of rats (Burke and Fenton, 1985; Bray et al., 1986).

In higher plants, similar effects of Zn on $O_2^{\cdot -}$ levels and $O_2^{\cdot -}$—generating NADPH oxidase have been found. Higher levels of $O_2^{\cdot -}$ were found in Zn-deficient plant roots (Cakmak and Marschner, 1988b,c) . The phenomena was thought to occur as a result of enhanced activity of an $O_2^{\cdot -}$ - generating NADPH oxidase as well as depressed Superoxide Dismutase (SOD) activity. SOD is the key enzyme involved in the scavenging of superoxide radicals in the plant. The most important plant SOD is a Cu-Zn metalloprotein in which Zn plays a key structural role affecting the enzyme stability. In Zn-deficient plants activity of catalase, an H_2O_2-scavenging enzyme, is also decreased (Cakmak and Marschner, 1988c), most probably as a result of inactivation of catalase by $O_2^{\cdot -}$ (Fridovich, 1986) or inhibited protein synthesis (see above). On the basis of these changes, it is suggested that Zn deficiency results in peroxidative damage to biomembranes by $O_2^{\cdot -}$ and $O_2^{\cdot}$ - derived toxic O_2 species. This damage can explain the well-known phenomena of enhanced leakage of organic and inorganic solutes from root cells (Welch et al., 1982; Cakmak and Marschner, 1988a). Thus the major role of Zn in membranes seems to be related to protection of membrane lipids and proteins from peroxidation.

4.4. Auxin metabolism:

Skoog (1940) first reported that IAA levels were reduced when tomato plants (*Lycosperscion esculentum*) were under Zn stress. Addition of Zn caused an increase in IAA and a corresponding stem elongation. He concluded that Zn was required for maintaining auxin in an active state and that the decreased IAA level was caused by enhanced oxidation. This conclusion is supported by the presence of indole carboxylic acid, an assumed oxidative product of IAA, in radish (*Raphanus sativus*) shoot (Domingo et al., 1990). However, Fujiwara and Tsutsumi (1954) reported that IAA oxidation in the Zn-deficient roots of barley was comparable with that of the control plant. Using the same species, Tsui (1948) confirmed Skoog's findings but also found that the decreased IAA levels were due to a reduced rate of synthesis of tryptophan, the precursor for the biosynthesis of IAA. He concluded that Zn is required directly for the synthesis of tryptophan and indirectly for the synthesis of IAA. This conclusion has been supported by Salami and Kenefick (1970), who demonstrated that addition of either Zn or L-tryptophan to a Zn-free nutrient solution eliminated Zn-deficiency symptoms in maize. In addition, tryptophan synthetase, an enzyme responsible for the synthesis of tryptophan from indole and serine, was found to require Zn for its maximum activity (Klein et al., 1962).

There is increasing evidence against a role for Zn in tryptophan synthesis. Takaki and Kushizaki (1970, 1976) found higher levels of free amino acid in Zn-deficient maize and tomato plants which increased as the deficiency became more severe. Free tryptophan content increased as Zn-deficiency symptoms became severe. This was also true for tryptamine in maize seedlings. These results have been confirmed in oat and barley (Takaki and Arita, 1985, 1986) and bean (Cakmak et al., 1989). Accumulation of tryptophan and tryptamine was not observed in plants with Fe, Cu, Mn, and B deficiency. These authors concluded that the reduction in IAA level in Zn-deficient plants is not brought about by impaired synthesis of tryptophan and more likely it is the conversion of tryptophan to IAA that is specifically inhibited. Thus Zn is necessary in the pathway of IAA biosynthesis from tryptophan.

The reason for this discrepancy in tryptophan levels is not known, though a different methodology was used in detecting these compounds (Cakmak et al., 1989). Increased concentration of tryptophan, and other amino acids, in Zn-deficient tissues is suggested to be more likely a result of inhibited protein synthesis and not inhibited conversion of tryptophan to IAA (Cakmak et al., 1989). Lower IAA concentrations in Zn-deficient tissues may be a result of enhanced oxidative degradation of IAA by oxygen radicals, especially by hydroxyl radicals, as shown *in vitro* by Cakmak (1988) confirming the results of Skoog (1940) and Domingo et al. (1990).

However, according to Law (1987), D-tryptophan is the precursor of IAA and not L-tryptophan, furthermore, the rate of transformation from L- into D- form is enhanced by gibberellin. No distinction between D- or L- forms of tryptophan nor determination of gibberellin levels was reported in the above experiments; thus a causal relationship between Zn and indole compounds was not established. Shkolnik et al. (1975) found that gibberellin-like substances decrease in the leaves of Zn-deficient bean. This GA reduction was confirmed by Suge et al. (1986) in barley, maize, and oat plants. Supplementation of Zn to Zn-stressed plant roots not only increased growth but also increased levels of gibberellin-like substances. Since gibberellic acid promoted the conversion of tryptophan and tryptamine to IAA (Muir and Lantican, 1968), Suge et al. (1986) suggest that it may be possible that GA is the primary candidate affected by Zn deficiency, rather than IAA. A causal relationship between Zn and indole compounds remains a topic for further study. The possible mechanisms of Zn deficiency effects on IAA metabolism are presented in Figure 2.

4.5. Defense mechanisms:

The role of Zn in the defense mechanisms of higher plants is controversial and will not be discussed here. Readers are referred to the detailed review of Graham (1983).

4.6. Reproduction:

Flowering and seed production are known to be severely depressed by Zn deficiency in beans, peas, and other plants (Reed, 1941; Hu and Sparks, 1990). Using solution culture, Ricemean and Jones (1959) found that dry matter of subterranean clover plants was influenced only slightly by raising Zn levels in the media, whereas the yields, were dramatically increased over the same range primarily because of increased number of inflorescences and seeds. The unchanged weight of individual seed regardless of Zn supply suggested that the effect of Zn was on seed formation rather than on seed growth.

Lower seed production under Zn deficiency can be attributed to (1) enhanced formation of abscissic acid in the plant, causing premature abscission of leaves and flower buds; (2) disruption of the development and physiology of anthers and pollen grains.

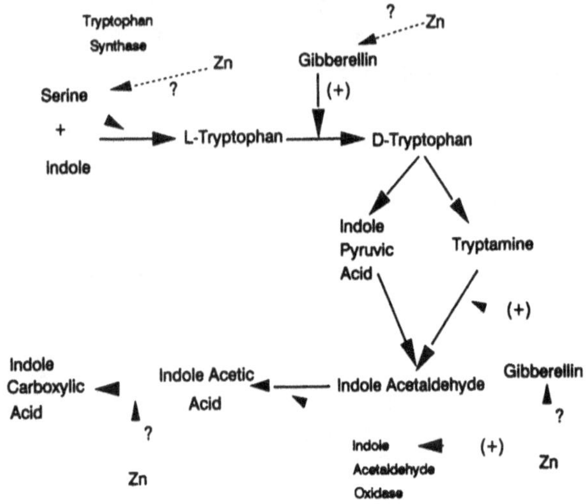

Figure 2. Proposed roles of zinc in indole acetic acid (IAA) synthesis.

Sharma et al. (1979) reported that Zn-deficient wheat developed small anthers with abnormal pollen grains; many anthers were found to be empty and completely devoid of pollen grains in acute Zn deficiency. Polar (1970) also observed higher Zn concentrations in reproductive organs and suggested that anther and pollen grain development in *Vicia faba* may have a higher requirement of Zn than vegetative growth.

5. Concluding remarks

Knowledge of the functional role of Zn is incomplete and in some instances remains controversial. The essentiality of Zn as a component of enzymes has been known for several decades, but it has been appreciated only recently that some of the functions of Zn may be replaced by other dissimilar elements. For example, Zn can be replaced by cobalt in carboxypeptidase, resulting in an increase in peptidase activity (Boardman and McGuire, 1990). Zn in superoxide dismutase (SOD) can also be replaced by cobalt without affecting the enzyme activity (Calabrese et al., 1972). Interactions between Zn and other elements such as P, Cu, and Cd have been reported widely, but they are still poorly understood (Suttle, 1975).

In general, the major portion of Zn in plant tissues exists as low molecular weight anionic complexes. Functional metalloproteins are only a small pool for Zn, but they represent the most important portion from a metabolic standpoint. The effects of Zn deficiency have not been directly attributed to enzyme malfunction, but to diminished activity of enzymes in certain tissues. This leads to a reduction of photosynthesis, decreased starch formation, reduced auxin level, accumulation of amino acids with a decrease in protein synthesis, an increase in permeability of bio-membranes, enhanced inorganic P content, and depression of male fertility.

The effects of Zn deficiency on physiological processes are unlikely to be uniform for all plant species and/or all tissues. For example, the effect of Zn deficiency is known to be most pronounced in meristematic tissue, where cell proliferation and differentiation occur. This may be due to the fact that Zn stress depresses synthesis of DNA, which suppresses gene replication and finally cell formation and cell division. Interactions with other elements and environmental conditions may also affect the manifestation of Zn deficiency. Many of the physiological effects are likely to be only indirect consequences of Zn deficiency and it has been suggested from animal studies that "the lack of Zn primarily restricts gene expression rather than enzyme activation" (Chester, 1992). This may also be true of Zn deficiency in plants.

Much remains to be done before the behavior of Zn in plants is fully understood. In the future, particular attention should be paid to the role of Zn in gene expression and regulation, the defense against oxygen-derived free radicals, and the prevention of photo-oxidation in chloroplasts.

References

Bettger W J and O'Dell B L A critical physiological role of Zn in the structure and function of biomembrane. Life Sci. 28, 1425-1438.

Boardman N K 1975 Trace elements in photosynthesis *In* Trace Elements in Soil-Plant-Animal Systems. Eds. D J D Nicholas and A R Efan. pp 199-212. Academy Press, London.

Boardman R and McGuire D O 1990 The role of zinc in forestry, I. Zinc in forest environments, ecosystems and tree nutrition. Forest Ecology and Management 37, 187-205.

Bowen H J M, Cause P A and Thick J 1962 The distribution of some inorganic elements in plant tissue extracts. J. Exp. Bot. 13, 257-267.

Bray T M, Kubow S and Bettger W J 1986 Effects of dietary Zn on endogenous free radical production in rat lung microsomes. J. Nutr. 116, 1054-1060.

Burke J P and Fenton M R 1985 Effects of zinc deficient diet on lipid peroxidation in liver and tumor subcellular membrane. Proc. Soc. Exp. Biol. Med. 179, 187-191.

Cakmak I 1988 Morphologische und physiologische veranglerungen bei zinkmangelpflanzen PhD Thesis. University of Hohenheim-Stuttgart, Germany.

Cakmak I and Marschner H 1987 Mechanism of phosphorous-induced zinc deficiency in cotton III. Changes in physiological availability of zinc in plants. Physiol. Plant. 70, 13-20.

Cakmak I and Marschner H 1988a Increase in membrane permeability and exudation in roots of Zn Deficient Plants. J. Plant Physiol. 132, 356-361.

Cakmak I and Marschner H 1988b Zinc-dependent changes in ESR signals, NADPH oxidase and plasma membrane permeability in cotton roots. Physiol. Plant. 73, 132-186.

Cakmak I and Marschner H 1988c Enhanced superoxide radical production in roots of zinc deficient plants. J. Exp. Bot. 39, 1449-1460.

Cakmak I, Marschner H and Bangerth F 1989 Effects of zinc nutritional status on growth, protein metabolism and levels of indole-3-acetic acid and other phytohormones in bean *(Phaseolus vulgaris* L.) J. Exper. Bot. 40, 405-412.

Calabrese L, Rotilio G and Mondovi B 1972 Cobalt erythrocauprein: Preparation and properties. Biochim. Biophys. Acta 263, 827-829.

Chandler W H 1937 Zinc as a nutrient for plants. Bot Gaz 98, 625-646

Chesters J K 1992 Trace element-gene interactions. Nutrition Reviews 50:217-223.

Chvapil M 1973 New aspects in the biological role of zinc: A stabilizer of macromolecules and biological membranes. Life Sci. 13, 1041-1049.

Czupryn M, Falchuk K H and Vallee B L 1987 Zinc deficiency and metabolism of histones and non-histone proteins in *Euglena gracilis*. Biochemistry 26, 8263-829.

Domingo A L, Nayatome Y, Tamai M and Takaki H 1990 Indole carboxylic acid in zinc-deficient radish shoots. Soil Sci. Plant Nutri. 36, 555-560.

Dwivedi R S and Takkar P N 1974 Ribonuclease activity as index of hidden hunger of zinc in crops. Plant and Soil 40, 170-181.

Everson R G and Slack C R 1968 Distribution of carbonic anhydrase in relation to the C_4 pathway of photosynthesis. Phytochemistry 7, 581-584.

Falchuk K H, Hardy C, Ulpino L and Vallee B L 1978 RNA metabolism, manganese, and RNA polymerases of zinc sufficient and zinc deficient *Euglena gracilis*. Proc. Nat. Acad. Sci. USA 75, 4175-4179.

Fellner S K 1963 Zinc free plant carbonic anhydrase; lack of inhibition by sulfonamides. Biochim Biophys Acta 77, 155-156.

Fridovich I 1986 Biological effects of the peroxide radical. Arch. Biochem. Biophys. 247, 1-11.

Fujiwara A. and Tsutsumi M 1954 Biochemical studies of micro-elements. Tohoku J. Agr. Res. 4, 47-52.

Giedroc D P, Keating K M, Williams K R, Konigsberg W H and Colemean J E 1986 Gene 32 protein the single stranded DNA binding protein from bacteriophage T4 is a zinc metalloprotein. Proc. Natl. Acad Sci. USA 83, 8452-8456.

Graham R D 1983 Effects of nutrient stress on susceptibility of plants to disease with particular reference to the trace elements. Adv. Bot. Res. 10, 221-276.

Grill E, Winnacker E and Zenk M H 1985 Phytochelatins: the principle heavy metal complexing peptides of higher plants . Science (Washington D.C.) 230, 674-676.

Groves J T and Dias R M 1979 Rapid amide hydrolysis mediated by copper and zinc. J. Am. Chem. Soc. 101, 1033-1035.

Hanas J S, Hazuda D J, Bogenhagen D F, Wu F Y H and Wu C W 1983 Xenopus transcription factor A requires zinc for binding to the 5S RNA gene. J. Biol. Chem. 258, 14120-14125.

Hatch M D and Burnell J N 1990 Carbon anhydrase activity in leaves and its role in the first step of C_4 photosynthesis. Plant Physiol. 93, 825-828.

Hu H and Sparks D 1990 Zinc deficiency inhibits reproductive development in 'Stuart' pecan. Hortscience 25, 1392-1396.

Jendrisak J and Burgess R R 1975 A new method for the large scale purification of wheat germ DNA-dependent RNA polymerase II. Biochemistry 14, 4639.

Johnson W J and Schrenk W G 1964 Nature of zinc containing substances in the alfalfa plant cell. Agr. Food Chem. 12, 210-213.

Jyung W H, Camp M E, Polson D E, Afams M W and Witter S H Differential response of two bean varieties to zinc as revealed by electrophoretic protein pattern. 1972 Crop Sci. 12, 26-29.

Jyung W H, Ehmann A, Schlender K K and Scala J 1975 Zinc nutrition and starch metabolism in *phaseolus vulgaris* L. Plant Physiol. 55, 414-420.

Kang Y J, Olson O J M and Busch H 1974 Phosphorylation of acid soluble proteins in isolated nucleoli of *Novikoff Hepatoma Ascites* cells. J. Biol. Chem. 258, 5580-5585.

Khan A and Weaver C M 1989 Pattern of Zn^{65} incorporation into soybean seeds by root absorption, stem injection and foliar application. J. Agric. Food Chem. 37, 855-860.

King S W and Fife T H 1983 Effects of metal ion substitution on carboxypeptidase A catalized hydrolysis of O-trans-cinnamoyl-L-ß-phenyllactate. Biochemistry 22, 3603.

Kitagishi K and Obata H 1986 Effects of zinc deficiency on the nitrogen metabolism of meristematic tissues of rice plants with reference to protein synthesis. Soil Sci. Plant Nutr. 32, 397-405.

Kitagishi K, Obata H and Kondo T 1987 Effect of zinc deficiency on 80S ribosome content of meristematic tissues of rice plant. Soil Sci. Plant Nutr. 33, 423-429.

Klein R M, Captuto E M and Witterhole B A 1962 The role of zinc in the growth of plant tissue culture. Am. J. Bot. 49, 323-327.

Law D M 1987 Gibberellin enhanced indole-3-acetic acid biosynthesis: D-tryptophan as the precusor of indole-3-acetic acid. Physiol. Plant. 70, 626-632.

Leece D R 1978 Distribution of physiologically inactive zinc in maize growing in a Black Earth Soil. Aust. J. Ag. Res. 29, 749-58.

Lindsay W L 1972 Zinc in soils and plant nutrition. Adv. Agron. 24, 147-186.

Marschner H and Cakmak I 1989 High light intensity enhances chlorosis and necrosis in leaves of zinc , phosphorous and magnesium deficient bean (*phaseolus vulgaris*) plant J. Plant Physiol. 134, 308-315.

Maze P 1914 Influences respective des éléments de la solution minérale sur le dévelopment du maïs. Ann. Inst. Pasteur 28, 21-68.

Muir R M and Lantican B P 1968 *In* Physiology and Biochemistry of Plant Growth Substances. Eds. G Settefield and F Wightman. pp 259-272. Range Press, Ottawa.

Ohki K 1976 Effect of zinc nutrition on photosynthesis and CA activity in cotton. Physiol. Plant. 38, 300-304.

Olsen S R 1972 Micronutrient interaction. In Micronutrients in Agriculture. Eds. J J Mortvedt, P M Giordano, W L Lindsay. pp 199-227. American Society of Agronomy, Madison.

Peterson P I 1969 The distribution of ^{65}Zn in *Agostis tenuis sibth* and *Astolonifera L* tissue. Physiol. Plant. 40, 130.

Polar E 1970 The distribution of Zn^{65} in the cotyledons of *Vivia faba* and its translocation during the growth and maturation of the plant. Plant and Soil 32, 1-17.

Prask J A and Plocke D J 1971 A role for zinc in the structural integrity of cytoplasmic ribosomes of *Euglena gracilis*. Plant Physiol. 48, 150-155.

Rahimi A and Schropp A 1984 Carbonic anhydrase activity and extractable zinc as indicators of the zinc supply of plants. Z. Pflanzenernahr. u. Bodenkund 147, 572-583.

Randall P J and Bouma D 1973 Zinc deficiency, carbonic anhydrase and photosynthesis in leaves of spinach. Plant Physiol. 52, 229-232.

Reed H S 1940 Effects of zinc deficiency on phosphate metabolism of the tomato plant. Am. J. Bot. 33, 778-784.

Reed H S 1941 The relation of zinc to seed production. J. Agr. Res. 4, 635-644.

Ricemean D C and Jones G B 1959 Distribution of zinc and copper in subterranean clover (*Trifolium subterraneum L.*) grown in culture solution supplied with graduated amount of zinc. Aust. J. Agr. Res. 9, 73-122.

Salami A U and Kenefick D C 1970 Stimulation of growth in zinc-deficient corn seedlings by the addition of tryptophan. Crop Sci. 10, 291-294.

Schmid W E, Haag H P and Epstein E 1965 Absorption of zinc by excised barley roots. Physiol. Plant. 18, 860-869.

Seethambaram Y and Das V S R 1984 RNA and RNase of rice (*Oryza sativa L.*) and pearl millet (*Pennisetum americanum L. Leeke*) under zinc deficiency. Plant Physiol. Biochem. 11, 91-94.

Seethambaram Y and Das V S R 1985 Photosynthesis and activities and C3 and C4 photosynthetic enzymes under zinc deficiency in *Oryza sativa* L. and *Pennisetum americanum*(L.) Leeke. Photosynthetica 19,72-79.

Sharma P N, Chatterjee C, Sharma C P, Nautiya N and Agarwala S C 1979 Effect of zinc deficiency on development and physiology of wheat pollen. J. India Bot. Soc 58,330-334.

Sharma C P, Sharma P N, Bisht S S and Nautiyal B D 1982 Zinc deficiency induced changes in cabbage [*Brassica oleracea var. Capitata*]. *In* Plant Nutrition 1982: Proceedings of the Ninth international Plant Nutrition Colloquium, Warwick University, England, August 22-27, 1982. Ed. A Scaife. pp 601-606.

Shkolnik M Y, Davydava V N and Mocheniat K I 1975 Effect of zinc on the content of gibberellin-like substances in *phaseolus* leaves. Fiziol. Rast. 22, 1021-1024 (*Russian*).

Shrotri C K, Tewari M N and Rathore V S 1978 Morphological and ultrastructural abnormalities in zinc deficient maize chloroplasts Nutritional deficiencies. Plant Biochem. J. 5, 89-96.

Shrotri C K, Tewari M N and Rathore V S 1980 Effects of zinc nutrition on sucrose biosynthesis in maize. Phytochemistry 19,139-140.

Singh R R and Gangwar M S 1974 Indian J. Agr. Sci. 43, 567.

Skoog F 1940 Relationship between zinc and auxin in the growth of higher plants. Am. J. Bot. 27, 939-950.

Sommer A L and Lipman C B 1926 Evidence on the indispensible nature of zinc and boron for higher green plant. Plant Physiol. 1, 231-249

Stiborova M, Ditrichova M and Brezinova A 1987 Mechanism of function of Cu^{2+} and Zn^{2+} on ribulose- 1,5-biphosphate carboxylase from barley (*Hordeum Vulgare L.*). Photosynthetica 21, 161-165.

Suge H, Takahashi H, Arita S and Takaki H 1986 Gibberellin relationships in zinc-deficient plants. Plant Cell Physiol. 27, 1010-1012.

Sukhijia P S, Randhawa V, Dhillon K S and Munshi S K 1987 The influence and zinc and sulphur deficiency on oil filling in peanut kernels. Plant and soil. 103:261-267.

Suttle N F 1975 Trace elements interaction in animals *In* Trace Elements in Soil-Plant-Animal Systems. Eds. D J D Nicholase and A R Egan. pp 273-286. Academy Press, London.

Takaki H and Arita S 1985 Free tryptophan in zinc deficient oat and barley. Bull. Fac. Agric. Miyazaki Univ. 32, 307-315.

Takaki H and Arita S 1986 Tryptamine in zinc-deficient barley. Soil Sci. Plant Nutr. 32, 433-442.

Takaki H and Kushizaki M 1970 Accumulation of free tryptophan and trytomine in zinc deficient maize seedling. Plant Cell Physiol. 11, 793-804.

Takaki H and Kushizaki M 1976 Zinc nutrition in higher plants. Bull. Nat. Inst. Agr. Sci. 28B, 75-118.

Thomson W W and Weier T E 1962 The fine structure of chloroplasts from mineral deficient leaves of *Phaseolus vulgaris*. Am. J. Bot. 49, 1047.

Tobin A J 1970 Carbonic anhydrase from parsley leaves. J. Biochem. 245, 2656-2666.

Torre M, Rodriguez A R and Saura-Calixto F 1991 Effect of diet on fiber and phytic acid on mineral availability. Crit. Rev. Food Sci. Nutr. 1, 1-22.

Tsui C 1948 The role of zinc in auxin synthesis in tomato plant. Am. J. Bot. 35, 172-179.

Turner R G 1970 The subcellular distribution of zinc and copper within the roots of metal-tolerant clones of

Agrostis tenuis sibth. New Phytol 69, 725-731

Turner R G and Marshall C 1971 The accumulation of ^{65}Zn by root homogenates of zinc-tolerant clones of *Agrostis tenuis sibth.* New Phytol. 70, 539-545

Vallee B L 1983 Zinc in biology and biochemistry. *In* Zinc Enzymes. Ed. T G Spiro. pp 1-24. John Wiley and Sons, New York.

Vodkin M H and Vodkin L O 1989 A conserved zinc finger domain in higher plants. Plant Mol. Biol. 12, 593-594.

Wainright S J and Woolhouse H W 1978 Inhibition by zinc of cell wall acid phosphatase from roots in zinc-tolerant and non-tolerant clones of Agrostis tenuis sibth. J. Exp. Bot. 29, 525-531.

Walker C D and Welch R M 1987 Low molecular weight complexes of zinc and other trace metals in lettuce leaf. J. Agr. Food Chem. 35, 721-727.

Welch R M, House W A and Allaway W H 1974 Availability of zinc from pea seeds to rats. J. Nutr. 104, 733.

Welch R M, House W A and Van Campen D 1976 Effects of oxalic acid on availability of zinc from spinach leaves and zinc sulfate to rats. J. Nutr. 107, 923-933.

Welch R M, Webb M J and Loneragan J F 1982 Zinc in membrane function and its role in phosphorus toxicity [Crops]. *In* Plant Nutrition 1982: Proceedings of the Ninth International Plant Nutrition

Colloquium, Warwick University, England, August 22-27, 1982. Ed. A Scaife. pp 710-715.

Williams R J P 1989 An introduction to the biochemistry of zinc. pp 15-32 *In* Zinc in Human Biology. Ed C F Mills, Springer Verlag, London

Chapter 8.

Genotypic Variation in Zinc Uptake and Utilization by Plants

ROBIN D. GRAHAM and ZDENKO RENGEL

1. Abstract

Genotypes of plants vary widely in their tolerance of Zn-deficient soils, both in Zn uptake and utilization. Tolerance, here termed Zn efficiency because it appears to involve in most cases more efficient extraction of soil Zn, is heritable and may therefore be exploited in the breeding of crop plants. This review considers the extent of variability in Zn eficiency in species of crop and natural plants, the physiological and biochemical nature of the mechanisms of efficiency, what is known of the genetics of inheritance, screening techniques for identifying Zn-efficient types in breeding programs, and the agronomic arguments for a breeding solution to the problem of Zn-deficient soils.

2. Introduction

Zinc-deficient soils are common all over the world in both tropical and temperate lands, but are most widespread in the Mediterranean region, including the cropping areas of Western and South Australia (Sillanpää and Vlek, 1985). Zinc deficiency is, unlike Fe deficiency, quite common in acid soils (e.g. in Cerrado soils in Brazil, Lopes and Cox, 1977). Genotypic differences in tolerance of crop plants to Zn deficiency are thus of universal interest.

Tolerance to Zn-deficient soils, as a genetic trait, is usually called Zn efficiency and defined as the ability of a cultivar (species etc.) to grow and yield well in soils too deficient in Zn for a standard cultivar (Graham, 1984). Zn efficiency, by this definition, does not imply a mechanism and may be determined in a simple field experiment. The agronomic significance of Zn efficiency in modern cultivars is emerging in current literature as an important adjunct to the use of Zn fertilizers in certain situations where topsoil drying, subsoil constraints or disease interactions limit the effectiveness of the latter. In addition, growing Zn-efficient genotypes of crop plants on Zn-deficient soils represents an environmentally-friendly approach which could reduce land degradation by limiting the use of machinery (see Thongbai et al., 1993) and by minimizing application of chemicals on agricultural land. A danger of exhausting soil micronutrient resources ("land mining") is negligible because the total supply of micronutrients in soils is sufficient for hundreds of years of sustainable cropping by new, efficient genotypes that are able to gain access to the micronutrient pools generally considered to be plant-unavailable.

Growing Zn-efficient plants on Zn-deficient soils represents the strategy of 'tailoring the plant to fit the soil' in contrast to the older strategy of 'tailoring the soil to fit the plant' (terminology according to Foy, 1983). The significance of such an approach should be assessed bearing in mind that, of all agricultural innovations, farmers most readily accept new cultivars because these can generally achieve improvement in yield without necessitating much change in agricultural practice (see Little, 1988). However, relatively slow progress in deciphering the genetics, physiology and biochemistry behind the mechanisms of Zn efficiency has hampered the development of genotypes of superior Zn efficiency through conscious breeding efforts geared specifically toward that purpose. Currently, the combination of both strategies mentioned (viz., growing more Zn-efficient genotypes and fertilizing them with smaller quantities of Zn, or less frequently) may be the most realistic approach to cropping on Zn-deficient soils.

3. Evidence for genetic variation

Crop species exhibit differential tolerance to Zn deficiency. Lentil, chickpea and pea were more sensitive to Zn deficiency than oilseeds and cereals (Tiwari and Dwivedi, 1990). Also, various species of *Eucalyptus* trees differed in tolerance to Zn deficiency (Dell and Wilson, 1985). Since there are well-documented differences in tolerance to Zn deficiency among genotypes within a species (see below), such comparisons between different species should be treated with caution because the apparent ranking will depend on the choice of cultivars to represent a particular species.

The most extensive literature on genotypic variation for Zn efficiency in crop plants comes from India where nearly half of $>10^5$ soil samples were considered low in Zn (Takkar, 1991). A review of the Indian literature was published recently (Takkar, 1992) covering 16 crops. Within a species, cultivars varied from tolerant (nil or little response to Zn fertilizer on deficient soil) to sensitive, where the response to Zn was several-fold.

Differential Zn efficiency among various genotypes has been reported for a number of other crops: spinach - *Spinacea oleracea* (Kohno, 1989; Willaert et al., 1990), potato - *Solanum tuberosum* (Sharma and Grewal, 1990), navy bean - *Phaseolus vulgaris* (Ambler and Brown, 1969; Jolley and Brown, 1991), tomato - *Lycopersicon esculentum* (Parker et al., 1992), pearl millet - *Pennisetum americanum* (Takkar et al., 1988), sorghum - *Sorghum vulgare* (Shukla et al., 1973; Ramani and Kannan, 1985), maize - *Zea mays* (Shukla and Raj, 1976; Clark, 1978; Ramani and Kannan, 1985), oats - *Avena sativa* (Brown and McDaniel, 1978), wheat - *Triticum aestivum* (Shukla and Raj, 1974; Mishra and Mehrotra, 1986; Shankar and Mehrotra, 1987; Solunke and Malewar, 1987; Graham, 1991), and others. An extensive screening of large numbers of genotypes was not performed for any of the crops mentioned, making it impossible to judge the extent of Zn efficiency available in a particular germplasm.

In southern Australia where Zn deficiency is also widespread (96% of crops had foliar Zn concentrations less than 20 mg/kg according to a survey of Mallee farms, Hannam, 1991), the range in sensitivity to Zn deficiency for wheat and barley was similar to that reported by Takkar and others in India: Zn-efficient cultivars yielded up to three times that of Zn-inefficient types (Graham, 1988; 1991). When a comparison was made under field conditions, the range in efficiency was relatively small (25-50%) in bread wheats, which have a 100-year history of breeding and adaptation in this area. However, the range was much greater (>100%) when poorly adapted durum wheats were included with the bread wheats (Graham et al., 1992). Large differences can be demonstrated in pots and in the field among the following cereals: ryes > triticales > bread wheats > durum wheats.

Most ryes are, on average, exceptionally well adapted to Zn-deficient soils. This tolerance is generally inherited in the triticales (Graham, 1984) though genomic interactions result in some attenuation of expression. Durum wheats, in our experience, are exceptionally poorly adapted to micronutrient-deficient soils (Graham and Pearce, 1979; Graham, 1988; 1991) and their Zn inefficiency (Table 1, cv Durati and Kamilaroi) is typical of their reaction to other deficiencies. Because of this poor adaptation, durum wheat production in South Australia is restricted to the most fertile soils. Curiously, Zn efficiency varies among durum wheats depending on the soil type (Table 2), a result which suggests that there are different mechanisms involved in handling simple deficiency in deep sands and a complex Zn deficiency in vertisols in which it is induced by exceptionally high levels of Mn and phosphate.

While studies of the cereals tend to dominate the literature on tolerance to Zn deficiency, it is a trait of considerable interest among grain and pasture legumes as well as trees (Nambiar, 1984). Notably, there have been recent attempts to breed for Zn efficiency in soybeans. Concentrations of Zn in fully developed trifoliate leaves differed by a factor of 2-3 between tolerant and intolerant lines, the latter giving a significant yield response to Zn on a sandy loam in Minnesota (Hartwig et al., 1991).

Table 1. Grain yields and grain Zn attributes of wheat genotypes grown in 1988 at Lameroo, S.A., a low rainfall, Zn-responsive sand-over-clay site. Zinc was applied at 25 kg/ha as Zn oxysulphate granules with the seed.

Genotype	Yield (t/ha)			Zn concentration (mg/kg)		Zn content (g/ha)		
	-Zn	+Zn	-Zn*100 / +Zn	-Zn	+Zn	-Zn	+Zn	-Zn*100 / +Zn
Excalibur	1.43	1.76	81	7.6	25	10.8	45	24
Schomburgk	1.11	1.32	84	7.5	24	8.3	31	27
Warigal 5RL	1.04	1.14	91	10.5	29	10.9	33	33
Warigal	1.00	1.22	82	13.0	26	13.0	31	27
Kite	0.95	1.37	69	9.5	24	9.0	33	27
TJB*MKR	0.63	1.30	48	12.1	30	7.6	38	20
Durati	0.45	1.12	40	9.6	25	4.3	28	15
Kamilaroi	0.45	1.29	35	8.8	23	4.0	30	13
LSD*(G×Zn)	0.31			3.4		4.9		

Source: Graham et al. (1992).

4. Genetics of Zn efficiency

Studies of addition lines have shown that Cu, Zn and Mn efficiency in rye were independent traits and carried on different chromosomes (Graham, 1984). Copper and Mn efficiency in rye and Mn efficiency in barley appear to be controlled by single major genes (Graham, 1984; McCarthy et al., 1988), just as Fe efficiency in soybeans (Weiss, 1943), B and Mg efficiency in celery (Pope and Munger, 1953a,b) and B efficiency in

Table 2. The grain yield (GY) of durum wheat cv Kamilaroi relative to that of its parent cv Durati, in contrasting soils in New South Wales and South Australia. Kamilaroi = Durati × Leeds.

	N.S.W. (black clay)		S.A. (light sand)	
	-Zn	+Zn	-Zn	+Zn
100 × GY Kamilaroi / GY Durati	130	105	95	110

Source of the N.S.W. data: Dr. Ray Hare, Agricultural Research Centre, Tamworth.

tomato (Wall and Andrus, 1962) likewise appear to be controlled at a single locus (all reported to be dominant). However, less is known of the genetics of Zn efficiency. The study of addition lines of rye (Graham, 1984) suggests several loci on as many different chromosomes are involved in Zn efficiency in rye; likewise, a few genes are involved in Zn efficiency in rice. The largest single screening exercise was of 3703 lines of paddy rice (Ponnamperuma, 1976; IRRI, 1979) where 388 lines were judged to be tolerant and a similar range of responses was observed. Following diallel analysis, a recent report suggested that the genetic effects responsible for the Zn efficiency trait in rice are mostly additive, and to a lesser extent dominant (Majumder et al., 1990).

Soybean varieties differ in their response to Zn fertilizer (Rao et al., 1977; Rose et al., 1981; Saxena and Chandel, 1992). Such a result is suggested to be a consequence of differential efficiency of Zn absorption; the distribution of F_3 lines from the cross between Zn-efficient and Zn-inefficient genotypes (330 F_3 lines tested) suggested that only a few genes control the Zn efficiency trait (Hartwig et al., 1991).

The various mechanisms of Zn efficiency are likely to be additive (as shown for rice, Majumder et al., 1990), putting great emphasis in a breeding program on stepwise compounding of genetic information (see Rengel and Jurkic, 1992). Such pyramiding into one locally adapted crop cultivar of a number of Zn efficiency mechanisms that are expressed at different levels of the plant organism (molecular, physiological, structural, or developmental, see Rengel, 1992) might follow the approaches of Yeo and Flowers (1986). In such a breeding program, genotypes having genes controlling a particular mechanism of Zn efficiency may be very important even though they themselves may not show phenotypically high overall Zn efficiency. Using genotypes from the local crop germplasm would be advantageous because development of cultivars with improved Zn efficiency may be expedited without severely disrupting the broad adaptation already achieved.

5. Mechanisms of Zn efficiency

A mechanistic explanation of differential Zn efficiency among genotypes of crop plants is still lacking. It may, however, be assumed that (i) efficiency mechanisms vary among crop species, (ii) more than one mechanism is often responsible for the level of Zn efficiency in a particular genotype, and (iii) increased efficiency of one genotype in comparison to another is due to involvement of additional mechanisms not present (expressed) in the less efficient genotype, or expressed at a lower intensity or rate.

Different mechanisms of Zn efficiency operate in Zn-deficient soil than in Zn-deficient nutrient solution. The wheat cv Excalibur was the most Zn-efficient genotype of those tested in Zn-deficient soil but was the least Zn-efficient in nutrient solution studies. In a Zn-deficient soil, the cv Excalibur has the ability to produce more roots of smaller diameter (<0.3 mm) (small-diameter roots have greater surface to volume ratio) than the cv. Gatcher (Dong Bei, Z. Rengel and R.D. Graham, unpublished, 1993). This would allow exploration of a larger volume of soil and hence more efficient scavenging of the small amounts of immobile Zn ions by the cv. Excalibur, rendering it Zn-efficient. Putative existence of this mechanism (root geometry differences) does not exclude the possibility of other mechanisms, including differential mycorrhizal infection, acting additively to make cv. Excalibur Zn-efficient when grown in Zn-deficient soil. Additional experiments are needed to ascertain that production of more roots of smaller diameter by the cv. Excalibur is the cause rather than the consequence of greater Zn efficiency.

Contrasting with the above, in nutrient solution where there is no restriction on Zn mobility in the root zone and where roots tend to be of greater diameter in all wheat cultivars screened, the cv. Excalibur ranked as the most inefficient genotype of 12 tested (the least amount of shoot dry matter produced, Z. Rengel and R.D. Graham, unpublished). It also had the highest root and shoot P concentrations (up to 23 g/kg of shoot dry matter at the 3-leaf stage which is well into the toxicity range, Reuter and Robinson, 1986) when grown in chelate-buffered nutrient solution containing 0.1 μM total Zn concentration (2 pM Zn^{2+} activity) and 0.1 mM P concentration. The poor performance of the cv. Excalibur under these conditions was apparently due to Zn deficiency-induced P toxicity, a phenomenon that has been well described in the literature (Webb and Loneragan, 1988 and references therein).

In addition to the importance of the Zn-P interaction in determining the level of Zn efficiency, an interaction between Fe and Zn may also be influential. A Zn-inefficient navy bean had higher Fe concentrations in tops under Zn deficiency than did a Zn-efficient genotype (Ambler and Brown, 1969). Zinc deficiency stimulated the initiation of the Fe-deficiency response in a Zn-inefficient navy bean by increasing the reduction and uptake of Fe which in turn enhanced Zn deficiency (Jolley and Brown, 1991). Since either Zn or Fe deficiency may stimulate higher production of phytosiderophores in Zn-efficient Aroona wheat than in Zn-inefficient Durati (Cakmak et al., 1993), the possibility that the effects resulted from impaired Fe utilization under Zn deficiency in Aroona but not in Durati has yet to be excluded. It remains to be established how effectively Zn-mobilizing phytosiderophores, as defined by Cakmak et al. (1993), would mobilize Zn from a mixture containing unavailable forms of both Fe and Zn.

Wheat genotypes differed in the rate of net Zn uptake from chelate-buffered nutrient solutions having various total Zn concentrations (0.1 to 10 μM) and corresponding Zn^{2+} activities (2 to 200 pM) (Z. Rengel and R.D. Graham, unpublished, 1993). However, no significant correlation was found between the rate of net Zn uptake and dry matter production for the genotypes tested. Moreover, cv. Durati (durum wheat) showed typical symptoms of Zn deficiency on fully expanded leaves when grown in solutions containing up to 0.5 μM total Zn, but had the same root and shoot concentrations of Zn as those genotypes that were free of visible symptoms of Zn deficiency. Similar results under different growing conditions were reported by Cakmak et al. (1993). It therefore appears that (i) total Zn concentrations in roots and leaves are a poor estimate of the Zn status of wheat (also maize, Gibson and Leece, 1981) and (ii) that differential compartmentalization and/or utilization of Zn within the plant cell may be one of the mechanisms of differential Zn efficiency.

Differences in Zn uptake kinetics were noted among genotypes of several crops. The maximum rate of Zn uptake was greater for Zn-inefficient rice IR26, while affinity for Zn was greater for Zn-efficient rice M101 (Bowen, 1986). The pattern of Zn absorption and Zn translocation from roots to shoots differed among sorghum hybrids; inheritance of the trait through the female parent was suggested (Ramani and Kannan, 1985). However, no ranking for Zn efficiency was given and maximum rates of uptake, even though apparently different among genotypes, were not statistically evaluated. Clearly, the data published to date do not allow for a proper assessment of the importance of the Zn uptake kinetics and the rate of transport of Zn to the plant tops in the expression of the Zn efficiency trait.

Seed Zn content may influence the performance of a particular genotype under Zn-deficient conditions. Genotypes with a higher seed Zn content produce more root (and shoot) material in the initial stages before seed reserves of Zn are exhausted. Greater root mass would be beneficial in the later stages when the plant has to rely on scavenging Zn from the environment. Our experiments show that while Excalibur wheat is more Zn-efficient than Gatcher at a low seed Zn level (around 250 ng Zn/seed), such a difference is dissipated when seed Zn content is around 700 ng Zn/seed (Z. Rengel and R.D. Graham, unpublished data). Clearly, higher seed Zn content is not the only mechanism responsible for good crop performance in a low-Zn environment because Zn-inefficient durum wheat cv Durati showed poorer growth than genotypes which had a seed Zn content about 2/3 of that in Durati seeds.

A possible mechanism of differential Zn efficiency about which no report was found in the literature is the difference in ability of genotypes to support mycorrhizal colonization of their roots. It has been shown that mycorrhizal roots of green gram (*Vigna radiata*) (Sharma and Srivastava, 1991), maize (Faber et al., 1990; Kothari et al., 1990; Sharma et al., 1992), and pigeonpea (*Cajanus cajan*) (Wellings et al., 1991) have higher tissue Zn concentrations than non-mycorrhizal plants. It remains to be established whether there is a correlation between mycorrhizal infection of different genotypes and their level of Zn efficiency.

In summary, a number of possible mechanisms may operate in different species and different genotypes of a species. These mechanisms are likely to be at several levels of plant organization (molecular, physiological, structural, or developmental). Some mechanisms (like differential root geometry, differences in the ability to sustain mycorrhizal infection) may operate in soil environments, while others (differential uptake kinetics, compartmentalization, transport, retranslocation and utilization of Zn, differences in production of Zn-mobilizing phytosiderophores, etc.) may be operational in both soil and nutrient solution environments. Zinc-P and Zn-Fe interactions may be expressed to a different extent in various genotypes and thus may variously influence the level of Zn efficiency.

6. Screening techniques for identifying Zn efficiency in breeding programs

Simple, fast and inexpensive techniques are needed in breeding programs to permit the assessment of hundreds or thousands of segregants from crosses of Zn-efficient × Zn-inefficient parents, without recourse to expensive field testing. The better we understand the mechanism of Zn-efficiency at a physiological level, the easier it is likely to be to develop an ideal screening technique.

The ultimate assessment of Zn efficiency is in a field experiment where the soil is deficient enough in Zn to seriously limit the yield of some lines. The Zn-efficiency of a genotype of interest is compared in yield with that of a standard cultivar. Often, such tests are conducted without any Zn treatments when there is a large number of genotypes to be screened. Genotypes under test may be located within an array of plots of a check genotype to provide a covariate on the heterogeneity of the site. Differences in general adaptation and yield potential are assumed to be small in comparison to the differences in Zn efficiency. If not, the introduction of a +Zn treatment (adequate Zn) will define the yield potential and allow the calculation of the Zn efficiency index:

$$\text{Zn efficiency} = \frac{\text{Yield (-Zn)}}{\text{Yield (+Zn)}} \times 100$$

This normalisation removes much of the background genetic effect but does not exclude Zn×G×E effects which can occasionally be important (Table 2; Graham, 1991). A paired-plot system of field screening of this type has been described by Graham (1991). Such a design, which is essentially a split-plot factorial, may be subjected to a nearest neighbour analysis of variance to quantify and correct for inherent soil variability (Wilkinson et al., 1983).

Some authors have stressed the importance of screening over a full range of stress, which necessitates a number of rates of Zn application, increasing the size and cost of the experiments (Boken, 1966; Blum, 1989; Paull, 1990). In studying the genetics of tolerance to B toxicity, Paull (1990) found that several loci were involved, each operating at a different level of stress. The presence or absence of tolerance alleles for a given locus could only be detected at an appropriate level of stress. Clearly, in this case multi-level testing is necessary, and as several genes may also be involved in Zn efficiency, full-range response curves may be required to make adequate progress with this trait. Graham et al. (1992) showed several different responses to Zn rates among five cultivars of wheat grown in pots, but multi-rate field experiments have been rare.

Screening in pots is common, being less expensive and faster than field work, and the problems of soil heterogeneity can be eliminated. However, the environment is less realistic, and we have found, like others, that low temperature, especially low soil temperature, is essential to inducing Zn deficiency stress. Moreover, root binding in small pots may be an independent limiting factor which can mask the full expression of the limitation of Zn.

Cakmak et al. (1993) demonstrated that under stress, efficient wheats synthesize phytosiderophores in the roots and release them to the environment. Since these organic acids are detectable by HPLC, a simple screening technique presents itself for use in solution culture-grown plants. The level of Zn stress can be efficiently controlled to exacting and reproducible limits by the chelate buffering technique (Norvell, 1991; Parker et al., 1992) and the level of phytosiderophores produced can be measured by the technique of Treeby et al. (1989). Another related approach is that of Zhang et al. (1989) in which the phytosiderophores released Zn bound to the heavy-metal-binding resin, Chelite 100. Such techniques show genuine promise. There is some evidence that the correlation between Zn stress-released phytosiderophores and field-assessed Zn efficiency is poor (H. Braun, pers. comm.), at least in some cases, and this must be fully assessed.

We anticipate that in the near future, molecular methods will dominate selection for Zn efficiency and many other traits. The use of RFLP mapping techniques or PCR leading to cDNA probes for regions of the genome segregating with the Zn efficiency trait will lead to rapid and efficient selection.

7. Agronomic considerations

Despite the fact that the efficacy of Zn fertilizers is considered to be generally good, there are clearly defined situations in which fertilizer Zn is sufficiently ineffective that a case can be made for breeding for Zn efficiency in crop plants (Graham et al., 1992). These situations include subclinical Zn deficiency which is otherwise untreated except through use of a Zn-efficient cultivar. The degree of Zn efficiency currently known to exist in germplasm banks is probably adequate to exploit for this purpose.

Zinc-deficient subsoil is difficult, expensive and/or impossible to fertilize. Yet nearly all soils, no matter how poor, have a sufficient content of micronutrients stored in the profile (Graham, 1984); the problem is usually one of availability, a problem which is half soil related and half plant (genotype) related. Responses to subsoil nutrients added in field trials (Graham et al., 1992) are immediate and often spectacular, and continue for a number of years (for example, 50-100% yield increases due to nutrients placed six seasons previously); the residual responses are principally to phosphorus and trace elements. We have observed that wheat grows poorly in potted subsoil even when fertilized with nitrogen and phosphorus. Although we are experimenting with deep injection of micronutrients through tubes welded down the back of deep-ripping tynes, we believe the better approach to this problem is to breed cereals with root systems which will penetrate subsoils of low phosphorus and micronutrient availability.

From physiological studies of roots (Welch et al., 1982; Loneragan et al., 1987), it appears that Zn is required in the external environment of the root for membrane function and cell integrity; Zn deficiency in the external environment promotes leakage of cell constituents such as sugars, amides and amino acids which are chemotaxic stimuli to pathogenic organisms. Moreover, a high internal Zn content did not prevent leakiness due to a deficiency of Zn external to the membrane (Welch et al., 1982); thus, mobility of Zn in the phloem from roots in soil zones high in Zn cannot fully compensate for the lack of Zn elsewhere in the soil profile. It follows that the roots of those genotypes which have a greater capacity to mobilize Zn strongly bound to soil particles in the rhizosphere will probably be better able to penetrate an infertile, calcareous subsoil. This view was confirmed by Nable and Webb (1993) in a pot experiment which compared wheat cultivars Excalibur and Gatcher in pots of Zn-deficient sand divided into upper and lower layers. Gatcher plants withdrew relatively more water than excalibur from the lower layer if that layer were treated with Zn. The advantage of Zn treatment in the lower layer was less, as expected, in the Zn-efficient cultivar, Excalibur.

Topsoil drying is another problem of wheat production on infertile soils of the southern Australia. When the topsoil dries as a result of a week or two of dry weather in spring, roots in the fertilizer zone become largely inactive and the plant must rely on deeper roots or retranslocation for further nutrient supply.

Table 3. Release of Zn bound to Chelite-N resin by root exudates collected from Zn-efficient Aroona and Zn-inefficient Durati wheat precultured in +/- Zn nutrient solutions (estimated from Fig. 3 of Cakmak et al., 1993).

Genotype	Zn mobilized (nmol/plant)	
	-Zn	+Zn
Aroona	110	10
Durati	30	11

Two further advantages accrue to Zn-efficient varieties if by virtue of their efficiency they also accumulate more Zn in the grain: firstly, better human nutrition may be achieved in regions where cereals form a major part of the diet if the normally low Zn concentration in cereal grains can be increased by breeding (Welch and House, 1983); and secondly, markedly better seedling vigour results where seed is kept for re-sowing in deficient soils. Seedling vigour is important in disease resistance, tolerance and avoidance, one of the strongest arguments for breeding for Zn efficiency.

Similarly, roots with leaky membranes, far from a fertilized layer, are at greater risk from pathogens. Recent studies have clearly linked Zn deficiency with decreased resistance of wheat to *Fusarium graminearum,* the crown rot fungus (Sparrow and Graham, 1988) and to *Rhizoctonia solani,* the bare patch fungus (Thongbai et al., 1993). The implication is that Zn-efficient genotypes in deficient soil, enjoying better nutrient status, should have greater resistance to such root pathogens. This hypothesis is supported by a strong negative association of susceptibility to crown rot (Burgess et al., 1984) with Zn efficiency (Graham et al., 1992) in all six wheats common to these two studies.

These agronomic arguments must be perceived by plant breeders as compelling if they are to allocate the resources to improving the Zn efficiency of current crop cultivars; that is, fertilizer and other agronomic strategies must be shown to be inadequate for overcoming the limitations of soils low in available Zn. Once this is accepted, the requirement is for fast and effective selection criteria, preferably applicable to seeds or seedlings.

References

Ambler J E and Brown J C 1969 Cause of differential susceptibility to zinc deficiency in two varieties of navy beans (*Phaseolus vulgaris* L.). Agron. J. 61, 41-43.

Blum A 1988 Plant Breeding for Stress Environments. CRC Press, Boca Raton, Fla. 223 pp.

Boken E 1966 Studies on methods of determining varietal utilization of nutrients. DSR Vorlag-Boghandel, Roy. Vet. Agric. Coll., Copenhagen.

Bowen J E 1986 Kinetics of zinc uptake by two rice cultivars. Plant Soil 94, 99-107.

Brown J C and McDaniel M E 1978 Factors associated with differential response of two oat cultivars to zinc and copper stress. Crop Sci. 18, 817-820.

Burgess L W, Klein T and Liddell C M 1984 Crown rot of wheat. *In* Research on Root and Crown Rots of Wheat in Australia. Proc. Workshop, Waite Institute, 14-15 February. Aust. Wheat Ind. Res. Council. pp 62-65.

Cakmak I, Gülüt K Y, Marschner H and Graham R D 1993 Effect of iron and zinc deficiency on phytosiderophore release in wheat genotypes differing in zinc efficiency. J. Plant Nutr. (*in press*).

Clark R B 1978 Differential response of maize inbreds to Zn. Agron. J. 70, 1057-1060.

Dell B and Wilson S A 1985 Effect of zinc supply on growth of three species of Eucalyptus seedlings and wheat. Plant Soil 88, 377-384.

Faber B A, Zasoski R J, Burau R G and Uriu K 1990 Zinc uptake by corn as affected by vesicular-arbuscular mycorrhizae. Plant Soil 129, 121-130.

Foy C D 1983 Plant adaptation to mineral stress in problem soils. Iowa State J. Res. 57, 355-391.

Gibson T S and Leece D R 1981 Estimation of physiologically active zinc in maize by biochemical assay. Plant Soil 63, 395-406.

Graham R D 1984 Breeding for nutritional characteristics in cereals. Adv. Plant Nutr. 1, 57-102.

Graham R D 1988 Development of wheats with enhanced nutrient efficiency: progress and potential. *In* Wheat Production Constraints in Tropical Environments. Ed. A R Klatt . pp 305-320. CIMMYT, Mexico D.F.

Graham R D 1991 Breeding wheats for tolerance to micronutrient deficient soil: present status and priorities. *In* Wheat for the Nontraditional Warm Areas. Ed. D A Saunders. pp. 315-332. CIMMYT, Mexico D.F.

Graham R D, Ascher J S and Hynes S C 1992 Selecting zinc-efficient cereal genotypes for soils of low zinc status. Plant Soil, 146, 241-250.

Graham R D and Pearce D T 1979 The sensitivity of hexaploid and octoploid triticales and their parent species to copper deficiency. Aust. J. Agric. Res. 30, 791-799.

Hannam R J 1991 Nutrition and zinc update. *In* Agronomy Technical Conference. South Australia Department of Agriculture. pp 90-94.

Hartwig E E, Jones W F and Kilen T C 1991 Identification and inheritance of inefficient zinc absorption in soybean. Crop Sci. 31, 61-63.

IRRI 1979 Annual Report, International Rice Research Institute, Los Baños, The Philippines.

Jolley V D and Brown J C 1991 Factors in iron-stress response mechanism enhanced by zinc-deficiency stress in Sanilac, but not Saginaw navy bean. J. Plant Nutr. 14, 257-266.

Kothari S K, Marschner H and Römheld V 1990 Direct and indirect effects of VA mycorrhizal fungi and rhizosphere microorganisms on acquisition of mineral nutrients by maize (*Zea mays* L.) in a calcareous soil. New Phytol. 16, 637-645.

Little R 1988 Plant-soil interactions at low pH. Problem-solving genetic approach. Commun. Soil Sci. Plant Anal. 19, 1239-1257.

Loneragan J F, Kirk G J and Webb M J 1987 Translocation and function of zinc in roots. J. Plant Nutr. 10, 1247-1254.

Lopes A S and Cox F R 1977 A survey of the fertility status of surface soils under 'Cerrado' vegetation in Brazil. Soil Sci. Soc. Am. J. 41, 742-747.

McCarthy K W, Longnecker N E, Sparrow D H B and Graham R D 1988 Inheritance of manganese efficiency in barley (*Hordeum vulgare* L.). *In* International Symposium on Manganese in Soils and Plants: Contributed Papers. Eds. M J Webb, R O Nable, R D Graham and R J Hannam. pp 121-122. Manganese Symposium, 1988 Inc., Adelaide.

Majumder N D, Rakshit S C and Borthakur D N 1990 Genetic effects on uptake of selected nutrients in some rice (*Oryza sativa* L.) varieties in phosphorus-deficient soil. Plant Soil 123, 117-120.

Mishra P H and Mehrotra O N 1986 Physiological studies on differential uptake of nitrogen and zinc by wheat varieties. I. Growth and yield aspects. Farm Sci. J. 1, 26-32.

Nable R O and Webb M J 1993 Further evidence that Zn is required throughout the root zone for optimal plant growth and development. Plant Soil (*in press*).

Nambiar E K S 1984 Increasing forest productivity through genetic improvement of nutritional characteristics. *In* Eds R Balland et al. Forest Potentials. Productivity and Value. Weyerhaueser Science Symposium, Tacoma, Wa. August 19-24, 1984. pp. 191-215

Norvell W A 1991 Reactions of metal chelates in soils and nutrient solutions. *In* Micronutrients in Agriculture. 2nd edition. Eds J J Mortvedt, F R Cox, L M Shuman and R M Welch. pp. 187-227. Soil Sci. Soc. Am., Madison, WI.

Paull J G 1990 Genetic studies on the tolerance of wheat to high concentrations of boron. Ph.D. thesis, University of Adelaide.

Parker D R, Aguilera J J and Thomason D N 1992 Zinc-phosphorus interactions in two cultivars of tomato (*Lycopersicon esculentum* L.) grown in chelator-buffered nutrient solutions. Plant Soil 143, 163-177.

Ponnamperuma F N 1976 Screening rice for tolerance to mineral stresses. *In* Plant Adaptation to Mineral Stress in Problem Soils. Ed. M J Wright. pp 341-353. Cornell Univ. Agric. Exp. Stn., Ithaca, New York.

Pope D T and Munger H M 1953a Heredity and nutrition in relation to magnesium deficiency chlorosis in celery. Proc. Am. Soc. Hort. Sci. 61, 472-480.

Pope D T and Munger H M 1953b The inheritance of susceptibility to boron deficiency in celery. Proc. Am. Soc. Hort. Sci. 61, 481-486.

Ramani S and Kannan S 1985 An examination of zinc uptake patterns by cultivars of sorghum and maize Differences amongst hybrids and their parents J Plant Nutr 8, 1199-1210
Rao V S, Gangwar M S and Rathore V S 1977 Genotypic variation in distribution of total and labelled zinc and availability of zinc (A and L values) to soybeans grown in mollisol J Agric Sci 8, 417-420
Rengel Z 1992 Role of calcium in aluminium toxicity New Phytol 121, 499-513
Rengel Z and Jurkic V 1992 Genotypic differences in wheat Al tolerance Euphytica 62, 111-117
Reuter D J and Robinson J B 1986 Plant Analysis An Interpretation Manual Inkata Press, Melbourne
Rose I A, Felton W L and Banke L W 1981 Response of four soybean varieties to foliar zinc fertilizer Aust J Exp Agric Anim Husb 21, 236-240
Saxena S C and Chandel A S 1992 Effect of zinc fertilization on different varieties of soybean (*Glycine max*) Indian J Agric Sci 62, 695-697
Shankar H and Mehrotra O N 1987 Response of wheat varieties to zinc application under varying levels of fertility Farm Sci J 2, 144-150
Sharma A K and Srivastava P C 1991 Effect of vesicular-arbuscular mycorrhizae and zinc application on dry matter and zinc uptake of green gram (*Vigna radiata* L Wilczek) Biol Fertil Soils 11, 52-56
Sharma A K, Srivastava P C, Johri B N and Rathore V S 1992 Kinetics of zinc uptake by mycorrhizal (VAM) and non-mycorhizal corn (*Zea mays* L) roots Biol Fertil Soils 13, 206-210
Sharma U C and Grewal J S 1990 Potato response to zinc as influenced by genetic variability J Indian Potato Assoc 17, 1-5
Shukla U C and Raj 1974 Influence of genetic variability on zinc response in wheat (*Triticum* ssp) Soil Sci Soc Am Proc 38, 477-479
Shukla U C, Arora S K, Singh Z, Prasad K G and Safaya N M 1973 Differential susceptibility of some sorghum (*Sorghum vulgare*) genotypes to zinc deficiency in soil Plant Soil 39, 423-427
Sillanpaa M and Vlek P L G 1985 Micronutrients and the agroecology of tropical and Mediterranean regions Fert Res 7, 151-167
Solunke S N and Malewar G U 1987 Differential response of wheat genotypes to zinc fertilization J Maharashtra Agricultural Universities 12, 382-383
Sparrow D H and Graham R D 1988 Susceptibility of zinc-deficient wheat plants to colonization by *Fusarium graminearum* Schw Group 1 Plant Soil 112, 261-266
Takkar PN 1991 Zinc deficiency in Indian soils and crops In Zinc in Crop Nutrition Lead Zinc Res Org Inc and Indian Lead Zinc Information Center, New Delhi pp 55-64
Takkar P N 1992 Requirement and response of crop cultivars to micronutrients in India - a review Plant Soil (*in press*)
Takkar P N, Bansal R L, Singh S P and Nayyar V K 1988 Response of pearl millet cultivars to zinc under field conditions Intern J Tropical Agric 6, 247-251
Thongbai P, Hannam R J, Graham R D and Webb M J 1993 Zn nutrition and *Rhizoctonia* root rot of cereals Plant Soil (*in press*)
Tiwari K N and Dwivedi B S 1990 Response of eight winter crops to zinc fertilizer on a Typic Ustochrept soil J Agric Sci (Camb) 115, 383-387
Treeby M, Marschner H and Romheld V 1989 Mobilization of iron and other micronutrient cations from a calcareous soil by plant-borne, microbial and synthetic metal chelators Plant Soil 114, 217-226
Wall J R and Andrus C F 1962 The inheritance and physiology of boron response in the tomato Am J Bot 49, 758-762
Webb M J and Loneragan J F 1988 Effect of zinc deficiency on growth, phosphorus concentration, and phosphorus toxicity of wheat plants Soil Sci Soc Am J 52, 1676-1680
Weiss M G 1943 Inheritance and physiology of efficiency in iron utilization in soybeans Genetics 28, 253-268
Welch R M and House W A 1983 Factors affecting the bioavailability of mineral nutrients in plant foods *In* Crops as Sources of Nutrients for Humans, Eds R M Welch and W H Gabelman pp 37-54 Am Soc Agron , Madison, WI
Welch R M, Webb M J and Loneragan J F 1982 Zinc in membrane function and its role in phosphorus toxicity *In* Plant Nutrition 1982 Proc 9th Int Coll Plant Nutrition Ed A Scaife pp 710-715
Wellings N P, Wearing A H and Thompson J P 1991 Vesicular-arbuscular mycorrhizae (VAM) improve phosphorus and zinc nutrition and growth of pigeonpea in a Vertisol Aust J Agric Res 42, 835-845
Wilkinson G N, Eckert S R, Hancock T W and Mayo O 1983 Nearest neighbour (NN) analysis of field experiments J Roy Stat Soc B45, 151-211
Willaert G, Verloo M and Sarrazyn R 1990 Influence of the cultivar on the accumulation of inorganic compounds in spinach Mededelingen van de Faculteit Landbouwwetenschappen Rijksuniversiteit Gent 155, 83-92

Yeo A R and Flowers T J 1986 Salinity resistance in rice (*Oryza sativa* L.) and a pyramiding approach to breeding varieties for saline soils. Aust. J. Plant Physiol. 13, 161-173.

Zhang F V, Römheld V and Marschner H 1989 Effect of zinc deficiency in wheat on the release of zinc and iron mobilizing root exudates. Z. Pflanzenernaehr. Bodenk. 152, 205-210.

Chapter 9.

Interactions Between Zinc and Other Nutrients Affecting the Growth of Plants

JACK F LONERAGAN and MICHAEL J. WEBB

1. Abstract

This paper examines the interactions between Zn and other nutrients in soil reactions, behaviour in plants and plant growth.

It stresses the need for identification of the factor responsible for any Zn response to the addition of another nutrient compound.

Of the many interactions of Zn with other nutrients, the most widespread and important to crop production are those with N and P fertilizers on soils with limiting supplies of both Zn and N or P. Similar interactions of Zn with other essential nutrients will also be important on soils with low supplies of both nutrients; such an interaction of Zn with Cu was strongly enhanced by an effect of Zn in depressing Cu absorption and almost eliminating grain production in wheat when Cu was not applied.

Other interactions with potential significance for crop production in specific situations include the enhancement of Zn deficiency through depression of Zn absorption by effects of high concentrations of Fe and Mn in flooded soils and of P in suppressing mycorrhizal infection of roots.

The many interactions of Zn with P are reviewed. Recent evidence that, when supplied high P at low Zn, plants accumulate high P in their leaves, precipitating Zn and increasing the plant's internal requirement for Zn, provides a new insight into the long puzzling phenomenon of "P enhanced Zn requirements".

2. Introduction

Other nutrients may interact with Zn by affecting its availability from soils and its status in the plant through the processes of growth or Zn absorption, distribution or utilization. In so doing, they may enhance or depress the response of plant growth to Zn. Conversely, Zn may affect other nutrients in the same ways.

Where an interaction does occur, it is important for diagnosis and effective treatment of Zn deficiency in crops to identify the factor responsible, and its site and mode of action. Where the nutrient is added as a salt, any of three factors other than the nutrient itself could be responsible for affecting a Zn response:-

- another ion in the salt
- Zn as a contaminant in the nutrient salt
- a change in the root environment.

Many reports of interactions of other nutrients with Zn must be rejected because of their failure to recognise and critically evaluate the possible role of each of these additional extraneous factors. Effects of nutrient additions on rhizosphere pH are especially important and will be considered briefly before reviewing the interactions of individual nutrients with Zn. The nature of the interactions between two nutrients in limiting supply will also be treated separately because of its general applicability to the interaction of Zn with all other nutrients.

2.1. pH-Zn interactions

Zinc absorption is particularly sensitive to changes in the pH of the rhizosphere which may respond directly to the addition of a nutrient salt or indirectly by changing the balance of anions and cations absorbed. In solution culture, decreasing pH below 7 progressively decreased Zn absorption in wheat by a mechanism of non-competitive inhibition by H^+ ions (Chaudhry and Loneragan, 1972b). But in soils, increasing pH generally depresses Zn absorption (Fig. 1) and induces or enhances Zn deficiency (Wear, 1956), apparently because its effects in enhancing Zn absorption on to soil components and increasing the formation of Zn-organic complexes (Chapter 1) dominate over its effect on the process of Zn absorption by roots. However, plant species may vary in their response to soil pH since increasing soil acidity while having no effect on plant growth enhanced Zn concentrations in the tops of subterranean clover plants but had no effect on those of oats grown on the same soil (Williams, 1977).

2.2. Interactions between two limiting nutrients

Of all the interactions between Zn and other nutrients, the most important and widespread are those which affect plant growth when the supplies of Zn and another nutrient are both limiting. Where another nutrient is very deficient, the plant may not respond to Zn alone but responds very strongly to the addition of Zn and the other nutrient together; severe Zn deficiency may affect the response to the other nutrient in the same way. Where the deficiencies are less severe, addition of the other nutrient alone may promote growth and induce or enhance Zn deficiency by diluting Zn in the plant; addition of Zn alone may affect the other nutrient in the same way; addition of Zn and the other nutrient together overcomes both deficiencies and promotes growth, usually by more than the sum of the increases from Zn and the other nutrient alone (eg Fig. 2a). All essential nutrients can be expected to interact with Zn in these ways as predicted by the extension to nutrients of Blackman's Law of Limiting Factors (Anderson and Thomas, 1946; Anderson, 1956).

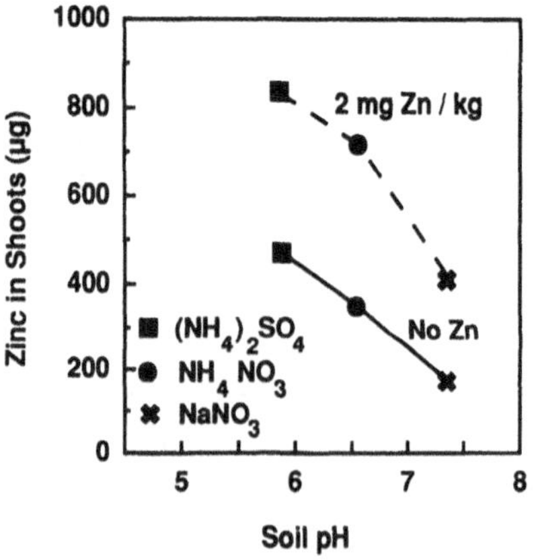

Figure 1. Effect of three N fertilizers on soil pH and its relationship to Zn content in shoots of *Sorghum vulgare*. Redrawn from Viets et al., (1957).

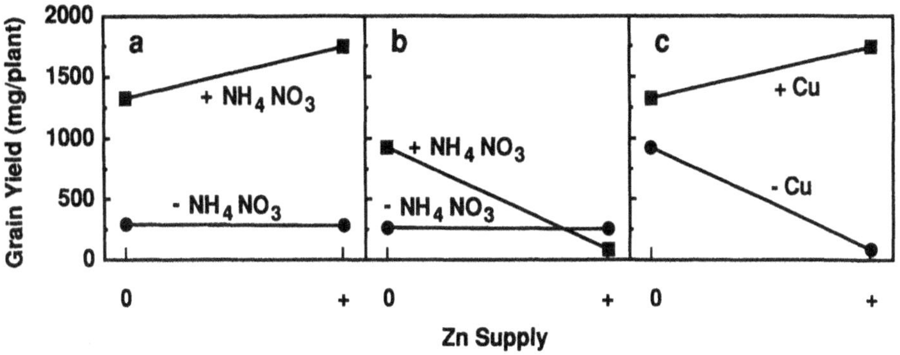

Figure 2. Responses and interactions of grain yield of wheat to NH_4NO_3, Cu and Zn on a N-deficient soil low in Cu and Zn; (a) interaction of NH_4NO_3 and Zn with Cu added, (b) interaction of NH_4NO_3 and Zn without Cu added, (c) interaction of Cu and Zn with NH_4NO_3 added. Redrawn from Chaudhry and Loneragan (1970).

3. P-Zn interactions

There is a voluminous and confusing literature on P-Zn interactions. Much of the confusion has arisen from workers who failed to identify the factor operative in an interaction or who used conditions irrelevant to Zn deficiency. Others have compounded the confusion by accepting conclusions without critical evaluation of the experimental conditions and data or by looking for a single magic phenomenon, 'the P-Zn interaction' or 'the P induced Zn deficiency', to explain the many phenomena now known to occur; some even look for an interaction under conditions where, since neither Zn nor P are limiting or excessive, there is no reason to expect any!

But even when doubtful findings are rejected, the subject of P-Zn interactions remains complex, involving many phenomena in both soils and plants. They fall into two categories according to whether increasing applications of P decrease or do not decrease Zn concentrations in plant shoots.

3.1. P decreases shoot Zn concentrations

The most common and important interaction in which P salts decrease plant Zn concentrations is that encountered when the supplies of both P and Zn are marginal or limiting and addition of P promotes growth sufficiently to dilute the concentration of Zn in the plant to levels which induce or enhance Zn deficiency (Boawn et al., 1954; Loneragan et al., 1979; Singh et al., 1988); plant growth responds to increasing supplies of P and Zn according to Blackman's Law of Limiting Factors as already discussed.

In other situations, increasing P has induced or enhanced Zn deficiency by depressing Zn concentrations in plant shoots more than can be explained by dilution from plant growth (Rogers and Wu, 1948; Loneragan 1951; Bingham et al., 1958; Sharma et al., 1968). Here, P must have depressed either absorption of Zn by roots or translocation of Zn from roots to shoots.

Zn absorption. There is good evidence for four mechanisms by which P salts can depress the absorption of Zn from soils:

- P suppresses root infection by vesicular arbuscular mycorrhizae

- cations added with P salts inhibit Zn absorption from solution
- H^+ ions generated by P salts inhibit Zn absorption from solution
- P enhances the sorption of Zn to soil components

Of these mechanisms, only the first has been shown unequivocally to induce Zn deficiency in plants grown in soils (Lambert et al., 1979). The second and third mechanisms are important in solution culture but not in soils for reasons discussed elsewhere (see *Micronutrient cation - Zn interactions* and *pH - Zn interactions*). The fourth has been demonstrated in experiments with hydrous oxides of Fe and Al (Stanton and Burger, 1967, 1970; Bolland et al., 1977) but the addition of P fertilizers to soils has been reported to have had variable effects on the retention of Zn (Saeed and Fox, 1979; Friesen et al., 1980). Barrow (1987) has suggested that some of this variability may be explained through the effects of Zn contamination and changing soil pH as already discussed. In addition, he has presented evidence that the process of P sorption on soils affects their retention of Zn by two mechanisms – one by changing pH and the other by changing the surface charge; he found no support that Zn formed links to the surface through P molecules as proposed earlier (Stanton and Burger, 1967, 1970). In chapter 2 Barrow discusses these reactions in detail.

Suggestions that P ions inhibit Zn absorption directly (Stukenholtz et al., 1966; Burleson and Page, 1967; Edwards and Kamprath, 1974; Safaya, 1976; Cogliatti et al., 1991) are not acceptable, as the conditions of the experiments on which the conclusions were based were unsatisfactory for various reasons (e.g. plants showed no response to Zn; Zn was precipitated by P; or, other factors which varied with P treatments such as cation concentration or pH were ignored). When nutrient anions were varied (Cl^-, $H_2PO_4^-$, NO_3^-, SO_4^{2-}) in solutions of otherwise constant composition, P had little effect on Zn absorption by wheat seedlings (Chaudhry and Loneragan, 1972a).

Zn transport. There is no definitive evidence that P induces Zn deficiency by inhibiting Zn transport from roots to shoots (see Chaudhry and Loneragan, 1970). However, under conditions of high Zn supply, P may immobilize Zn in roots through formation of Zn phytate which has been shown to occur in a wide range of crop plants (Van Steveninck et al., 1993).

3.2. P does not decrease shoot Zn concentrations

More puzzling than the P-Zn interactions in which increasing P depresses Zn concentrations in plant shoots, are those interactions where increasing P in low Zn treatments induces symptoms of Zn deficiency and depresses plant growth while having no effect on Zn concentrations in plant shoots or even increasing them; yet application of Zn eliminates the symptoms and restores plant growth (Millikan, 1951, 1963; Boawn and Leggett, 1964; Boawn and Brown, 1968; Millikan et al., 1968; Loneragan et al., 1979, 1982; Christensen and Jackson, 1981). This syndrome has been attributed to an effect of increasing P within the plant which enhances the plant's internal Zn requirement i.e. a "P enhanced Zn requirement".

Factors which increase Zn concentrations. Before considering "P enhanced Zn requirements", two effects of P salts which sometimes enhance Zn concentrations in plants need to be mentioned, as they have caused confusion. They are an increase in the acidity of the root environment (Terman et al., 1966) and an inadvertent increase in Zn supply from Zn contamination of the P salt (Piper and Walkley, 1943; Bingham, 1959;

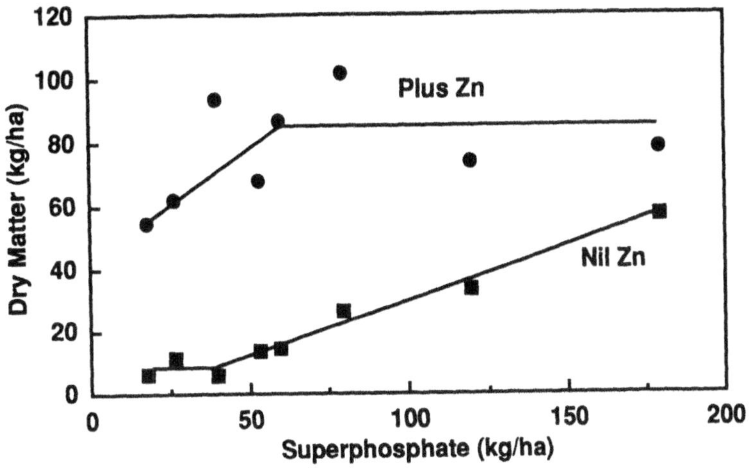

Figure 3. Effect of superphosphate (contaminated with Zn) and $ZnSO_4$ (7.8 kg/ha) on dry matter production of *Phalaris tuberosa*. Data from Anderson (1946).

Ozanne et al., 1965). Where P salts are contaminated with Zn, the response of plant growth to P and the interaction between P and Zn can be unusual and complex as Anderson (1946, 1956, Fig. 3) has shown. Contamination of superphosphate with Zn has also explained a linear relation between P and Zn concentrations in shoots of wheat plants given increasing levels of superphosphate (Piper and Walkley, 1943) which had previously been interpreted as support for a P enhanced Zn requirement (Millikan, 1940).

P enhanced Zn requirements. Early observations that mottle leaf of citrus was corrected by adding Zn but was intensified by increasing P, led to suggestions that high P in the plant might render its Zn less available (Chapman et al., 1937), perhaps by precipitating Zn in the leaf (West, 1938). The suggestions were supported by evidence that symptoms induced by high P treatments and corrected by Zn correlated with the ratio of concentrations of P/Zn but not with Zn (Millikan, 1951).

A plague of papers on P/Zn ratios followed but few have contributed to any further understanding of the phenomenon. Many have confirmed the correlation between symptoms and P/Zn ratios and have advocated "critical" P/Zn ratios for diagnosis of Zn deficiency. However, while "critical" P/Zn ratios could be established for individual experiments, they varied widely with species and conditions as, for example, in grape leaves with critical values of around 150 for plants in soil and over 1000 for plants in nutrient solution (Marschner and Schropp, 1977). Moreover, leaf symptoms correlated as well with leaf P concentrations as with the ratio of concentrations of P/Zn(Webb, 1987), leading to examination of the relation between Zn deficiency and P concentrations in plants.

Zn deficiency enhances P toxicity in shoots. Under conditions of high P supply such as may occur in solution cultures or siliceous sands, Zn deficiency has enhanced P concentrations in leaves or shoots of many species to levels known to be toxic to plants (see Table 7 of Loneragan et al., 1982). In subterranean clover (Loneragan et al., 1979), potato (Christensen and Jackson, 1981), okra (Loneragan et al., 1982), cotton (Cakmak and Marschner, 1986) and wheat (Webb and Loneragan, 1988), the effects and interactions of Zn and P on the intensity of leaf symptoms bore little relation to their Zn concentration but paralleled toxic P concentrations in leaves so closely that uncontrolled

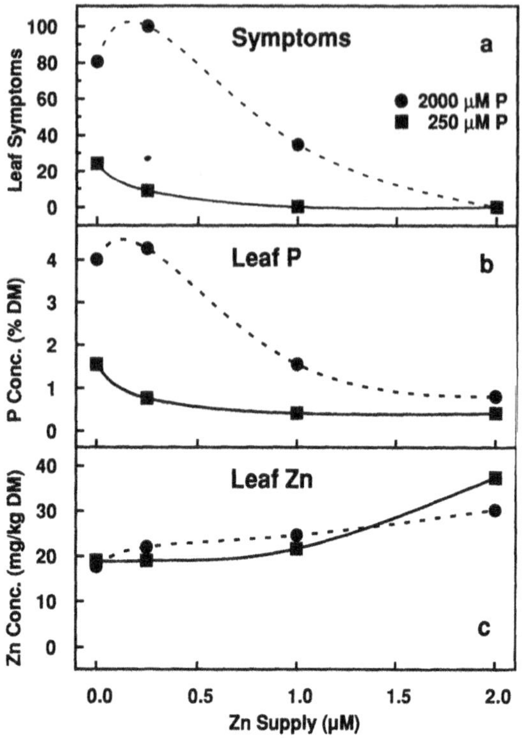

Figure 4. Effect of Zn supply on (a) symptoms resembling Zn deficiency, (b) P concentration, and (c) Zn concentration in mature leaves of okra grown in nutrient solution at two levels of P. Redrawn from Loneragan et al. (1982).

accumulation of P at low Zn was suggested as responsible for the syndrome of "P enhanced Zn requirements" (Fig. 4; Loneragan et al., 1979, 1982).

In potato, okra and cotton, low Zn treatments combined with high P supply induced P toxicity by enhancing the rate of P absorption into the plant and by causing P to accumulate preferentially in leaves, apparently by depressing the export of P from the leaves (Marschner and Cakmak, 1986). In wheat, Zn deficiency had only a transitory effect in stimulating P absorption and induced P toxicity primarily through preferential accumulation of P in older leaves by depressing P export (Webb, 1987; Webb and Loneragan, 1990).

P inactivation of plant Zn. While the role of Zn in preventing accumulation of P to toxic concentrations in leaves appears to explain the development in them of Zn deficiency symptoms induced by high P without any decrease in their Zn concentrations, there is now compelling evidence that precipitation of Zn by high P within the plant is the primary mechanism responsible for the syndrome.

In cotton, increasing P supply which enhanced symptoms of Zn deficiency in leaves, had no effect on their total Zn concentrations but depressed the fraction extracted by water (Fig. 5; Cakmak and Marschner, 1987). At all levels of Zn supply from severely deficient to adequate, increasing P supply depressed the proportion of Zn extracted from roots, stems and leaves from around 60% to nearly 30%. The concentration of water soluble Zn in leaves was closely correlated with visual Zn deficiency symptoms, and levels of chlorophyll, superoxide dismutase and membrane permeability.

Relevance to crop production. The syndrome of "P enhanced Zn requirements" has been produced many times in plants grown in siliceous sand and water culture experimental

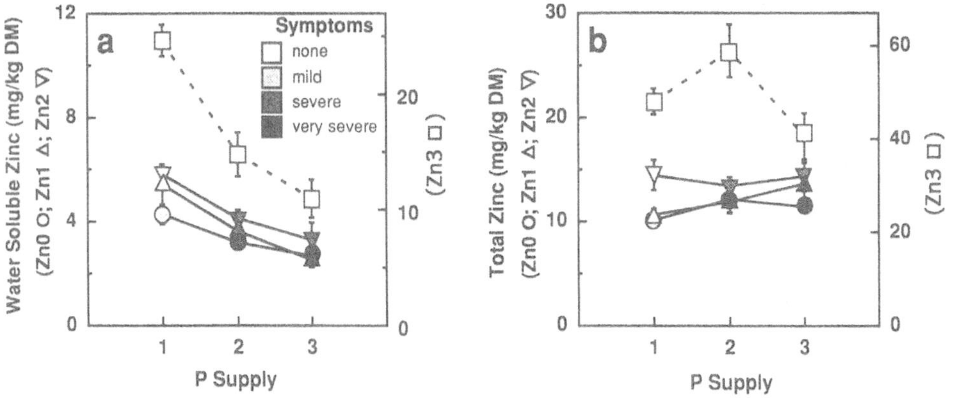

Figure 5. Effect of P supply (P1 = $2.5 \times 10^{-5}M$, P2 = $2.0 \times 10^{-4}M$, P3 = $6.0 \times 10^{-4}M$) on (a) water soluble and (b) total Zn concentrations in mature leaves of cotton grown at four Zn levels (Zn0 = 0, Zn1 = $2.0 \times 10^{-9}M$, Zn2 = $2.0 \times 10^{-8}M$, Zn3 = $1.0 \times 10^{-6}M$). The severity of Zn deficiency symptoms is represented by the intensity of shading within data symbols. Redrawn from Cakmak and Marschner (1987).

systems with their very high levels of P supply and has also been recently reported in water culture systems with P maintained at 10 μM (Parker, 1993). But there are no valid reports that it occurs at the very much lower concentrations of P commonly encountered in soil solutions. Moreover, the very high plant P concentrations associated with the enhancement of Zn deficiency symptoms in this phenomenon are seldom encountered in crop plants. It therefore appears likely that "P enhanced Zn requirements" are an artifact of glasshouse experimental procedures and have little relevance to crop production.

4. N-Zn interactions

Nitrogen salts have both ameliorated and severely enhanced Zn deficiencies. The strongest interactions have resulted from the effects of N in promoting plant growth and, to a lesser extent, in changing the pH of the root environment.

Since N limits plant growth on many soils, it is not surprising to find many reports of positive interactions between increasing levels of N and Zn fertilizers. For example, when grown on a N deficient soil with adequate levels of all nutrients other than N and Zn, grain yield of wheat did not respond to Zn in the absence of NH_4NO_3 fertilizer but responded strongly in its presence (Fig. 2a; Chaudhry and Loneragan, 1970). The response was attributed to the dilution of Zn in the plant resulting from the strong promotion of plant growth by the correction of N deficiency.

By contrast, a strong negative N-Zn interaction was observed in the same experiment when Cu was omitted from the basal fertilizer (Fig. 2b). In this case, the addition of NH_4NO_3 fertilizer promoted grain yield in the absence of Zn and induced moderate Cu deficiency primarily by dilution of Cu through shoot growth: adding Zn together with NH_4NO_3 enhanced the Cu deficiency so severely that grain yield dropped below the level in plants without NH_4NO_3. This unusual N-Zn interaction resulted from a Cu-Zn interaction discussed below.

On other low Zn soils of higher fertility, N fertilizers have ameliorated or intensified Zn deficiency by affecting Zn absorption through changing soil pH (eg. Fig. 1; Viets et al., 1957). Acidifying effects of ammonium may explain why concurrent dressings of N with $ZnSO_4$ were effective in controlling Zn deficiency in crops on soils where $ZnSO_4$ alone was unsatisfactory (Viets et al., 1953).

Two other effects of N salts on Zn nutrition have been observed. In short term studies with wheat seedlings, NH_4^+ salts inhibited Zn absorption from low concentrations of Zn^{2+} (Chaudhry and Loneragan, 1972a,c). The inhibition by NH_4^+ ions was stronger than by alkali and alkaline earth cations but was competitive with them, so that its effects would be diminished by the relatively high concentrations of competing ions in soil and culture solutions; in soils, any direct effect of NH_4^+ ions on Zn absorption would also rapidly disappear with nitrification and consequent soil acidification.

A suggestion that N may induce Zn deficiency by restricting transport to the shoots (Ozanne, 1955) was not supported by the distribution of Zn in the Zn deficient wheat plants of the experiment discussed above (Chaudhry and Loneragan, 1970) in which NH_4NO_3 strongly enhanced the proportion of Zn in plant shoots. Nevertheless, some effect of N supply on the distribution of Zn within the plant seems likely in view of the retention of Zn in non-senescing leaves and its export with N during senescence (Hill et al., 1979).

5. Macronutrient cation-Zn interactions

There is good evidence that the macronutrient cations Ca, Mg and K, as well as other alkali and alkaline earth cations, inhibit the absorption of Zn by plants from solutions. They need to be considered when interpreting the results of solution culture experiments involving Zn nutrition. But in soils, their inhibitory activity seems much less important than the effects of their salts on soil pH.

For six legumes grown in solution culture at constant pH, Zn concentrations were highest in the shoots of plants at the lowest Ca level where plants were not Ca deficient and progressively decreased with increasing Ca concentrations in solution (Bell et al., 1989; Fig. 6). The finding that Ca inhibited Zn absorption is in accord with short term studies with wheat seedlings in which increasing concentrations of $Ca(NO_3)_2$ from 0 to 40 m*M* progressively inhibited the rate of Zn absorption by a non-competitive mechanism with no additional effect to 100 m*M*; the inhibition was attributed to Ca as varying anions had little effect on Zn absorption whereas substituting other cations for Ca did (Chaudhry and Loneragan, 1972a,c; Fig. 6).

The macronutrient cations K, NH_4 and Mg and other alkali and alkaline earth cations all inhibited the rate of Zn absorption strongly from solutions of low Ca concentration; with increasing Ca concentrations, their inhibitory effects weakened and, in the case of the two ions (K, Mg) tested at sufficiently high Ca concentration (2.5 - 10 m*M*), eventually disappeared, suggesting that they operated through the same mechanism as Ca (Chaudhry and Loneragan, 1972a,c). However, since H^+ ions above pH 4 and Cu^{2+} inhibited the rate of Zn absorption by mechanisms in which Ca^{2+} did not compete (Chaudhry and Loneragan, 1972b), their inhibitory activities would add to that of Ca^{2+}.

In soils, the effects of Ca compounds on Zn nutrition are variable, apparently due to effects of the salts on soil pH and hence on Zn in the soil solution as discussed earlier. Thus, Zn concentrations in plants growing in soil increased slightly when treated with

Figure 6. Effect of competing cations (Ca (■, ●) and Mg(▲)) on Zn concentration (■) of six legumes (Bell et al., 1989); and Zn absorption rate (●, ▲) by wheat seedlings (Chaudhry and Loneragan, 1972c).

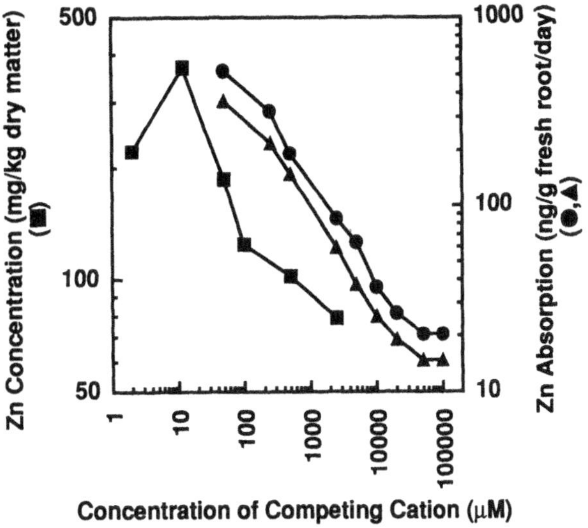

$CaSO_4$ which decreased soil pH from 5.6 to 4.8, but decreased strongly when treated with an equivalent amount of $CaCO_3$ which increased soil pH from 5.7 to 6.6 (Wear, 1956).

6. Micronutrient-Zn interactions

6.1. Cu-Zn interactions

In addition to the normal interaction between two limiting nutrients, Cu and Zn may interact in several other ways -

- Zn strongly depresses grain yield of wheat by depressing Cu absorption
- Cu competitively inhibits Zn absorption
- Cu nutrition affects the redistribution of Zn within plants.

A very strong Cu-Zn interaction has been observed in the grain yield of wheat crops growing on soils deficient in both Cu and Zn (Toms, 1958; Chaudhry and Loneragan, 1970; Kausar et al., 1976). In the presence of NH_4NO_3 in the N-Cu-Zn experiment discussed above (Fig. 2; Chaudhry and Loneragan, 1970), it appears to have two components - the usual positive interaction between increasing supply of two nutrients when both are limiting and an abnormal depression of grain yield with addition of Zn in the absence of Cu. In this latter treatment, Zn intensified Cu deficiency in the plants by depressing Cu uptake.

The effect of Zn in depressing Cu uptake may result from the competitive inhibition of Zn on Cu absorption as suggested by Bowen (1969) from short term studies with excised leaf discs. While Bowen's data are not definitive, a converse competitive inhibition of Cu^{2+} ions on Zn^{2+} absorption has been established in short term, water culture studies with excised tissues (Schmid et al., 1965; Hawf and Schmid, 1967; Bowen, 1969) and seedlings (Chaudhry and Loneragan, 1972b; Giordano et al., 1974). But while Zn severely depressed Cu uptake by wheat growing in soil as already described, Cu did not depress Zn absorption in the same experiment. Nor are we aware of any study showing a depression in plant growth or yield through an effect of Cu in depressing Zn uptake.

The reason for this apparent conflict between results in soil and solution culture, may lie in the effects of complex formation on the activities of the divalent Cu^{2+} and Zn^{2+} ions which are thought to be the dominant forms of Cu and Zn absorbed (Kochian, 1991; Chapter 4). In the solution studies, Cu and Zn were present as divalent ions whereas in most soils they are predominantly present in complexed forms; in soils, a much higher proportion of Cu is complexed than Zn (Hodgson et al., 1965, 1966; Geering and Hodgson, 1969; Chapter 1) so that Zn^{2+} activity would be much higher than Cu^{2+} activity at the absorbing sites, making it an effective competitor in Cu absorption and making its absorption less sensitive to competition from Cu.

The effect of complex formation on the activities of Cu^{2+} and Zn^{2+} probably also accounts for the lack of Cu-Zn interactions in plants grown for long periods in water culture solutions using chelate buffering techniques; when Cu and Zn were present in adequate supply and predominantly in chelated forms, increasing Cu^{2+} activity had little effect on Zn concentrations in roots and shoots of maize (Bell et al., 1991) and increasing Zn^{2+} activity had little effect on Cu concentrations in roots and shoots of barley (Norvell and Welch, 1993).

Copper nutrition of plants has also been shown to affect the redistribution of Zn from wheat leaves. In Cu deficient wheat plants, the senescence of the oldest leaf and the export from it of N, Cu and Zn was delayed compared with plants given adequate Cu (Hill et al., 1979). The effect of Cu on Zn was probably indirect through its effect on senescence.

6.2. Fe-Zn interactions

While the interactions between Zn and Fe have not caused as much controversy as the interaction between Zn and P, they appear to be as complex. Increasing Zn supply to plant roots has been shown to increase (Watanabe et al., 1965; Jolley and Brown, 1991), have little effect on (Norvell and Welch, 1993) or decrease (Safaya, 1976; Jolley and Brown, 1991) Fe concentration in shoots. The reciprocal effect of increasing Fe, generally has only a depressive effect on Zn concentration in plant tissues (Watanabe et al., 1965; Zhang et al., 1991b) although it has been shown to increase (Giordano et al., 1974), have no effect on (Chaudhry and Loneragan, 1972b) or decrease (Giordano et al., 1974; Rashid et al., 1976; Zhang et al., 1991b) rates of Zn absorption by plant roots.

The apparently conflicting results are probably due to differences in experimental details, especially in plant species and the concentration, ionic state and complexation of Fe. Few of the reported effects are likely to affect crop yields in farming situations with the possible exception of unusually high concentrations of Fe^{2+} in flooded soils suppressing Zn absorption and enhancing Zn deficiency in rice. Even this interaction will be strongly modified in soils by complex effects of flooding on soil reactions affecting the concentration of Zn and other ions in the soil solution (Beckwith et al., 1975; Forno et al., 1975).

Effects of Fe on Zn absorption. When present at 10 μM, Fe^{2+} had no effect on the rate of Zn absorption by wheat seedlings from solutions containing 1 or 10 μM Zn and 50 mM $Ca(NO_3)_2$ (Chaudhry and Loneragan, 1972b). But when added as 100 μM Fe^{2+}, a form and concentration likely to occur in flooded rice paddies, Fe completely suppressed Zn absorption by rice seedlings from solutions of 0.05 μM $ZnCl_2$ and no Ca (Giordano et al., 1974).

Effects of Fe deficiency on Zn absorption. Fe deficiency has increased Zn concentrations

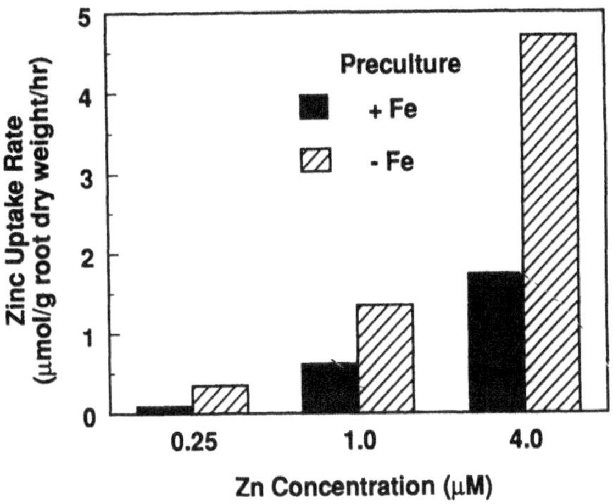

Figure 7. Effect of preculture with or without Fe on subsequent Zn uptake by wheat. Redrawn from Zhang et al. (1991b).

in the shoots of dicotyledonous species (Agarwala et al., 1979) and rates of Zn absorption in both dicotyledonous (Römheld et al., 1982) and grass (Fig. 7; Zhang et al., 1991b) species. For dicotyledonous plants, the mechanism for increasing Zn absorption is probably that of acidification of the rhizosphere resulting from the Strategy I reaction to Fe deficiency (Marschner et al., 1989). For grasses, phytosiderophores released in the Strategy II mechanism of reaction to Fe deficiency (Zhang et al., 1991b) have enhanced the mobilisation of Zn from a calcareous soil (Treeby et al., 1989). However, they did not enhance Zn absorption into the root in the same way as they did for Fe (Marschner et al., 1989). The effects of these mechanisms on Zn absorption would be additional to any effect Fe might have in directly inhibiting Zn absorption as discussed above.

Effects of Zn deficiency on absorption of Fe. Zn deficiency has increased Fe concentrations in the shoots of both Strategy I (sugar beet - Rosell and Ulrich, 1964; navy beans - Ambler and Brown, 1969) and Strategy II (corn - Jackson et al., 1967) plants, possibly by mechanisms involving acidification in the rhizosphere, and release of reductants and phytosiderophores.

As with Fe deficiency, Zn deficiency in cotton and sunflower acidified nutrient solutions (Cakmak and Marschner, 1990) when supplied with nitrate-N. While this effect might play some role in the enhancement of Fe concentrations in shoots of Strategy I plants, more emphasis has been placed on the role of Zn deficiency in enhancing the excretion of reductants and other exudates from roots. In navy beans, Brown (1979) showed that reduction of Fe^{III} to Fe^{II} was enhanced by Zn deficiency and that the effect was greater in a Zn-inefficient variety (Sanilac) than in a Zn-efficient variety (Saginaw). More recently, Jolley and Brown (1991) confirmed the differences in Fe^{III} reduction between cultivars and further showed that greater amounts of reductant were released in the Zn-inefficient variety (Sanilac) than the Zn-efficient variety (Saginaw) under Zn deficiency stress. They attributed the increased Fe concentration in Sanilac compared to Saginaw to increased reduction at the root surface and increased release of reductants, and suggested that the greater availability of Fe depressed Zn absorption by competition. Using a range of strategy I plants, Zhang et al. (1991a) showed that exudates from Zn-deficient plants were capable of mobilizing more Fe from Fe^{III} hydroxides than were exudates from Zn adequate plants. Surprisingly, there was little difference in their ability to mobilize Zn from a Zn chelate or calcareous soil!

Like Fe deficiency, Zn deficiency of Strategy II plants also enhanced mobilization, absorption and accumulation of Fe by production and release of phytosiderophores (Zhang et al., 1989; 1991a). Acidification of the rhizosphere may also have been involved since, unlike Fe deficiency which does not acidify the rhizosphere of Strategy II plants, Zn deficiency acidified nutrient solutions in which wheat was grown (Loneragan et al., 1987).

Effects of Zn on translocation of Fe. When Zn and Fe (as ^{59}Fe) were supplied to different root systems of healthy watercress, increasing Zn to 100 μM or more enhanced the retention of Fe in stems and roots to which Fe had been applied and decreased its translocation to other roots and leaves (Cumbus et al., 1977). That the response to Mn was somewhat different, having little effect on distribution, indicates some specificity in this interaction. It is possible that the effect is a competitive inhibition at the site of unloading from the xylem.

6.3. *Co, Mn, Na-Zn interactions*

Under some conditions, Co, Mn and Na may all inhibit Zn absorption. But none of these effects is likely to affect growth of crops except perhaps for high concentrations of Mn in combination with Fe depressing absorption of Zn by rice in flooded soils.

When present at the same concentration as Zn^{2+} in solutions containing Ca^{2+}, Co^{2+} had no effect on the absorption of Zn by sugar cane leaf discs (Bowen, 1969) or wheat seedlings (Chaudhry and Loneragan, 1972b). But when present in 10-fold excess, Co^{2+} depressed the rate of Zn absorption by 10% in wheat seedlings (Chaudhry and Loneragan, 1972b).

Under similar conditions and even at 10-fold excess over Zn^{2+}, Mn^{2+} had no effect on the absorption of Zn by excised roots (Schmid et al., 1965), intact seedlings (Hawf and Schmid, 1967; Chaudhry and Loneragan, 1972b) or leaf discs (Bowen, 1969). Nor did Mn have much effect on Zn concentrations of roots or shoots of barley when total Mn concentration and Mn^{2+} activity were varied 10,000-fold from deficient to near toxic levels in a complete nutrient solution with the activities of the transition metals buffered by an excess of HEDTA (Webb et al., 1993). By contrast, when present at 2,000-fold excess in the absence of Ca^{2+}, Mn^{2+} depressed the rate of Zn absorption by about 50%, leading to the suggestion that the high concentrations of reduced Mn and Fe which develop in paddy culture may enhance Zn deficiency in rice (Giordano et al., 1974).

There is evidence that Na^+ inhibits the absorption of Zn by the same mechanism as, but more weakly than, K^+ (Chaudhry and Loneragan, 1972a).

6.4. *B-Zn interactions*

In solution culture, low Zn treatments, which had no effect on growth, enhanced B concentrations to toxic levels in barley (Graham et al., 1987). The resulting B-Zn interaction was similar to the P-Zn interaction discussed earlier in which Zn deficiency enhanced P toxicity in several species. The interesting suggestion that low Zn availability in alkaline soils with high B in semi-arid regions might depress crop yields by enhancing B toxicity has yet to be established. However, the suggestion is viable since, in soils, Zn deficiency enhanced B concentration of wheat (Singh et al., 1990) while decreasing wheat dry matter.

7. Conclusions

While Zn interacts with other nutrients in many ways, few, other than those involving correction of deficiencies of both Zn and another nutrient, appear to be important in crop production. Where interactions do occur, they sometimes result, not from the nutrient to which they are attributed, but from other factors associated with the addition of the nutrient compound. For diagnosis and effective treatment of Zn deficiency, it is essential to identify the operative factor.

The value of careful, critical research in the understanding of nutrient interactions is well illustrated by recent research which has provided a new insight into the long puzzling phenomenon of "P enhanced Zn requirements". It contrasts with the huge volume of uncritical and unproductive research correlating P/Zn ratios with Zn deficiency and serves as a warning against the current trend of producing "desktop" interactions from the uncritical development of relationships between every possible combination of nutrients resulting from the availability of multi-element analyzers and high capacity computers.

References

Agarwala S C, Mehrotra S C, Bisht S S and Sharma C P 1979 Mineral nutrient element composition of three varieties of chickpea grown at normal and deficient levels of iron supply. J. Indian Bot. Soc. 58,153-162.

Ambler J E and Brown J C 1969 Cause of differential susceptibility to zinc deficiency in two varieties of navy beans (*Phaseolus vulgaris* L.). Agron. J. 61, 41-43.

Anderson A J 1946 Fertilizers in pasture development on peat soils in the lower south east of South Australia. J. Counc. Sci. Ind. Res. Aust. 19, 394-403.

Anderson A J and Thomas M P 1946 Molybdenum and symbiotic nitrogen fixation. Bull. Counc. Sci. Industr. Res. Aust. 198, 7-24.

Anderson A J 1956 Effects of fertilizer treatments on pasture growth. Proc. Seventh Int. Grassl. Congr., Massey Agricultural College, NZ, pp. 323-333.

Barrow N J 1987 The effects of phosphate on zinc sorption by a soil. J. Soil Sci. 38, 453-459.

Beckwith R S, Tiller K G and Suwadji E 1975 The effects of flooding on the availability of trace metals to rice in soils of differing organic matter status. *In* Trace Elements in Soil-Plant-Animal Systems. Edited by Nicholas D J D and Egan A R pp 135-149. Academic Press, New York.

Bell P F, Chaney R L and Angle J S 1991 Determination of the copper^{2+} activity required by maize using chelator-buffered nutrient solutions. Soil Sci. Soc. Amer. J. 55, 1366-1374.

Bell R W, Edwards D G and Asher C J 1989 Effects of calcium supply on uptake of calcium and selected mineral nutrients by tropical food legumes in solution culture. Aust. J. Agric. Res. 40, 1003-1014.

Bingham F T 1959 Micronutrient content of phosphorus fertilizers. Soil Sci. 88, 7-10.

Bingham F T, Martin J P and Chastain J A 1958 Effects of phosphorus fertilization of California soils on minor elements nutrition of citrus. Soil Sci. 86, 24-31.

Boawn L C, Viets F G and Crawford C L 1954 Effect of phosphate fertilizers on zinc nutrition of field beans. Soil Sci. 78, 1-7.

Boawn L C and Brown J C 1968 Further evidence for a P-Zn imbalance in plants. Soil Sci. Soc. Amer. Proc. 32, 94-97.

Boawn L C and Legget G E 1964 Phosphorus and zinc concentrations in Russet Burbank potato tissue in relation to development of zinc deficiency symptoms. Soil Sci. Soc. Amer. Proc. 28, 229-232.

Bolland M D A, Posner A M and Quirk J P 1977 Zinc adsorption by goethite in the absence and presence of phosphate. Aust. J. Soil Res. 15, 279-286.

Bowen J E 1969 Absorption of copper, zinc and manganese by sugar cane tissue. Plant Physiol 44, 255-261.

Brown J C 1979 Effect of zinc stress factors affecting iron uptake in navy bean. J. Plant Nutr. 1, 171-183.

Burleson C A and Page N R 1967 Phosphorus and zinc interactions in flax. Soil Sci. Soc. Amer. Proc. 31, 510-513.

Cakmak I and Marschner H 1986 Mechanism of phosphorus-induced zinc deficiency in cotton. I. Zinc deficiency-enhanced uptake rate of phosphorus. Physiol. Plant. 68, 483-490.

Cakmak I and Marschner H 1987 Mechanism of phosphorus-induced zinc deficiency in cotton. III. Changes in physiological availability of zinc in plants. Physiol. Plant. 70, 13-20.
Cakmak I and Marschner H 1990 Decrease in nitrate uptake and increase in proton release in zinc deficient cotton, sunflower and buckwheat plants. Plant Soil 129, 261-268.
Chapman H D, Vanselow A P and Liebig G F 1937 The production of citrus mottle-leaf in controlled nutrient cultures. J. Agric. Res. 55, 365-379.
Chaudhry F M and Loneragan J F 1970 Effects of nitrogen, copper, and zinc fertilizers on the copper and zinc nutrition of wheat plants. Aust. J. Agric. Res. 21, 865-879.
Chaudhry F M and Loneragan J F 1972a Zinc absorption by wheat seedlings: I. Inhibition by macronutrient ions in short-term experiments and its relevance to long-term zinc nutrition. Soil Sci. Soc. Amer. Proc. 36, 323-327.
Chaudhry F M and Loneragan J F 1972b Zinc absorption by wheat seedlings: II. Inhibition by hydrogen ions and by micronutrient cations. Soil Sci. Soc. Amer. Proc. 36, 327-331.
Chaudhry F M and Loneragan J F 1972c Zinc absorption by wheat seedlings and the nature of its inhibition by alkaline earth cations. J. Exp. Bot. 23, 552-560.
Christensen N W and Jackson T L 1981 Potential for phosphorus toxicity in zinc-stressed corn and potato. Soil Sci. Soc. Amer. J. 45, 904-909.
Cogliatti D H, Alcocer N and Santa Maria G E 1991 Effect of phosphorus concentration on zinc-65 uptake in *Gaudinia fragilis*. J. Plant Nutr. 14, 443-452.
Cumbus I P, Hornsey D J and Robinson L W 1977 The influence of phosphorus zinc, and manganese on absorption and translocation of iron in watercress. Plant Soil 48, 651-660.
Edwards J H and Kamprath E J 1974 Zinc accumulation by corn seedlings as influenced by phosphorus, temperature, and light intensity. Agron. J. 66, 479-482.
Forno D A, Yoshida S and Asher C J 1975 Zinc deficiency in rice. I. Soil factors associated with the deficiency. Plant Soil 42, 537-550.
Friesen D K, Juo A S R and Miller M H 1980 Liming and lime-phosphorus-zinc interaction in two Nigerian ultisols: I. Interactions in the soil. Soil Sci. Soc. Amer. J. 44, 1221-1226.
Geering H R and Hodgson J F 1969 Micronutrient cation complexes in soil solution: III. Characterization of soil solution ligands and their complexes with Zn^{2+} and Cu^{2+}. Soil Sci. Soc. Amer. Proc. 33, 54-59.
Giordano M, Noggle J C and Mortvedt J J 1974 Zinc uptake by rice, as affected by metabolic inhibitors and competing cations. Plant Soil 41, 637-646.
Graham R D, Welch R M, Grunes D L, Cary E E and Norvell W A 1987 Effect of zinc deficiency on the accumulation of boron and other mineral nutrients in barley. Soil Sci. Soc. Amer. J. 51, 652-657.
Hawf L R and Schmid W E 1967 Uptake and translocation of zinc by intact plants. Plant Soil 27, 249-260.
Hill J, Robson A D and Loneragan J F 1979 The effect of copper supply on the senescence and the retranslocation of nutrients of the oldest leaf of wheat. Ann. Bot. 44, 279-287.
Hodgson J F, Geering H R and Norvell WA 1965 Micronutrient cation complexes in soil solution: Partition between complexed and uncomplexed forms by solvent extraction. Soil Sci. Soc. Amer. Proc. 29, 665-669.
Hodgson J F, Lindsay W L and Trierweiler J F 1966 Micronutrient cation complexing in soil solution: III. Complexing of zinc and copper in displaced solution from calcareous soils. Soil Sci. Soc. Amer. Proc. 30, 723-726.
Jackson T L, Hay J and Moore D P 1967 The effects of zinc on yield and chemical composition of sweet corn in Willamette Valley. Amer. Soc. Hort. Sci. 91, 462-471.
Jolley V D and Brown J C 1991 Factors in iron-stress response mechanism enhanced by Zn-deficiency stress in Sanilac, but not Saginaw navy beans. J. Plant Nutr. 14, 257-265.
Kauser M A, Chaudhry F M, Rashid A, Latif A and Alam S M 1976 Micronutrient availability to cereals from calcareous soils. I. Comparative Zn and Cu deficiency and their mutual interaction in rice and wheat. Plant Soil 45, 397-410.
Kochian L V 1991 Mechanisms of micronutrient uptake and translocation in plants. *In* Micronutrients in Agriculture. 2nd edition. Edited by Mortvedt J J, Cox F R, Shuman L M and Welch R M pp 229-296. Soil Science Society of America, Madison.
Lambert D H, Baker D E and Cole H 1979 The role of mycorrhizae in the interaction of phosphorus with zinc, copper, and other elements. Soil Sci. Soc. Amer. J. 43, 976-980.
Loneragan J F 1951 Effect of applied phosphate on the uptake of zinc by flax. Aust. J. Sci. Res. B4, 108-114.
Loneragan J F, Grove T S, Robson A D and Snowball K 1979 Phosphorus toxicity as a factor in zinc-phosphorus interaction in plants. Soil Sci. Soc. Amer. J. 43, 966-972.
Loneragan J F, Grunes D L, Welch R M, Aduayi E A, Tengah A, Lazar V A and Cary E E 1982 Phosphorus

accumulation and toxicity in leaves in relation to zinc supply. Soil Sci. Soc. Amer. J. 46, 345-352.

Loneragan J F, Kirk G J and Webb M J 1987 Translocation and function of zinc in roots. J. Plant Nutr. 10, 1247-1254.

Marschner H and Cakmak I 1986 Mechanism of phosphorus-induced zinc deficiency in cotton. II. Evidence for impaired shoot control of phosphorus uptake and translocation under zinc deficiency. Physiol. Plant. 68, 491-496.

Marschner H and Schropp A 1977 Vergeichende Untersuchungen über die Empfindlichkeit von 6 Unterlagensorten der Weinrebe gegenüber P-induziertem Zn-Mangel. Vitis 16, 79-88.

Marschner H, Treeby M and Römheld V 1989 Role of root-induced changes in the rhizosphere for iron acquisition in higher plants. Z. Pflanzenernähr. Bodenk. 152, 197-204.

Millikan C R 1940 Zinc Requirement of wheat. The effect of superphosphate. J. Dept. Agric., Victoria. 38, 135-136.

Millikan C R 1951 Diseases of flax and linseed. Zinc deficiency. Dept. Agric. Vict., Aust. Tech. Bull. No. 9, 90-108.

Millikan C R 1963 Effects of different levels of zinc and phosphorus on the growth of subterranean clover (*Trifolium subterraneum* L.). Aust. J. Agric. Res. 14, 180-205.

Millikan C R, Hanger B C and Bjarnason E N 1968 Effect of phosphorus and zinc levels in the substrate on ^{65}Zn distribution in subterranean clover and flax. Aust. J. biol. Sci. 21, 619-640.

Norvell W A and Welch R M 1993 Growth and nutrient uptake by barley (*Hordeum vulgare* L. cv Herta): Studies using an N-(2-Hydroxyethyl)ethylenedinitrilotriacetic acid buffered nutrient solution technique. I. Zinc ion requirements. Plant Physiol. 101, 619-625.

Ozanne P G 1955 Effect of nitrogen on zinc deficiency in subterranean clover. Aust. J. Biol. Sci. 8, 47-55.

Ozanne P G, Shaw T C and Kirton D J 1965 Pasture responses to traces of zinc in phosphate fertilizers. Aust .J. Exp. Agric. Anim. Husb. 5, 29-33.

Parker D R 1993 Novel nutrient solutions for zinc nutrition research: buffering free zinc^{2+} with synthetic chelators and P with hydroxyapatite. *In* Proceedings of the XII International Plant Nutrition Colloquium, Perth, Australia, 21-24 September, 1993. Edited by Barrow N J. In Press. Kluwer Academic Publishers, Dordrecht, The Netherlands.

Piper C S and Walkley A 1943 Copper, zinc and manganese in some plants of agricultural interest. J. Counc. Sci. Ind. Res. Aust. 16, 217-234.

Rashid A, Chaudhry F M and Sharif M 1976 Micronutrient availability to cereals from calcareous soils. III. Zinc absorption by rice and its inhibition by important ions of submerged soils. Plant Soil 45, 613-623.

Rogers L H and Wu C 1948 Zinc uptake by oats as influenced by application of lime and phosphate. J. Amer. Soc. Agron. 40, 563-566.

Römheld V, Marschner H and Kramer D 1982 Responses to Fe deficiency in roots of "Fe-efficient" plant species. J. Plant Nutr. 5, 489-498.

Rosell R A and Ulrich A 1964 Critical zinc concentrations and leaf minerals of sugar beet plants. Soil Sci. 97, 152-167.

Saeed M and Fox R L 1979 Influence of phosphate fertilization on zinc adsorption by tropical soils. Soil Sci. Soc. Amer. J. 43, 683-686.

Safaya N M 1976 Phosphorus-zinc interaction in relation to absorption rates of phosphorus, zinc, copper, manganese, and iron in corn. Soil Sci. Soc. Amer. J. 40, 719-722.

Schmid W E, Haag H P and Epstein E 1965 Absorption of zinc by excised barley roots. Physiol. Planta. 18, 860-869.

Sharma K C, Krantz B A, Brown A L and Quick J 1968 Interaction of Zn and P in top and root of corn and tomato. Agron. J. 60, 453-456.

Singh J P, Karamanos R E and Stewart J W B 1988 The mechanism of phosphorus-induced zinc deficiency in bean (*Phaseolus vulgaris* L.). Can J. Soil Sci. 68, 345-358.

Singh J P, Dahiya D J and Narwal R P 1990 Boron uptake and toxicity in wheat in relation to zinc supply. Fert. Res. 24, 105-110.

Stanton D A and Burger R du T 1967 Availability to plants of zinc sorbed by soil and hydrous iron oxides. Geoderma 1, 13-17.

Stanton D A and Burger R du T 1970 Studies on zinc in selected orange free state soils: V. Mechanisms for the reaction of zinc with iron and aluminium oxides. Agrochemophysica 2, 65-76.

Stukenholtz D D, Olsen R J, Gogan G and Olson R A 1966 On the mechanism of phosphorus-zinc interaction in corn nutrition. Soil Sci. Soc. Amer. Proc. 30, 759-763.

Terman G L, Allen S E and Bradford B N 1966 Response of corn to zinc as affected by nitrogen and phosphorus fertilizers. Soil Sci. Soc. Amer. Proc. 30, 119-124.

Toms J 1958 The use of copper and zinc in the cereal-growing districts of Western Australia. J. Dept. Agric. Western Australia, series 3, vol 7, 197-203.

Treeby M, Marschner H and Römheld V 1989 Mobilization of iron and other micronutrient cations from a calcareous soil by plant-borne, microbial, and synthetic metal chelators. Plant Soil 114, 217-226.

Van Steveninck R F M, Barbare A, Fernando D R and Van Steveninck M E 1993 The binding of zinc in root cells of crop plants by phytic acid. *In* Proceedings of the XII International Plant Nutrition Colloquium, Perth, Australia, 21-24 September, 1993. Edited by Barrow N J. In Press. Kluwer Academic Publishers, Dordrecht, The Netherlands.

Viets F G, Jnr, Boawn L C, Crawford C L and Nelson C E 1953 Zinc deficiency in corn in central Washington. Agron. J. 45, 559-565.

Viets F G, Jnr, Boawn L C and Crawford C L 1957 The effect of nitrogen and types of nitrogen carrier on plant uptake of indigenous and applied zinc. Soil Sci. Soc. Amer. Proc. 21, 197-201.

Watanabe F S, Lindsay W L and Olsen S R 1965 Nutrient balance involving phosphorus, iron, and zinc. Soil Sci. Soc. Amer. Proc. 29, 562-565.

Wear J I 1956 Effect of soil pH and calcium on uptake of zinc by plants. Soil Sci. 81, 311-315.

Webb M J 1987 Zn function in P accumulation by plants. PhD thesis. Murdoch University, Western Australia.

Webb M J and Loneragan J F 1988 Effect of zinc deficiency on growth, phosphorus concentration, and phosphorus toxicity of wheat plants. Soil Sci. Soc. Amer. J. 52, 1676-1680.

Webb M J and Loneragan J F 1990 Zinc translocation to wheat roots and its implications for a phosphorus/zinc interaction in wheat plants. J. Plant Nutr. 13, 1499-1512.

Webb M J, Norvell W A, Welch R M and Graham R D 1993 Using a chelate-buffered nutrient solution to establish the solution activity of Mn^{2+} required by barley (*Hordeum vulgare* L. cv Herta). Plant Soil 153, 195-205.

West E S 1938 Zinc-cured mottle-leaf in citrus induced by excess phosphate. J. Counc. Sci. Ind. Res. 11, 182-184.

Williams C H 1977 Trace metals and superphosphate: toxicity problems. Aust. Inst. Agric. Sci. J. 43, 99-109.

Zhang F, Römheld V and Marschner H 1989 Effect of zinc deficiency in wheat on the release of zinc and iron mobilizing root exudates. Z. Pflanzenernähr. Bodenk. 152, 205-210.

Zhang F, Römheld V and Marschner H 1991a Release of zinc mobilizing root exudates in different plant species as affected by zinc nutritional status. J. Plant Nutr. 14, 675-686.

Zhang F, Römheld V and Marschner H 1991b Diurnal rhythm of release of phytosiderophores and uptake rate of zinc in iron-deficient wheat. Soil Sci. Plant Nutr. 37, 671-678.

Chapter 10.

Zinc Phytotoxicity

R. L. CHANEY

1. Abstract

After "natural" phytotoxicity from Al or Mn in strongly acidic soil, Zn phytotoxicity is the most extensive microelement phytotoxicity, far more important than Cu, Ni, Co, Cd, or other metals. Zn has been extensively dispersed, and has reached phytotoxic concentrations in many soils due to anthropic contamination from many sources (fertilizers, pesticides, manures, sewage sludges, smelters, incinerators, mines, galvanized products). As soil pH falls, Zn solubility and uptake increase and potential for phytotoxicity increases. When plant leaves reach about 300-1000 mg Zn/kg DW (typical phytotoxic level is 500 mg/kg DW in diagnostic leaves), yield is reduced. At least in acidic soils, phytotoxicity is indicated by Zn-induced Fe-deficiency-chlorosis.

The physiology of Zn phytotoxicity in leaves is complicated, resulting from Zn interference in chlorophyll biosynthesis, and other biochemical reactions. In acidic soils, Zn usually causes severe Fe-deficiency chlorosis in dicots. Crops such as lettuce, mustard, and beet are highly susceptible to excessive soil Zn. In strongly acidic soils, grasses are usually much more Zn tolerant than dicots. However, in neutral or alkaline soils, *Poaceae* species are more sensitive to soil Zn than are dicots, apparently due to the interference of Zn in phytosiderophore function. Zn and other strongly chelated metal ions are able to displace Fe from mugineic acid and cause severe phytotoxicity. The natural increased secretion of phytosiderophores at alkaline pH increases the dissolved Zn in the soil, increases convective and diffusive movement of Zn to the root, and causes relatively greater susceptibility to soil Zn in grasses than other species.

Plant tolerance of Zn is an inheritable physiological property in many species. "Ecotypic" tolerance to Zn has been observed as soon as 20 years after Zn contamination of acidic soils. Highly Zn-tolerant individuals exist in wild type seed for these species. Some species tolerate soil Zn by excluding Zn by the roots (e.g., 'Merlin' red fescue [*Festuca rubra* L.]). Others tolerate higher foliar concentrations of Zn. Still others transport Zn rapidly to the shoots, and tolerate very high foliar Zn (up to 40,000 mg/kg DW in alpine pennycress [*Thlaspi caerulescens* J.and C. Presl.]). Compartmentalization in the vacuole and strong chelation (by malate, citrate, glutathione and possibly phytochelatins) in the cytoplasm apparently provide the high tolerance seen in most tolerant genotypes. Researchers are presently studying Zn and Cd metabolism in species such as *Thlaspi* in order to develop a Phyto-Remediation crop which can be used to "de-pollute" contaminated soils, allowing the shoot Zn to be recycled as an ore.

2. Introduction

Phytotoxicity of soil Zn has been reviewed by several authors during the last 20 years (Chaney and Giordano, 1977; Foy et al., 1978; Chang et al., 1992; Bingham et al., 1986; Berry and Wallace, 1989; Beckett et al., 1979; Collins, 1982; Cummings and Tomsett, 1992; Patterson, 1971; Rauser, 1990; Wagner and Krotz, 1989; Woolhouse, 1983). As research on both practical and fundamental aspects of Zn phytotoxicity has progressed, we have much more understanding of how to correct Zn phytotoxicity by practical agronomic measures, and to regulate environmental contamination processes to avoid future problems with excessive soil Zn. Because both soil management practices and plant tolerance differences affect whether Zn phytotoxicity is observed, it has been difficult to develop regulatory practices for deliberate addition of Zn in waste materials (e.g. sewage sludge), or to determine when a phytotoxic hazardous soil contamination has occurred which requires remediation. As will be discussed below, the overwhelming effect of soil pH on Zn phytotoxicity makes the decision of causality of Zn phytotoxicity more difficult. Did a problem develop because the farmer failed to maintain soil pH at levels recommended for normal productive use of the soil, or was the contamination so extreme that poor pH management is less important than the contamination process. Because Zn phytotoxicity is the most important limitation on soil Zn to protect humans, livestock, plants, wildlife, or the environment, detailed evaluation of Zn phytotoxicity was made during the recent work to develop standards for the beneficial utilization of sewage sludge on cropland in the US (Chaney and Ryan, 1993; Chang et al., 1992).

3. Sources and significance of Zn contamination.

High soil Zn is a potential limiting phytotoxic heavy metal in many contaminated soils. In contrast with the extensive Ni-rich serpentinite-derived soils which can cause "natural" Ni phytotoxicity if strongly acidic, few natural soils contain high levels of Zn. Nearly all surface soils which now contain excessive Zn are a result of human activities. Smelter (Zn, Pb, Cu) and incinerator emissions, excessive applications of Zn fertilizers and pesticides, use of highly contaminated sewage sludges and some livestock manures, dispersal of Zn (and Pb and Cu) mine and beneficiation wastes, and release of Zn from galvanized (Zn plated) surfaces can cause Zn contamination of soils sufficient to cause Zn phytotoxicity, at least at low soil pH. As soil pH falls, Zn concentration in the soil solution and plant uptake increase, and the potential for Zn phytotoxicity is more severe (Chaney et al., 1975; White et al., 1979a; 1979c; Francis et al., 1985). Thus, Zn phytotoxicity is a potential problem whenever Zn-rich soils become acidic. Rainfall leaching and N-fertilizers acidify soils with time, causing present Zn contamination to comprise future Zn phytotoxicity risk if soil pH is not managed reasonably. The recent large body of research on land application of sewage sludge offers important information about potential Zn phytotoxicity because sludges will be one of the more important long-term sources of soil Zn enrichment.

Zinc phytotoxicity has been observed in acidic soils as a result of Zn contamination from a variety of sources. Agricultural soils are seldom contaminated enough to cause Zn phytotoxicity (Holmgren et al., 1993). A few soils are enriched in Zn and other elements because they developed on an outcropping of a metal deposit, while others were contaminated by seepage from ore bodies (e.g., Cannon, 1955; Staker, 1942; Rosen et al., 1978), or dispersal of ore or tailings or smelter emissions (Takijima and Katsumi, 1973; Boon and Soltanpour, 1991; Baker and Bowers, 1988). In traditional agriculture, Zn phytotoxicity has been observed in cotton and soybean after accumulation of Zn from pesticide sprays used in peach production (Lee and Craddock, 1969), and in peanut after Zn pesticide sprays on pecans (Keisling et al., 1977). Utilization of high cumulative application rates of sewage sludge rich in Zn has caused phytotoxicity in many species (Chaney and Giordano, 1977; Logan and Chaney, 1983; Marks et al., 1980; Williams, 1980; Berrow and Burridge, 1990). Mixing ground rubber (contains 0.5-4% Zn) in potting media or using rubber as a mulch, or mixing ash from burning rubber materials into soils has caused phytotoxicity (Milbocker, 1974; Patterson, 1971). Accumulation of Zn from galvanized containers or below galvanized fences or electricity transmission towers has caused Zn toxicity, and even caused selection of Zn tolerant ecotypes of several grasses (Al-Hiyaly et al., 1988, 1990; Antonovics et al., 1971; Bradshaw, 1977; Foy et al., 1978; Baker, 1981, 1987; Jones, 1983). Perhaps the definitive case for selection of Zn-tolerant ecotypes was the finding that Zn-tolerant ecotypes of grasses were selected in a narrow line immediately below a galvanized fence when soils were acidic (Antonovics et al., 1971).

Accumulation of Zn from emissions of Zn, Pb, or Cu smelters has caused severe phytotoxicity in many plant species and caused ecosystem disruption at many locations. For example, emissions of the Zn smelters at Palmerton, PA, have killed or prevent regrowth of natural forests (Beyer, 1988; Buchauer, 1973; Oyler, 1988), and caused Zn toxicity in garden crops, lawn grasses, deer, and grazing horses (Beyer, 1988; Chaney et al., 1988). Zinc phytotoxicity to grasses is so severe in the Borough of Palmerton that many homeowners have covered their lawns with stones or other mulch materials. These lawn and garden soils contain as high as 10,000 mg total Zn/kg and 100 mg total Cd/kg, while Cu and other elements are not present at excessive levels. Even addition of limestone to raise soil pH to 7 does not correct Zn toxicity enough to allow Kentucky bluegrass to persist on these soils. Baker and Bowers (1988) evaluated growth of lettuce on Palmerton area soils, noting that high soil Zn from long-term smelter pollution was much less phytotoxic or bioavailable than freshly added Zn salts. After adjusting soil pH and adding fertilizers to smelter polluted soils, time is required for metal fixation processes to proceed in the soil (e.g., Brümmer et al., 1986). Severe soil Zn contamination has been found to have occurred at essentially every Zn smelter or mine which operated before 1950. The older the technology, the more severe the soil contamination. For example, Davies and Roberts (1979) described the residual soil contamination from prehistoric mining in Wales.

Thus, Zn contamination of soils is very widespread wherever humans mined, smelted, or used Zn. Because Zn is a highly volatile metal, and an important metal of commerce, it has been widely dispersed by stack emissions. The combination of high soil Zn enrichment and strongly acidic soil pH is all that is required for Zn phytotoxicity to be observed.

4. Effect of soil pH and other soil properties on Zn phytotoxicity.

The solubility of Zn increases as soil pH is lowered. Protons compete with Zn for adsorption on the metal binding sites of soils. Soil pH is the most important variable in soils affecting Zn phytotoxicity, (eg Williams, 1980; Chaney et al., 1975;1978;1982) Sanders and Adams (1987), and Sanders et al. (1986, 1987) show this relationship of pH and Zn solubility. Chaney, White and Simon (1975) showed the interaction of soil pH and total soil Zn (added $ZnSO_4$) in a pot study with soybeans. As soil Zn was increased, the critical pH for alleviation for Zn phytotoxicity rose.

As soil pH becomes lower, other metals also become more soluble and potentially phytotoxic. Soil Mn can be reduced to Mn^{2+} and become phytotoxic at even neutral soil pH, but soil Mn normally becomes phytotoxic when soil pH falls below about 5.5. Below this pH, soil microbes which oxidize Mn^{2+} to MnO_2 are inhibited and Mn^{2+} accumulates in the soil. In several studies in which Zn phytotoxicity occurred in agricultural fields, the soil pH had fallen to well below 5, and both Mn and Zn contributed to the observed yield reduction (Lee and Craddock, 1969; Lutrick et al., 1982; King and Morris, 1972). In our early studies of Zn phytotoxicity, it appeared that added Zn displaced soil Mn and the situation might better be described as a Zn-induced Mn-phytotoxicity (White et al., 1979a; White and Chaney, 1980). However, we now believe this was an artifact of the methods used in the research. High applications of soluble Zn salts can cause effects different from those of Zn which accumulates in soils over time, especially when the soils are air-dried during the study. Others have discussed the plethora of artifacts which result from air drying of soils, and have called dried soils "lab dirt". Drying reduces MnO_2 and kills Mn^{2+} oxidizers such that simply air drying acidic Mn rich soils can induce Mn phytotoxicity.

Pot studies in greenhouses give significantly different Zn uptake results than do field studies (e.g., deVries and Tiller, 1978). Part of the difference is due to restriction of all plant roots to the contaminated soil. Restriction of root length decreases plant availability of other nutrients obtained by diffusion from soil particles. Soil temperature is usually higher in greenhouse pots than in the field, and transpiration is greater in the greenhouse. These differences caused Zn uptake by lettuce to be about 4-times higher than found in the field. No comparisons have been reported for phytotoxic levels of soil Zn.

Adding Zn salts such as $ZnSO_4$ normally strongly acidify soils as the Zn displaces protons from the adsorption surfaces (e.g., White et al., 1979c). However, even when soil pH was corrected by addition of sufficient $CaCO_3$ to bring all Zn treatments to the same pH, added Zn salts increased Mn uptake by soybeans so that the resulting phytotoxicity was a result of both Zn and Mn (White and Chaney, 1980). However, in field studies where pH stays at 5.5 or above, and is equal among treatments, added sludge Zn did not increase Mn uptake by plants. In more recent pot studies, Chaney et al. (1990) used soils maintained at field moisture, and added a sewage sludge containing 5% Zn, and adjusted pH equally across treatments. They found no increase in plant Mn with increasing added Zn, at least in the range of pH > 5.5 (see Table 1). As seen in the field studies noted above (Lee and Craddock, 1969; Lutrick et al., 1982), when soil pH falls to well below 5.5, natural increases in soil Mn availability can allow significant interactions of Zn with Mn.

The specific Zn adsorption capacity of the soil is also important in Zn phytotoxicity. At equal pH and total Zn, Zn phytotoxicity is more severe on light textured soils than on heavy textured soils. Research on Zn applied by sewage sludge has added further information about Zn phytotoxicity in relation to added soil Zn. As originally reported by Chaney et al. (1982), and recently reviewed by Corey et al. (1987), plant metal concentration approaches a plateau with increasing sludge application rate for a sludge. This happens because the sludge adds not only metals, but also specific-metal-adsorption capacity. At low sludge application rates, specific-metal-binding-sites in the soil hold metals more strongly than the sludge metal-adsorption-sites. However, as metal concentration in the soil-sludge mixture increases, the soil metal-adsorption-sites become saturated. Above low sludge application rates, the sludge metal adsorption sites control metals phyto-availability in the soil-sludge mixture. Another implication of this model is that plant metal concentration is an increasing curvilinear function of increased sludge metal concentration at equal metal application rates (other factors unchanged). Supporting this model, Zn toxicity was induced in all species at higher levels of soil Zn in the Chaney et al. (1990) study using a sludge very high in Zn, but Zn phytotoxicity has not resulted at pH≥ 5.5 with even high loading rates of median quality sewage sludges (≤1500 mg Zn/kg).

5. Physiological aspects of Zn phytoxicity.

The fundamental biochemical mechanism of Zn phytotoxicity has not been identified for any plant. However, different responses have been observed. In strongly acidic soils, most species become chlorotic when they are exposed to excessive soil Zn. Spraying $FeSO_4$ or chelated Fe on the leaves corrects the chlorosis, indicating that Zn had interfered with Fe uptake, translocation, or utilization in the leaves.

Ambler et al. (1971) studied the effect of Zn on uptake and translocation of Fe by soybean in nutrient solutions. They found that increased solution Zn interfered with the Fe-stress-response, and inhibited Fe uptake and translocation. However, in most soil systems leaves do not contain deficient levels of Fe even when they are severely chlorotic due to excessive soil Zn (Rosen et al., 1977; White et al., 1979a). Thus, it appears that Zn in leaf cells interferes with utilization of Fe, perhaps in chlorophyll biosynthesis.

In further experiments of White et al. (1979b), reciprocal grafts were made between soybean cultivars with lower or higher tolerance to soil Zn. Tolerance (resistance to yield reduction or chlorosis) was a character of the shoots, not the roots. This indicates that Fe uptake and translocation (a root characteristic) did not control Zn tolerance. However, rootstock genotype controlled the concentration of Zn in the leaves. Thus, physiological tolerance of Zn is a characteristic of plant shoots, at least in economic plant species.

Boawn and Rasmussen (1971) also noted that plants suffering Zn phytotoxicity had lower shoot P levels. This might result from inhibited root length, which would reduce the uptake of P by roots, or it might result from the insolubility of Zn-phosphate in root cells. If phosphate is precipitated in the root cells, it is not available for transport to the shoots.

Many other physiological studies have been reported. Some careful studies have been conducted on the inhibition of enzymes from plants with different levels of Zn tolerance (Mathys, 1990; Woolhouse, 1983). However, it is not clear that Zn tolerance or susceptibility to Zn phytotoxicity is related to any particular enzymic activity in any species.

6. Crop differences in susceptibility to Zn phytotoxicity .

Crops differ remarkably in susceptibility to phytotoxicity from soil Zn. In acidic soils, most grasses are much more tolerant than most dicots, but this order is reversed in alkaline soils as discussed above. Chaney et al. (1978) and Chaney and Giordano (1977) have reviewed the available literature on crop differences in tolerance of sludge applied metals. Leafy vegetable crops and the beet family are highly sensitive to Zn phytotoxicity compared to most other dicots. Many legumes are also very sensitive.

Cultivars of plant species have been observed to differ somewhat in susceptibility to Zn phytotoxicity. Polson and Adams (1970) found bean cultivars differed, and White et al. (1979a; 1979b; 1979c) found that soybean cultivars differed in susceptibility, but not by as much as two-fold. Boawn and Rasmussen (1971) compared Zn uptake and phytotoxicity in a range of economic species (Table 1), and Boawn (1971) reported on a field study with leafy vegetable crops grown on neutral pH soil amended with $ZnSO_4$. Among the leafy vegetables, spinach and chard accumulated Zn more strongly and were more susceptible than other crops.

Table 1. Comparison of Zn uptake and tolerance by different crop species on neutral pH soil amended with $ZnSO_4$ (Boawn and Rasmussen, 1971).

Crop Group	Crop	Zn in shoots, mg/kg mg Zn/kg soil= 300	 500	Yield Decrease at P < 0.05
		—mg/kg DW—		%
Poaceae	Field Corn	484	763	26
	Sweet Corn	475	713	29
	Sorghum	748	1030	15
	Barley	910	2110	15
	Wheat	522	909	10
Dicots	Field Bean	151	257	-
	Snapbean	111	213	-
	Alfalfa	142	345	13
	Clover	161	252	-
	Pea	285	489	-
	Lettuce	250	665	27
	Spinach	640	945	10
	Potato	163	327	-
	Sugarbeet	509	1070	22
	Tomato	316	514	24

In the discussion of methods to conduct research on Zn phytotoxicity in soils, data from Chaney et al. (1990) were discussed. This study used excellent methods for study of soil Zn phytotoxicity, using non-dried soils, Zn adsorbed on sewage sludge, pH equivalence of treatments, etc. However, it was a pot study. The differences among crop species are similar to many other reports in the literature with lettuce and dicots being much more sensitive than grasses (Chaney et al., 1978; Chaney and Giordano, 1977). In these studies, many cultivars of 4 grass species were studied. Only 'Merlin', a genotype selected for metal tolerance, had good resistance to excessive soil Zn. These data and other unpublished studies using chelator-buffered nutrient solutions clearly showed that 'Merlin' excludes Zn from the roots and shoots compared to other red fescue cultivars. The process of selection of cultivars by modern breeding methods generally excludes the unique Zn tolerance genes selected at metal contamination sites. Williams (1986) did find some difference among wheat and barley cultivars in susceptiblity to excessive sludge-applied Zn. Small differences were also noted by Carlton-Smith and Davis (1983).

Table 2. Difference among crops in tolerance of high soil Zn concentrations at pH 5.5. Means of yield, chlorosis, and shoot composition for three treatments (870, 1740, and 3480 mg Zn/kg soil; from 2, 4, and 8% of a sewage sludge containing 5% Zn) added to non-dried Sassafras sandy loam. Normalized yield is % of control for that species; chlorosis score is Green=1 to Severe Chlorosis=5 (Chaney, Lee, and Murray, 1990).

CROP	N	Geometric Mean Yield	Geo. Norm. Yield	Geometric Shoot Zn	Shoot Mn	Chlorosis Score	Height
		g/pot	%-Control	—mg/kg DW—			cm
Red Fescue	61	0.46 e	17.2 ef	965 cd	71.0 d	1.8 ef	12.8 e
Merlin	9	3.73 ab	83.0 a	183 g	39.5 fg	1.2 h	15.8 d
K. Bluegrass	68	1.06 d	29.6 de	645 ef	35.3 g	1.6 fgh	8.7 f
Tall Fescue	72	2.09 bc	69.8 ab	1060 cd	55.5 d-g	1.7 fg	25.5 b
P. ryegrass	63	1.80 cd	42.9 bcd	1040 cd	58.3 def	2.2 de	22.5 c
C. bluegrass	9	1.38 cd	16.4 f	898 cd	53.4 d-g	2.3 d	8.0 f
Little bluestem	9	1.11 cd	39.5 bcd	804 de	153. c	.	.
Soybean	9	0.18 f	11.4 f	1120 c	196. b	4.2 b	.
Lettuce	9	0.01 g	1.2 g	3620 a	244. a	4.6 a	.
Cyperus	9	5.70 a	33.6 cd	580 f	60.4 de	3.1 c	57.3 a
Barley	9	3.87 ab	57.6 abc	1660 b	40.8 efg	1.4 gh	57.8 a

Means in a column followed by the same letter are not significantly different at the 5% level according to the Waller-Duncan K-ratio t test.

7. Tolerance of high soil Zn.

Some plants are extremely tolerant of excessive soil Zn. Generally, these genotypes were selected on mine waste contamination or smelter contamination sites in many countries. The genes apparently were selected because of the severe "natural selection" for Zn tolerance on the contaminated soils. Zn tolerant "ecotypes" of several species have been selected at Zn mine waste sites in several nations (Antonovics et al., 1971; Baker, 1987). Interestingly, genes for Zn tolerance exist in natural populations of many forage species (Walley et al., 1974).

Metal resistance mechanisms remain elusive (Wagner and Krotz, 1989; Wainwright and Woolhouse, 1975; Woolhouse, 1983; Baker, 1987; Baker and Brookes, 1989; Bradshaw, 1977). A number of approaches have been investigated, but little "proof" has been accumulated that a single mechanism works for all species. Some ecotypes appear to exclude metals. Others appear to detoxify metals in the cytoplasm by chelation or precipitation. Citrate, malate, glutathione, and phytochelatins appear to chelate Zn in tolerant species (e.g., Brookes et al., 1981; Mathys, 1980; Wagner and Krotz, 1989). After the discovery of phytochelatins (see Rauser, 1990), it was believed that the long awaited "tolerance" mechanism had been identified. However, by exposing cells or roots to an inhibitor of glutathione biosynthesis, the precursors of phytochelatins are depleted. Reese and Wagner (1987) found that inhibition of phytochelatin biosynthesis did not change the Zn tolerance of tobacco cells, suggesting that Zn tolerance is not mediated by phytochelatins.

The strongest model for Zn tolerance today is that of vacuolar compartmentalization.

Vogeli-Lange and Wagner (1990) found that Cd and Zn were accumulated in vacuoles of tabacco leaves. Van Steveninck et al. (1987) reported that the Zn-tolerant grass *Deschampsia caespitosa* formed specialized vacuoles in root cortex cells which accumulated high levels of Zn and appear to explain Zn tolerance by this ecotype. The possible role of phytochelatins in metal tolerance by plants remains unclear because these compounds chelate many metals, but ecotypes are often (but not always) selected with specific tolerance to a single metal (Baker, 1987).

Although genetic tolerance of metals is theoretically interesting, and has been a persistent challenge to science, it has only been of practical significance in revegetation of highly metal contaminated soils (Williamson and Johnson, 1981). The very requirement for metal tolerance indicates that serious adverse effects are being manifested in ecosystems and the environment.

8. Use of chelator buffering to study Zn at phytotoxic levels.

Many problems have been experienced in study of Zn phytotoxicity, and have lead to mis-interpretation of the literature on Zn phytotoxicity. Two problems have commonly occurred, both related to displacement of Fe from FeEDTA and related Fe-chelates. As Chaney, Bell, and Coulombe (1989), Norvell (1991), and Parker, Chaney, and Norvell (1993) have discussed, chelation equilibria for the relatively non-Fe^{3+}-specific chelators such as EDTA allow Zn to displace Fe from the Fe^{3+}-chelate at normal nutrient solution pH. This has two consequences: 1) little or no FeEDTA remains in the solution to supply required soluble Fe to the root, causing Fe-stress; and 2) the Zn^{2+} activity is far below the expected level because the Zn is chelated by the EDTA. Figure 1a shows the loss of Fe from several chelators commonly used in plant nutrition research in a 0.5 Hoagland solution containing equimolar amounts of Fe-chelate and Zn. Fe is substantially displaced from FeEDTA, FeHEDTA, and FeDTPA at practical pH levels. However, Fe is not displaced from FeHBED. HBED is a relative of EDDHA; both have higher specificity for Fe^{3+} than the EDTA family of chelators because HBED and EDDHA contain phenolic ligands. HBED was used in this modelling because of the complexity of chelation by EDDHA; modelling of EDDHA is no longer considered valid since Bannochie and Martell (1989) showed that EDDHA is comprised of several optical isomers with different

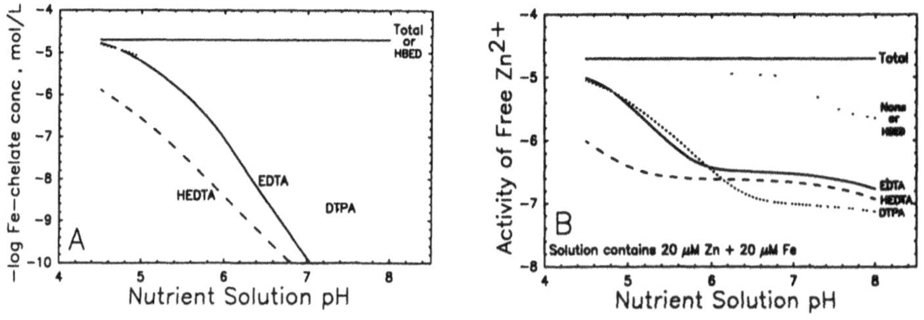

Figure 1. Equilibria involving Zn and Fe^{3+}-chelates in 0.5 Hoagland solutions: 1A: Effect of added Zn on Fe^{3+} remaining soluble with different Fe^{3+}-chelates; 1B: Effect of added Zn on activity of Zn^{2+} with different Fe^{3+}-chelates.

metal chelate formation constants. HBED is simpler, and the formation constant database is more complete for HBED than for EDDHA (see Parker et al., 1993). Figure 1b shows the effect of solution pH on activity of Zn^{2+} in these same solutions. Chelators which bind Zn relatively stronger than Fe^{3+} allow displacement of Fe at low pH, decreasing Zn^{2+} activity. Clearly, if one wants to supply controlled Zn^{2+} activity in study of Zn phytotoxicity with non-*Poaceae* plant species, one should use a highly selective chelator such as FeHBED or FeEDDHA.

Even when the proper Fe-chelate is used, plant uptake of Zn and sorption of Zn on roots reduces the activity of solution Zn^{2+} over time. Buffering of Zn^{2+} can prevent this problem, as can use of flowing nutrient solutions with Zn concentration maintained. Chaney et al. (1989) described the use of chelator-buffering to control Zn^{2+} activity in nutrient solutions independent of solution pH. Sufficient chelator is added to prevent precipitation of Fe. If the "excess" chelator (the concentration of chelator in excess of the strongly chelated microelement cations) is kept constant while Zn is varied, the activities of other microelement cations is held constant and pH has little effect on activity of any of the microelement cations. Further, microelement cation activities similar to those in soil solution can be maintained in chelator-buffered media. Parker et al. (1993) describe this more completely. These chelator-buffered nutrient solutions are very useful for study of Zn deficiency (e.g., Parker et al., 1992; Norvell and Welch, 1993). However, the chelators used to provide deficient levels of Zn cannot be used to study phytotoxic levels of Zn. Because the concentration of total Zn in nutrient solutions required to cause phytotoxicity is 10μM or higher, there are two approaches to attain "adequately" controlled Zn^{2+} activity. The first is to use a selective Fe-chelate so that Fe is not displaced to confound the Zn toxicity study, and replace the solutions frequently. The second is to use a weak Zn chelator such as EGTA [ethylene-bis-(oxyethylenenitrilo)-tetraacetate] to buffer all elements except Fe, coupled with use of a selective Fe^{3+}-chelate. Figure 2 shows leaf and root Zn for sunflower grown with EGTA buffered nutrient solution and FeEDDHA. In a preliminary experiment with higher Zn^{2+} activity, young leaves became chlorotic when Zn toxicity reduced yields by over 25%. Some yield reduction occurred without apparent Fe-chlorosis, and leaf Zn exceeded 400 mg/kg DW.

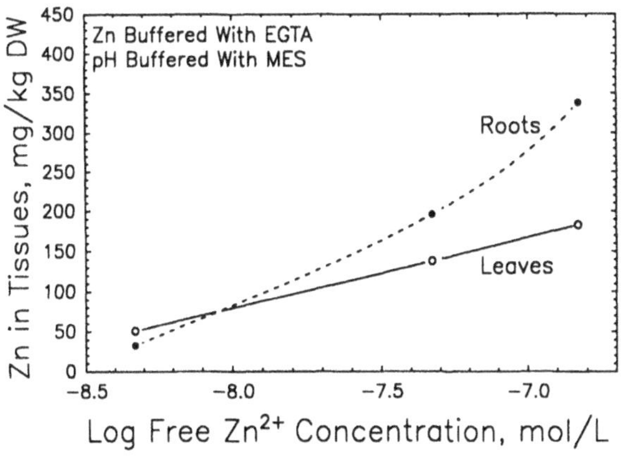

Figure 2. Leaf and root Zn concentration for Hybrid 954 sunflower grown with EGTA buffered (100 μM excess EGTA) nutrient solution and FeEDDHA.

These methods are not relevant to study of *Poaceae* species because these species use the phytosiderophore system to obtain Fe. Provision of toxic levels of buffered Zn with adequate levels of Fe^{3+} is not possible. It is possible to study deficient levels of Zn with adequate Fe^{3+} for *Poaceae* (e.g., Norvell and Welch, 1993). But if excess chelator is not included, added Zn displaces the Fe (see also Bell et al., 1990). An alternative method was developed to supply phytotoxic levels of Zn to *Poaceae* using Fe^{2+} buffered with Ferrozine (FZ), and a weak Zn chelator to buffer Zn, Cd, etc (Chaney et al., 1990). If pFe^{2+} (activity) is maintained at 8.5, rice, wheat and barley obtain adequate Fe for normal growth, and little Fe precipitates on the roots (*Poaceae* species grown with normal 1:1 molar ratio FeHEDTA added to 0.5 Hoagland solution have very high Fe contamination of roots [Chaney and Bell, 1987]). It appears that the high EGTA-buffered Zn (5-50 μM) does not interfere with Fe^{2+} equilibria enough to be a problem, but the Zn-ferrozine chelate formation constant is not available to allow complete calculation. In this system, 16.1 μM Fe^{2+} is buffered with 80.0 μM FZ (3 mol FZ/Fe = 48.3; "excess" FZ = 31.7 μM) to provide pFe^{2+} activity of 8.5, and Zn is buffered with EGTA (5 μM total Zn with 100 μM "excess" EGTA to give pZn^{2+} activity of 6.08). The levels of Mn and Cu were 2 and 0.5 μM, respectively. This Zn activity gave wheat and rice shoots with about 200 mg Zn/kg DW with adequate Fe, and about 400-600 mg Zn/kg with deficient Fe ($pFe^{2+} < 9.0$). Normal foliar Zn would be achieved with 0.5-1.0 μM total Zn in this system, but deficient Zn could not be practically obtained with EGTA.

The GEOCHEM-PC computer program is very convenient for calculating nutrient solution equilibria important in microelement research. Recent papers by Parker et al. (1993a; 1993b) discuss the program and use of the program in nutrient solution studies.

9. Prognosis for avoiding or preventing Zn phytotoxocity.

Regulations to Prevent Zn Phytotoxicity. Environmental regulations are one way society can prevent future occurrence of Zn phytotoxicity. Future additions of Zn to soils will be considered very differently than the historic situations which have lead to present Zn phytotoxicity. Most Zn contamination sources have come under different management or regulation, reducing the likelihood of excessive Zn accumulating in soils. Few Zn pesticide sprays are used in agriculture, although some remain approved. Smelter and other stack emissions are strongly regulated in developed countries today, although use of tall stacks to avoid local phytotoxicity and ecosystem toxicity allows some Zn to be widely dispersed and accumulate in soils.

Farm practices continue to apply Zn to soils, but most of this Zn is hidden in manures. When high levels of Cu are added to feed of swine and poultry, diet Zn must be increased to avoid Cu toxicity to the livestock. Thus, these manures are enriched in both Cu and Zn, usually about 1000 mg/kg DW for both metals. This manure Zn represents a higher quantity of soil Zn enrichment than does application of urban sewage sludges and MSW composts. As in the case of sewage sludge, manure Zn phyto-availability might be reduced by addition of Fe or Mn to the manure to increase the specific metal adsorption capacity of the manure before application. Maintenance of soil pH near 7 on fields which receive high cumulative applications of Cu- and Zn-rich manures (or even equivalent metal salts) can prevent the expected metal phytotoxicity (Anderson et al., 1991). However, if the pH of these soils is allowed to fall to below 5, phytotoxicity would be expected based on many research studies.

Zn fertilizers are regularly applied to soils where deficiency has been important in the past, or where soils tests indicate present need for Zn fertilizer. Because Zn fertilizers cost money, over-application is uncommon. Unnecessary Zn fertilizer application remains a concern of agricultural managers. Manure is a disposal problem for intensive animal production farms, and Zn and Cu will be applied to soils in these manures. The EEC is considering regulations on feed Cu and Zn levels to avoid these problems, but little progress has been made, partially because sheep did not suffer Cu toxicity when grazing pastures to which the Cu- and Zn-rich manures were land applied (see Chaney and Ryan, 1993).

Remediation of Zn contaminated soils. Some soils have become so Zn contaminated (and usually so Cd contaminated) that some actions are needed to prevent dispersal of the contaminated soils, and environmental toxicity from the soil Zn. The Zn smelter at Palmerton, Pennsylvania, is an example which represents many smelter and mine waste contaminated soils around the world. In the US, these sites become "Superfund" hazardous waste sites, and some method is considered to reverse the environmental hazard. These practices are called "remediation" of hazardous sites. Mine waste sites in the UK and other European countries have received similar public attention. Because of the high cost of soil removal and replacement, alternative *in situ* technologies are sought to alleviate the environmental risk. Although Zn is usually severely phytotoxic at these locations, the environmental risk assessment process is driven by protection of humans from the co-pollutant, Cd. Only more recently has the prevention of toxicity to plants and wildlife become a significant part of this process.

It has been clear for decades that some method is required to prevent erosive dispersal of the barren contaminated soils at these locations. Bradshaw and colleagues introduced the use of metal tolerant grasses (and minimal fertilization and limestone application) to revegetate these soils (Bradshaw, 1977; Smith and Bradshaw, 1979; Johnson et al., 1977; Williamson and Johnson, 1981; Oyler, 1988). 'Merlin' red fescue is very effective at providing erosion control at these sites, and this tolerant plant excludes Zn and Cd very effectively (Chaney et al., 1990). However, even this plant suffers Zn phytotoxicity when soil pH falls on highly contaminated soils, and it tends to fail when grazed or walked on. Fertilization may be required to maintain effective vegetative cover (McNeilly and Johnson, 1981); required N cannot be supplied by N-fixation because legumes with ecotypic Zn tolerance similar to the grasses have not been found.

The metal tolerant red fescue ecotype (and commercial cultivar), 'Merlin', has provided vegetative stabilization of soil on the Zn toxic Blue Mountain at Palmerton if some limestone and fertilizer nutrients are provided (Oyler, 1988; Chaney et al., 1988). Oyler (1988) also reported some success with tall fescue and some warm season grasses when he added limestone to the soil, and surface applied a mixture of sewage sludge and fly ash to these Zn toxic soils. Chaney et al. (1990) evaluated tolerance of Zn by cultivars of 4 grass species. Although little tolerance was found in Kentucky bluegrass or perennial ryegrass, several turf-type tall fescue cultivars and 'Merlin' red fescue offer much promise for grasses which can grow on sites with potentially Zn toxic soils. Improved Zn-tolerant species, cultivars, or ecotypes may be a practical solution to revegetation of extensive Zn toxic soils in the vicinity of Zn smelters and other Zn-polluted soil.

Potential for phyto-remediation of Zn contaminated soils. A new approach to decontaminate these soils is based on the unusual ability of some plant species to "hyperaccumulate" Zn. Although most economic plants suffer significant yield reduction when foliar Zn exceeds 500 mg/kg DW, a few rare plants are Zn tolerant because they transport Zn rapidly to the shoots, and tolerate very high foliar Zn (up to 40,000 mg/kg DW in alpine pennycress [*Thlaspi caerulescens* J.and C. Presl.]). Chaney (1983) concluded that this unusual Zn accumulation in plants might be useful. Baker and Brooks (1992) have further discussed this possibility. Although soil Cd risk to the environment will usually drive the public desire to decontaminate Zn + Cd contaminated soils, the removal of both corrects the potential for ecosystem injury from soil metals. The ash of these plants contains sufficient Zn (20-40%) to support recycling the Zn and Cd into commercial channels. We have been conducting research to develop a technology to decontaminate Zn + Cd contaminated soils, and worked to understand the Zn physiology of these plants (Brown et al., 1993a; 1993b). Because nearly all Zn hyperaccumulators have low biomass, the genes may have to be transferred to an agronomically adapted plant to allow development of a practical phyto-remediation technology. Soil pH can be adjusted to achieve plant Zn and Cd levels at the beginning of phytotoxicity and made more acidic with time as the soil is decontaminated.

Acknowledgement

I gratefully acknowledge the cooperation and assistance of many colleagues in my research on heavy metals in soils, plants, animals, and the environment. J.S. Angle, A.J.M. Baker, P.F. Bell, J.C. Brown, S.L. Brown, Y. Chen, R.B. Corey, C.D. Foy, C.E. Green, F.A. Homer, A.L. Page, W.A. Norvell, D.R. Parker, J.A. Ryan, L.O. Tiffin, M.C. White, and R.J. Wright have cooperated in my research on Zn, and/or discussed many of these concepts with me, and made it possible to reach these understandings.

References

Al-Hiyaly, S.A.K., T. McNeilly, and A.D. Bradshaw. 1988. The effect of zinc contamination from electricity pylons — Evolution in a replicated situation. New Phytol. 110:571-580.

Al-Hiyaly, S.A.K., T. McNeilly, and A.D. Bradshaw. 1990. The effect of zinc contamination from electricity pylons. Contrasting patterns of evolution in five grass species. New Phytol. 114:183-190.

Ambler, J.E., J.C. Brown and H.G. Gauch. 1970. Effect of zinc on translocation of iron in soybean plants. Plant Physiol. 46:320-323.

Anderson, M.A., J.R. McKenna, D.C. Martens, S.J. Donohue, E.T. Korngay and M.D. Lindemann. 1991. Long-term effects of copper rich swine manure application on continuous corn production. Commun. Soil Sci. Plant Anal. 22:993-1002.

Antonovics, J., A.D. Bradshaw, and R.G. Turner. 1971. Heavy metal tolerance in plants. Adv. Ecol. Res. 7:1-85.

Baker, A.J.M. 1987. Metal tolerance. New Phytol. 106(Suppl.):93-111.

Baker, A.J.M. and R.R. Brooks. 1989. Terrestrial higher plants which hyperaccumulate metal elements — A review of their distribution, ecology, and phytochemistry. Biorecovery 1:81-126.

Baker, D.E. and M.E. Bowers. 1988. Human health effects of cadmium predicted from growth and composition of lettuce in gardens contaminated by emissions from zinc smelters. Trace Subst. Environ. Health 22:281-295.

Bannochie, C.J. and A.E. Martell. 1989. Affinities of racemic and meso forms of N,N'-ethylenebis-[2-(*o*-hydroxyphenyl)glycine] for divalent and trivalent metal ions. J. Amer. Chem. Soc. 111:4735-4742.

Beckett, P.H.T., R.D. Davis, and P. Brindley. 1979. The disposal of sewage sludge onto farmland: The scope of the problems of toxic elements. Water Pollut. Contr. 78:419-445.

Bell, P.F., R.L. Chaney and J.S. Angle. 1991a. Free metal ion and total metal concentration as indices of metal availability for barley. Plant Soil 130:51-62.

Bell, P.F., R.L. Chaney and J.S. Angle. 1991b. Determination of the free Cu^{2+} activity required by corn (*Zea mays* L.) using chelator-buffered nutrient solutions. Soil Sci. Soc. Am. J. 55:1366-1374.

Berrow, M.L. and J.C. Burridge. 1990. Persistence of metal residues in sewage sludge treated soils over seventeen years. Intern. J. Environ. Anal. Chem. 39:173-177.

Berry, W.L. and A. Wallace. 1989. Zinc phytotoxicity: Physiological responses and diagnostic criteria for tissues and solutions. Soil Sci. 147:390-397.

Beyer, W.N. 1988. Damage to the forest ecosystem on Blue Mountain from zinc smelting. Trace Subst. Environ. Health 22:249-262.

Bingham, F.T., F.J. Peryea and W.M. Jarrell. 1986. Metal toxicity to agricultural crops. Metal Ions in Biological Systems 20:119-156.

Boawn, L.C. 1971. Zinc accumulation characteristics of some leafy vegetables. Soil Sci. Plant Anal. 2:31-36.

Boawn, L.C., and P.E. Rasmussen. 1971. Crop response to excessive zinc fertilization of alkaline soil. Agron. J. 63:874-876.

Boon, D.Y. and P.N. Soltanpour. (1992). Leed, cadmium and zinc contamination of Aspen garden soils and vegetation J Environ Qual 21, 82-86

Bradshaw, A.D. 1977. The evolution of metal tolerance and its significance for vegetation establishment on metal contaminated sites. Proc. Intern. Conf. Heavy Metals in the Environment 2(II):299-322.

Brookes, A., J.C. Collins, and D.A. Thurman. 1981. The mechanism of zinc tolerance in grasses. J. Plant Nutr. 3:695-705.

Brown, S.L., R.L. Chaney, J.S. Angle and A.J.M. Baker. 1993. Zinc and cadmium uptake of *Thlaspi caerulescens* grown in nutrient solution. Soil Sci. Soc. Am. J. (*submitted*).

Brown, S.L., R.L. Chaney, J.S. Angle and A.J.M. Baker. 1993. Zinc and cadmium uptake by *Thlaspi caerulescens* and *Silene cucubalis* in relation to soil metals and soil pH. J. Environ. Qual. (*submitted*).

Brümmer, G.W., J. Gerth, and U. Herms. 1986. Heavy metal species, mobility and availability in soils. Z. Pflanzenernähr. Bodenk. 149:382-398.

Buchauer, M.J. 1973. Contamination of soil and vegetation near a zinc smelter by zinc, cadmium, copper, and lead. Environ. Sci. Technol. 7:131-135.

Cannon, H.L. 1955. Geochemical relations of zinc-bearing peat to the Lockport dolomite, Orleans County, New York. Geol. Surv. Bull. 1000-D: 119-185.

Carlton-Smith, C.H., and R.D. Davis. 1983. Comparative uptake of heavy metals by forage crops grown on sludge-treated soil. pp 393-396. *In* Proc. Internat. Conf. Heavy Metals in the Environment-Heidelberg. CEP Consultants, Edinburgh.

Chaney, R.L. 1983. Plant uptake of inorganic waste constituents. pp 50-76. *In* J.F. Parr, P.B. Marsh and J.M. Kla (eds.) Land Treatment of Hazardous Wastes. Noyes Data Corp., Park Ridge, NJ.

Chaney, R.L. and P.F. Bell. 1987. The complexity of iron nutrition: Lessons for plant-soil interaction research. J. Plant Nutr. 10:963-994.

Chaney, R.L., P.F. Bell and B.A. Coulombe. 1989. Screening strategies for improved nutrient uptake and use by plants. HortSci. 24:565-572

Chaney, R.L., W.N. Beyer, C.H. Gifford, and L. Sileo. 1988. Effects of zinc smelter emissions on farms and gardens at Palmerton, PA. Trace Subst. Environ. Health 22:263-280.

Chaney, R.L., Y. Chen, P.F. Bell and J.S. Angle. 1990. Using chelator-buffered nutrient solutions to determine the pFe^{2+} requirement of tomato and soybean. Agron. Abstr. 1990:225.

Chaney, R.L. and P.M. Giordano. 1977. Microelements as related to plant deficiencies and toxicities. pp. 234-279. *In* L.F. Elliott and F.J. Stevenson (eds.). Soils for Management of Organic Wastes and Waste Waters. American Society of Agronomy, Madison, WI.

Chaney, R.L., P.T. Hundemann, W.T. Palmer, R.J. Small, M.C. White, and A.M. Decker. 1978. Plant accumulation of heavy metals and phytotoxicity resulting from utilization of sewage sludge and sludge composts on cropland. pp. 86-97. *In* Proc. Natl. Conf. on Composting Municipal Residues and Sludges. Information Transfer, Inc., Silver Spring, MD.

Chaney, R.L., M.-H. Lee and J.J. Murray. 1990. Response of yellow nutsedge, barley, lettuce, soybean, little bluestem, Canada bluegrass, and cultivars of tall fescue, red fescue, Kentucky bluegrass, and perennial ryegrass to excessive sewage-sludge applied soil zinc in an acidic soil. Final Report. Army Engineers, Waterway Experiment Station.

Chaney, R.L., S.B. Sterrett, M.C. Morella, and C.A. Lloyd. 1982. Effect of sludge quality and rate, soil pH, and time on heavy metal residues in leafy vegetables. pp 444-458. *In* Proc. Fifth Annu. Madison Conf. Appl.

Res. Pract. Municipal and Industrial Waste. Univ. Wisconsin-Extension, Madison, Wisconsin.
Chaney, R.L., M.C. White, and P.W. Simon. 1975. Plant uptake of heavy metals from sludge use on land. pp. 169-178. *In* Proc. 2nd Natl. Conf. on Management of Municipal Wastewater Sludges. Information Transfer Inc., Silver Spring, MD.
Chang, A.C., T.C. Granato and A.L. Page. 1992. A methodology for establishing phytotoxicity criteria for chromium, copper, nickel, and zinc in agricultural land application of municipal sewage sludges. J. Environ. Qual. 21:521-536.
Collins, S.C. 1982. Zinc. pp. 145-169. *In* W.W. Lepp (ed.). Effect of Heavy Metal Pollution on Plants. Vol 1. Effect of Trace Metals on Plant Function. Applied Science Publishers, NJ.
Corey, R.B., L.D. King, C. Lue-Hing, D.S Fanning, J.J. Street, and J.M. Walker. 1987. Effects of sludge properties on accumulation of trace elements by crops. pp. 25-51. *In* A.L.Page T.J.Logan and J.A. Ryan (eds.) Land Application of Sludge. Lewis Publishers Inc., Ann Arbor, MI.
Cumming, J.R. and A.B. Tomsett. 1992. Metal tolerance in plants: signal transduction and acclimation mechanisms. pp. 329-364. *In* D.C. Adriano (ed.). Biogeochemistry of trace metals. Lewis Publishers.
Davies, B.E., and L.J. Roberts. 1978. The distribution of heavy metal contaminated soils in Northeast Clwyd, Wales. Water, Air, Soil Pollut. 9:507-518.
Davis, R.D., and P.H.T. Beckett. 1978. Critical levels of twenty potentially toxic elements in young spring barley. Plant Soil 49:395-408.
deVries, M.P.C. and K.G. Tiller. 1978. Sewage sludge as a soil amendment, with special reference to Cd. Cu, Mn, Ni, Pb, and Zn - Comparison of results from experiments conducted inside and outside a greenhouse. Environ. Pollut. 16:213-240.
Foy, C.D., R.L. Chaney, and M.C. White. 1978. The physiology of metal toxicity in plants. Annu. Rev. Plant Physiol. 29:511-566.
Francis, C.W., E.C. Davis, and J.C. Goyert. 1985. Plant uptake of trace elements from coal gasification ashes. J. Environ. Qual. 14:561-569.
Holmgren, G.G.S., M.W. Meyer, R.L. Chaney and R.B. Daniels. 1993. Cadmium, lead, zinc, copper, and nickel in agricultural soils of the United States of America. J. Environ. Qual. 22:335-348.
Johnson, M.S. and A.D. Bradshaw. 1977. Prevention of heavy metal pollution from derelict mine sites by vegetative stabilization. Trans. Inst. Min. Metall. 864:47-55.
Johnson, M.S., T. McNeilly, and P.O. Putwain. 1977. Revegetation of metalliferous mine spoil contaminated by lead and zinc. Environ. Pollut. 12:261-277.
Johnson, N.B., P.H.T. Beckett, and C.J. Waters. 1983. Limits of zinc and copper toxicity from digested sludge applied to agricultural land. pp. 75-81. *In* R.D. Davis, G. Hucker, and P. L'Hermite (eds.). Environmental Effects of Organic and Inorganic Contaminants in Sewage Sludge. D. Reidel Publ., Dordrecht.
Jones, R. 1983. Zinc and cadmium in lettuce and radish grown in soils collected near electrical transmission (hydro) towers. Water, Air, soil Pollut. 19:389-395.
Keisling, T.C., D.A. Lauer, M.E. Walker and R.J. Henning. 1977. Visual, tissue, and soil factors associated with Zn toxicity of peanuts. Agron. J. 69:765-769.
King, L.D. and H.D. Morris. 1972. Land disposal of liquid sludge: II. The effect on soil pH, manganese, zinc, and growth and chemical composition of rye (*Secale cereale* L.). J. environ Qual. 1:425-429.
Lee, C.R., and G.R. Craddock. 1969. Factors affecting growth in high-zinc medium: Influence of soil treatments on growth of soybeans on strongly acid soil containing zinc from peach sprays. Agron. J. 61:565-567.
Lutrick, M.C., W.K. Robertson, and J.A. Cornell. 1982. Heavy applications of liquid-digested sludge on three ultisols: II. Effects on mineral uptake and crop yield. J. Environ. Qual. 11:283-287.
Marks, M.J., J.H. Williams, and C.G. Chumbley. 1980. Field experiments testing the effects of metal contaminated sewage sludges on some vegetable crops. pp 235-251. *In* Inorganic Pollution and agriculture. Min. Agr. Fish. Food Reference Book 326, HMSO, London.
Mathys, W. 1980. Zinc tolerance by plants. pp. 415-437. *In* J.O. Nriagu (ed.) Zinc in the Environment. Part 2: Health Effects. Wiley-Interscience, New York.
Milbocker, D.C. 1974. Zinc toxicity to plants grown in media containing polyrubber. HortSci. 9:545-546.
Morrey, D.R., M.S. Johnson, and J.A. Cooke. 1984. A comparison of metal tolerant and non-tolerant varieties of *Festuca rubra* for use in the direct hydraulic seedings of metalliferous fluorspar mine tailings. J. Environ. Management 19:99-105.
Norvell, W.A. 1991. Reactions of metal chelates in soils and nutrient solutions. pp. 187-227. *In* J.J. Mortvedt et al. (eds.) Micronutrients in Agriculture. 2nd Edition. Soil Sci. Soc. Am., Madison,WI.
Norvell, W.A. and R.M. Welch. 1993. Growth and nutrient uptake by barley (*Hordeum vulgare* L. cv. Herta):

Studies using an N-(2-hydroxyethyl)ethylenedinitrilotriacetic acid-buffered nutrient solution technique. 1. Zinc ion requirements. Plant Physiol. 101:619-625.

Oyler, J. 1988. Revegetation of metals-contaminated site near a zinc smelter using sludge/fly ash amendments: Herbaceous species. Trace Subst. Environ. Health. 22:306-320.

Parker, D.R., J.J. Aguilera and D.N. Thomason. 1992. Zinc-phosphorus interactions in two cultivars of tomato (*Lycopersicon esculentum* L.) grown in chelator-buffered nutrient solutions. Plant Soil 143:163-177.

Patterson, J.B.E. 1971. Metal toxicities arising from industry. *In* Trace Elements in Soils and Crops. Min. Agric. Fish. Food, Tech. Bull. 21:193-207.

Polson, D.E. and M.W. Adams. 1970. Differential response of navy beans (*Phaseolus vulgaris* L.) to zinc. I. Differential growth and elemental composition at excessive zinc levels. Agron. J. 62:557-560.

Rauser, W.E. 1990. Phytochelatins. Annu. Rev. Biochem. 59:61-86.

Reese, R.N. and G.J. Wagner. 1987. Effects of buthionine sulfoxamine on Cd-binding peptide levels in suspension-cultured tobacco cells treated with Cd, Zn, or Cu. Plant Physiol. 84:574-577.

Rosen, J.A., C.S. Pike and M.L. Golden. 1977. Zinc, iron, and chlorophyll metabolism in zinc-toxic corn. Plant Physiol. 59:1085-1087.

Rosen, J.A., C.S. Pike, M.L. Golden and J. Freedman. 1978. Zinc toxicity in corn as a result of a geochemical anomaly. Plant Soil 50:151-159.

Sanders, J.R. and T.M. Adams. 1987. The effects of pH and soil type on concentration of zinc, copper and nickel extracted by calcium chloride from sewage sludge-treated soils. Environ. Pollut. A43:219-228.

Sanders, J.R., S.P. McGrath, and T.M. Adams. 1986. Zinc, copper, and nickel concentrations in ryegrass grown on sewage sludge-contaminated soils of different pH. J. Sci. Food Agric. 37:961-968.

Sanders, J.R., S.P. McGrath, and T.M. Adams. 1987. Zinc, copper, and nickel concentrations in soil extracts and crops grown on four soils treated with metal-loaded sewage sludges. Environ. Pollut. A44:193-210.

Smith, R.A.H. and A.D. Bradshaw. 1979. The use of metal tolerant plant populations for the reclamation of metalliferous wastes. J. Appl. Ecol. 16:595-612.

Staker, E.V. 1942. Progress report on the control of zinc toxicity in peat soils. Soil Sci. Soc. Am. Proc. 7:387-392.

Takijima, Y., and F. Katsumi. 1973. Cadmium contamination of soils and rice plants caused by zinc mining. 1. Production of high-cadmium rice on the paddy fields in lower reaches of the mine station. Soil Sci. Plant Nutr. 19:29-38.

Van Steveninck, R.F.M., M.E. Van Steveninck, D.R. Fernando, D.L. Godbold, W.J. Horst, and H. Marschner. 1987. Identification of zinc-containing glogules in roots of a zinc-tolerant ecotype of *Deschampsia caespitosa*. J. Plant Nutr. 10:1239-1246.

Vogeli-Lange, R. and G.J. Wagner. 1990. Subcellular localization of cadmium and cadmium-binding peptides in tobacco leaves: Implication of a transport function for cadmium binding peptides. Plant Physiol. 92:1086-1093.

Wagner, G.J. and R.M. Krotz. 1989. Perspectives on cadmium and zinc accumulation, accommodation, and tolerance in plant cells: The role of cadmium-binding peptide versus other mechanisms. pp. 325-336. *In* D.H. Hamer and D.R. Winge (eds.). Metal Ion Homeostasis: Molecular Biology and Chemistry. A.R. Liss. New York.

Wainwright, S.J. and H.W. Woolhouse. 1975. Physiological mechanisms of heavy metal tolerance in plants. pp. 231-257. *In* M.J. Chadwick and G.T. Goodman (eds.) The Ecology of Resource Degradation and Renewal. Blackwell Sci. Publ., Oxford.

Walley, K.A., M.S.I. Khan, and A.D. Bradshaw. 1974. The potential for evolution of heavy metal tolerance in plants. I. Copper and zinc tolerance in *Agrostis tenuis*. Heredity 32:309-319.

White, M.C., and R.L. Chaney. 1980. Zinc, cadmium, and manganese uptake by soybean from two zinc- and cadmium-amended coastal plain soils. Soil Sci. Soc. Am. J. 44:308-313.

White, M.C., R.L. Chaney, and A.M. Decker. 1979a. Differential cultivar tolerance of soybean to phytotoxic levels of soil Zn. II. Range of soil Zn additions and the uptake and translocation of Zn, Mn, Fe, and P. Agron. J. 71:126-131.

White, M.C., R.L. Chaney, and A.M. Decker. 1979b. Role of roots and shoots of soybean in tolerance to excess soil zinc. Crop Sci. 19:126-128.

White, M.C., A.M. Decker, and R.L. Chaney. 1979c. Differential cultivar tolerance to phytotoxic levels of soil Zn. I. Range of cultivar response. Agron. J. 71:121-126.

Williams, J.H., 1980. Effect of soil pH on the toxicity of zinc and nickel to vegetable crops. pp 211-218. *In* Inorganic Pollution and Agriculture. Reference Book 326. Min. Agr. Fish. Food. HMSO, London.

Williams, J.H. 1986. Varietal tolerance in cereals to metal contamination in a sewage treated soil. pp. 537-542.

In P. L'Hermite (ed.) Processing and Use of Organic Sludge and Liquid Agricultural Wastes. Reidel Publ., Dordrecht.

Williamson, A. and M.S. Johnson. 1981. Reclamation of metalliferous mine wastes. pp. 185-212. *In* N.W. Lepp (ed.) Effect of Heavy Metal Pollution on Plants. Vol. 2. Metals in the Environment. Applied Science Publ., London.

Woolhouse, H.W. 1983. Toxicity and tolerance in the responses of plants to metals. pp. 246-300. *In* O.L. Lange et al. (eds.) Physiological Plant Ecology III: Responses to the Chemical and Biological Environment. Springer-Verlag, New York.

Chapter 11.

The Distribution and Correction of Zinc Deficiency.

P.N. TAKKAR and COLIN D. WALKER

1. Abstract

Zinc deficiency is widely reported in agricultural production. Field trial results indicating Zn deficiency have been reported on most major soil types. The distribution of Zn deficiency is commonly assessed in terms of proxies for yield such as soil or plant testing.

The correction of Zn deficiency has focussed on obtaining the maximum improvement of yield. Correction has been demonstrated to have a residual effect benefiting subsequent crops, however the duration of this effect varies with the nature of the soil and cropping system. Depending on the application (usually somewhere between 2.5 and 25 kg Zn ha^{-1} when inorganic Zn salts are applied to the soil), a Zn application usually ameliorates Zn deficiency for around seven subsequent crops, and on lighter acidic soils it may be for considerably longer.

2. Introduction

Zinc deficiency is a serious micronutrient deficiency, threatening world food production. Therefore knowledge and identification of Zn deficient areas would help develop appropriate strategies to combat its deficiency. Although use of Zn in many areas is receiving due emphasis, knowledge of efficient and economical methods to correct Zn deficiency on a long term basis and in specific soil-cropping systems is desirable. This review attempts to bring in focus the distribution of Zn deficient areas, as well as efficient methods to correct Zn deficiency.

3. Distribution of zinc deficiency

3.1. Maps, field surveys, and reports

Australia and New Zealand: Very extensive Zn deficient areas occur in Australia (Stephens and Donald, 1958; Williams and Andrew, 1970; Anderson, 1970). For example widespread Zn deficiency is found in the famous Ninety Mile Desert on the border of Victoria and South Australia (Riceman, 1948). The most extensive Zn and Cu deficient province, comprising 8 million ha, exists in the south-west of Western Australia (Gartrell, 1974; Donald and Prescott, 1975). In Western and Southern Australia widespread Zn deficiency occurred in several million hectares of clover-grass pastures growing on calcareous or siliceous sands and loams, as well as on calcareous or strongly acidic soils of coastal areas and offshore islands in Tasmania (Riceman, 1948). Zinc deficiency in citrus was first recorded on podsolic soils of Western Australia by Pittman and Owen

(1936). In Queensland and northern New South Wales, Zn deficiency in wheat (Duncan, 1967a), linseed (Duncan, 1967b) and maize (Weir and Holland, 1980) have been observed in vertisol soils, often where calcareous subsoils are exposed. In New Zealand, Zn deficiency resulted in poor growth of many forage and other crops on shallow Niue soils (Widdowson, 1966).

United States of America: Extensive Zn deficiency has also been observed in the USA. As early as 1937, Chandler reported that Zn deficiencies occurred over an area more than 650 km long in California. Many pecan orchards were abandoned in certain valleys of Arizona because of Zn deficiency (Finich and Kinnison, 1933). Surveys indicated scattered Zn deficient areas in eight states in 1956 (Beeson, 1957) and in 1965 in 40 states (SSSA, 1965). Nutrient deficiency tends to be assessed in terms of plant analysis in the USA (Hodgson et al., 1971). The deficiency was widespread in corn in the mid-west (Lingle and Holmberg, 1957), and in calcareous, cut, heavily irrigated sandy soils of the west coast, and the poor sandy soils of Florida.

Central and South America: In tropical America, it is reported that Zn deficiency could limit production over much of the region (Sanchez and Cochrane, 1980). Zinc deficiency is widespread in the highly weathered soils (oxisols, ultisols and latosols). in the Campo Cerrado region of the central plateau of Brazil, in the Lalanos Orientales of Colombia and Venequelan, as well as in Costa Rica, Gautemala, Mexico and Peru (Stiles, 1961; Igue and Bornemisza, 1967; Larez and Sanchez, 1971; Dantas, 1971; Sanchez, 1977; CIAT, 1978; Galrao and Felho, 1981; Galrao, 1990). In north-west Puerto Rico (Badillo-Feliciano and Lopez, 1980) and in pineapple plantations of French Guinea (Malavolta et al., 1962), Zn deficiencies are most striking problems and are associated with low levels of Zn in the parent material (Marinho and Igue, 1972; Peralta et al., 1981).

Europe: Zinc deficiency has been recorded in leached chernozem in Bulgaria (Stratieva et al., 1990), in red Mediterranean and alluvial soils of Cyprus and Greece, reddish-brown and alluvial soils of Spain, shale derived soils of Ireland (Macneidhe et al., 1986), heavily-limed soils of Norway (Myhr, 1988) and in some soils of France, Netherlands, Poland, Sweden, Switzerland and United Kingdom (Ryan et al., 1967).

India, Pakistan, Bangladesh, Japan and Philippines: Zinc deficiency was first recognised by Nene (1966) as a problem of rice on an alkali soil in India. Later it was reported in rice in Pakistan, Japan and the Philippines (Yoshida et al., 1973; Yoshida and Tanaka, 1969). Nearly 0.5 million ha of irrigated rice in the Philippines (Castro, 1977), and over 8 million ha in South-East Asia (Ponnamperuma, 1982), suffer from Zn deficiency. The distribution of Zn deficiency in India has been reviewed by many workers (Kanwar and Randhawa, 1974; Randhawa and Takkar, 1975; Takkar and Randhawa, 1978; Katyal and Sharma, 1979; Takkar et al., 1989; Takkar, 1991b). Analysis of more than 113,000 soil samples for available Zn (mostly DTPA extractable) and about 20,000 plant analyses from 15 states and territories indicated 44-46% are deficient in Zn. The deficiency ranged between 60-70% in the states of Haryana, Madhya Pradesh and Uttar Pradesh; 50-59% in Andhra Pradesh and Punjab; 30-49% in Kerala, Bihar and Tamil Nadu; 20-29% in Delhi, Gujrat, Karnataka and Rajasthan, and less than 20% in Jammu and Kashmir and Pondichery. The incidence of Zn deficiency was greater in calcareous, coarse textured sands; in high water table "Flood Plain" soils; and in saline and sodic or low organic matter soils (Takkar et al., 1989; Takkar, 1991a). Based on these analyses and on thousands of fertilizer responsive experiments on farmers' fields, a map showing Zn deficient areas has been compiled (Takkar, 1991b). Currently in India about 8,000 tonnes of Zn a year is used as a plant nutrient, equivalent to 1.6 million ha receiving 5 kg Zn $ha^{-1}yr^{-1}$. In Bangladesh, more than 2 million ha of the wet cultivable area is deficient in Zn (Karim et al., 1991).

China: In China, Zn deficiency is reported in calcareous soils including loessial soils, calcareous alluvial soils along the Yellow river, brown "dab" soils, chestnut soils, rendzina, calcareous purplish soils, and desert soils. The calcareous and neutral paddy soils along the Yangtze river and throughout northern China are more deficient than the acid rice soils of southern China (Liu Zheng, 1991). The total cultivable area to which Zn has been applied is more than one million ha (Li-Shu-fan et al., 1991).

USSR: In USSR, Zn deficiency has been recorded in rice on meadow chernozem (solonetzic soil), saline sodic soils (Sheudzhen et al., 1991; Servetuik et al., 1991) and in cotton soils (Ponomorev and Ponomareva, 1989).

Africa: Zinc deficiency in the alfisols and ultisols of west Africa are common (Cottenie et al., 1981). Liming and continuous cropping are further increasing Zn deficiency in these soils (Juo and Uzu, 1977). Zinc deficiency has been observed in maize on some of the soils of south-west Nigeria (Kayode and Agboola, 1983), on sands in Zimbabwe (Tanner and Grant, 1974) and ferrallitic soils derived from gneiss in Malagasy, (Velly et al., 1974), in oil palm on leached sandy soils at Etoumbi in the Congo, (Ferrand et al., 1951), in cacao in high pH and P soils in Ghana (Greenwood and Hayfron, 1951) and western Nigeria (Egbe and Omotoso, 1972), and in flooded soils in Sierra Leone (Haque and Kamara, 1976).

Zinc deficiency is common in rice grown in vertisols in north-east Nigeria (Chad Basin) (Kang and Okoro, 1976) and the Gezira region of Sudan (Magar and Babakir 1965), in brownish-red and brownish-yellow sand, brown muck and loess-derived soils of Israel (Ryan et al., 1967), and arid calcareous soils of Iran (Maftoun and Karimian, 1989). In South Africa, Zn deficiency is widespread in pears, prunes, guava and apricot on calcareous soils of south-west Cape Province.

3.2. Soil factors associated with zinc deficiency

Zinc deficiency has been reported to be associated with high pH soils, low pH soil with low total or extractable Zn content, limed acid soils, calcareous soils, cut (ie, levelled) soils, sodic soils, soils with very low or very high organic matter, sandy soils, wetland or ill-drained soils, and those with high available P concentrations. These associations of Zn deficiency with soil properties are described elsewhere in this volume.

3.3. Environmental factors associated with zinc deficiency

Plants are more susceptible to Zn deficiency under adverse climatic conditions, and this results from effects on both plant and soil. Zinc deficiency is more frequently associated with flooded soils than with dryland soils as a result of reaction of Zn with free sulfide (Mikkelsen and Shiou, 1977) or with sesquioxides (Sajwan and Lindsay, 1988). For example, under submerged conditions of rice cultivation, native or fertilizer Zn moves into amorphous sesquioxide precipitates (Takkar and Sidhu, 1979; Singh and Abrol, 1986) or franklinite as the final form of applied Zn (Sajwan and Lindsay, 1988). The delay of Zn fertilizer applications until well after flooding of lateritic soils for rice minimises its absorption to sesquioxides (Mandal et al., 1992). However, wetting and drying cycles did not affect the levels of exchangeable Zn in a sandy loam soil (Nambiar, 1975), probably due to low Fe content.

Poor aeration was also one of the factors for Zn deficiency in crops grown on orchards, camping sites and old corrals because of trampling and puddling of the soil (Lucas and Knezek, 1972). Anaerobic conditions appear to act in part through the need for alcohol dehydrogenase to produce energy in roots under anoxic conditions.

Zinc deficiency is more prevalent in cool and wet seasons. Bauer and Lindsay (1965) found that warm, moist pre-incubations of soil increased the subsequent uptake of Zn by maize. In controlled experiments, temperatures below 16°C during growth were associated with decreased Zn uptake to the tops of maize (Ellis et al., 1965), linseed (Moraghan, 1980), tomato (Fawusi and Ormrod, 1975), and barley (Schwartz et al., 1987). Soil temperature effects appear to be largely on rate of mineralisation of Zn. Barley grown at a root temperature of 10°C had decreased root growth with thicker, shorter roots, but accumulated higher concentrations of Zn in the roots when supplied at adequate levels (Schwartz et al., 1987).

High light intensity and long day-length have been shown to be major factors in the development of foliar Zn deficiency symptoms (Ozanne, 1955; Marschner and Cakmak, 1989). This effect may be mediated through a number of roles that Zn has in photosynthesis, such as in superoxide dismutase and carbonic anhydrase, and in protein synthesis.

4. Correction of zinc deficiency.

4.1. Effectiveness of different applications and forms of zinc

Zinc deficiency can be corrected by treating the soil in various ways; the crop seed or foliage; or the roots of transplanted crops by dipping or by enriching the beds of nursery seedlings with Zn. For economic effectiveness the lowest rate of the most effective source giving maximum yield is desired. The cost/unit of Zn should also be considered when selecting sources. Despite the greater effectiveness of chelates (ZnEDTA), other Zn sources might be more economical to apply, even if higher Zn rates are required (Mortvedt, 1991; Takkar, 1991b).

Rate and Method of Soil Application: Many field experiments have indicated that 2.5-25 kg ha^{-1} Zn as $ZnSO_4$ or 0.3-6 kg ha^{-1} Zn as chelates applied as broadcast and mixed, or banded below the seed proved most effective in correcting Zn deficiency in field and vegetable crops. These methods proved superior to top dressing, side dressing or side banding before sowing or planting (Whitney and Murphy, 1969; Walsh and Schulte, 1970; Murphy and Walsh, 1972; Takkar et al., 1974; Mann et al., 1978; Takkar and Singh, 1979; Wier and Holland, 1980; Takkar and Bansal, 1987).

These large variations in the rate of Zn application have emanated from the sensitivity of crops to soil type and deficiency status, soil environment, Zn sources and their residual effects, and methods of application. Thus many aspects need to be considered when interpreting the results of studies. For example, the optimum rate of Zn application to rice was high (22 kg ha^{-1}) in highly sodic (pH>10.0) and in floodplain soils (Takkar and Nayyar, 1981) compared to 11 kg ha^{-1} in moderately alkaline soils (pH 9.4-9.7; Takkar and Singh, 1978), and 2.5 kg ha^{-1} in sandy alkaline soil. In Australia, the rates of Zn application to cereals and pasture on light soils were lower, at 0.6 to 2.4 kg ha^{-1}

and progressed to 1.8 to 3.9 kg ha^{-1} on heavy soils (Gartrell and Glencross, 1969). For the sensitive crop linseed on heavy soils the rate of Zn was markedly higher, at 7-29 kg ha^{-1} (Duncan, 1967b). The rate also differed with the method of application of Zn to corn, snap bean, onions and potatoes. It was much higher with broadcast, 0.7 to 1.3 kg ha^{-1}, compared to 0.3 to 0.7 kg ha^{-1} when banded or 0.17 kg ha^{-1} when applied as chelate to the foliage (Walsh and Schulte, 1970).

Zinc Carriers: Many Zn sources (eg., inorganic or organic compounds, industrial byproducts, Zn mixed or incorporated in macronutrient fertilizer both in solid and liquid forms) have been tested for their effectiveness to correct Zn deficiency in a large number of crops. A very characteristic response is that the relative effectiveness varies markedly with Zn rate.

For upland crops, field experiments using at least four Zn rates, including the control, of each source should be done to determine the relative effectiveness of Zn carriers. Often the chelate, Zn EDTA, is more effective than ZnO at low Zn rates while all sources are equally effective at the highest Zn rate. Chelated Zn sources have been shown to be more effective than $ZnSO_4$ for maize (Prasad and Sinha, 1981; Hergert et al., 1984), flax (Welch et al., 1967), and field beans (Boawn et al., 1957). In low croplands that flood it is important to consider the performance of Zn fertilizers against fixation, causing loss of residual effectiveness which is discussed later.

The effectiveness of Zn carriers varies with method and form of application. Crop yields with $ZnSO_4$, ZnEDTA and a Zn polyflavonoid were equal when broadcast, but when banded the yield with Zn EDTA was superior to that with $ZnSO_4$ (Brown and Krantz, 1966). Granulation reduced the effectiveness of these sources except for ZnEDTA. The soluble $ZnSO_4$ and sparingly soluble ZnO and Zn-frits were equally effective for maize on Cerrado soils of Brazil; for grain sorghum and pea beans on Ritzville fine sandy loam (Boawn et al., 1957); and for wheat, sorghum, Bengal gram (chickpeas), soybean, maize, and barley on diverse soils (Table 1). But for wheat and rice on alluvial soils, $ZnSO_4$ was most effective (Table 1). Granule size of Zn-frits determines effectiveness, eg. Zn-frits compared favourably with $ZnSO_4$ for alfalfa only when its granule size was at least 200 mesh (Holden and Brown, 1965). The ineffectiveness of a physical mixture of single super and $ZnSO_4$ may have resulted from segregation of $ZnSO_4$ during application.

Coating of Zn onto granular fertilizers eliminates the possibility of its being segregated in dry blending of fertilizers. But Ellis et al. (1964) showed that pea bean yields on a Zn deficient soil were similar with $ZnSO_4$ or ZnO when these were blended with, incorporated within, or coated onto a granular fertilizer. Irrespective of the method of application, $ZnSO_4$ exceeded ZnO in increasing the plant Zn uptake.

Foliar Applications: Zinc may be applied as a foliar spray, as with other nutrients such as Fe and Mn. Foliar fertilization has particular application where soil fixation, pollution or salinity render nutrient uptake by the root system less effective. Trials with low concentrations (0.5-1.0% of $ZnSO_4.7H_2O$ in water) achieved comparable biomass yields to seedling dips or soil Zn treatments, but with greater grain yield in most of the cases from the latter methods (Table 2). In first attempts, yields obtained with foliar applications of lime-neutralised sprays were usually lower than with soil applied Zn (Mann et al., 1978; Takkar and Singh, 1978; Nayyar et al., 1990; Katyal and Friesen, 1988; Takkar et al., 1989; Bansal and Nayyar, 1989). Also un-neutralised foliar sprays (1-2%) to rice on severely Zn deficient sodic soils were inferior to a Zn root dip (2-4% ZnO slurry) or a

Table 1. Effects of Zn Carrier (applied at 10-11 kg Zn ha^{-1}) on Grain Yields (t ha^{-1}) of Crops on Zn Deficient Soils in India.

Zn Carrier	Wheat	Bengal Gram	Soybean	Sorghum	Maize	Wheat	Rice	Rice*	Barley
Control	3.4	1.6	1.1	2.1	2.3	4.0	6.4	3.6	4.2
$ZnSO_4.7H_2O$	3.9	2.2	1.4	3.0	3.2	5.6	7.6	6.0	4.8
ZnO	4.0	2.1	1.3	2.9	3.3	4.8	6.8	5.1	4.8
$Zn_3(PO_4)_2$ *or* $ZnCl_2{}^b$	3.5	2.1	1.4	-	-	5.1^b	7.2	-	-
Zn-frits *or* $SSP^{\dagger}+ZnSO_4{}^C$	-	-	1.6	-	34	49c	-	5.2	4.4
LSD(0.05)	0.2	0.2	0.2	-	-	0.2	0.4	0.4	0.5
Soil Type	Medium Black		Deep Black	Black	Red	Alluvial			
Location	1	2	3	4	5	6	7	8	9
Texture	C	C	C	C	S	SL	L	L-LS	Sil-L
DTPA-Zn:	-	0.42	0.23	0.83	0.80	<0.6	<0.6	<0.6	0.52

Notes: 1 to 3 = Madhya Pradesh; 4-5 = Tamil Nadu; 6-7 = Haryana; 8 = Punjab; 9 = Bihar States.
S=Sand(y); C=Clay; L=Loam; Sil-L = Silty loam; LS = Loamy sand.
* Average values of three experiments † Single superphosphate
Source: Takkar et al. (1989)

$ZnSO_4$ soil application (Sadana and Takkar, 1983; Table 2).

Lower responses to foliar sprays than with soil Zn may result from a delayed cure of the disorder. However, issues of appropriate rates and timing, the use of humectants to prolong nutrient absorption, the number of applications, new ameliorants to avoid burning leaves, and variations in foliar applications for particular environments are being examined (Alexander, 1986; Alexander and Schroeder, 1987; Brennan, 1991). No doubt foliar sprays make more efficient use of Zn fertilizer than soil application, if additional operating costs are discounted. They are effective as emergency or supplementary treatments, often with other sprayed chemicals.

The advantage of foliar fertilization has long been recognised for fruit trees, the technique has been strongly advocated in horticulture (Thorne, 1957), and a recent study confirms its importance (Littlemore et al., 1991). Foliar sprays often employ chelated forms of Zn (ZnEDTA) with more success than with inorganic salts (Brennan, 1991). Chelates are more expensive but also more compatible with other sprayed agricultural chemicals than $ZnSO_4$ (Alexander and Schroeder, 1987). The need for chelates for Zn in sprays is not as critical as for Fe (Thorne, 1957).

Root Dressing and Seed Treatments: These fertilizer strategies seek to provide Zn to the plant early in its life (Scott, 1989). They can be quite labour intensive but again make efficient use of Zn fertilizer. Since the efficiency of soil-applied Zn is very low, efforts have been made to develop better, inexpensive and reliable methods to enhance its efficiency while still correcting Zn deficiency. As early as 1970, Yoshida et al. reported that dipping roots of rice in 1% ZnO slurry (0.1 kg ha^{-1} Zn) was as effective as broadcast application of 10 kg ha^{-1} Zn as $ZnSO_4$ after puddling or as a preplanting application.

Table 2. Some Typical Comparisons of Crop Yields in Response to Zinc Applied to Foliage, Roots, or Soil. Foliage applications are of 0.2-2% Zn solutions, often not stating a rate per unit area. Root applications are by dipping seedlings in 2-4% ZnO slurry. Soil applications are of 5-23 kg Zn ha^{-1}. Yields of controls are not shown, but were all normal.

Plant Species	% Response, over control.			LSD	Source
	Foliar	Root	Soil	(P=0.05)	
Barley	114	70		18	Macneidhe and Fleming,1988
Eggplant	17	42	4-25	12	Iyengar and Raja, 1988
Maize	20		32	8	Takkar et al., 1989
Rice	13-27		21-38	13	Sharma et al., 1982
Rice	39-56	78-83	139	22	Sadana and Takkar, 1983
Sorghum	8		28	8	Takkar et al., 1989
Wheat	2-29		22	10	Duncan, 1967a
Wheat	16		36	10	Takkar et al., 1989

Seedling dips of ZnO may also be used with transplanted vegetables (Iyengar and Raja, 1988) and sugarcane sets, although insufficient for full Zn supply with the latter (Nayyar et al., 1984). In the Philippines treating rice seedlings with 2% ZnO has become popular (Katyal and Ponnamperuma, 1974; Castro, 1976). This practice did not catch on with the farmers in India because of a certain limitation, specifically: standing the treated seedlings in the water of the rice field prior to planting was robbing the roots of loosely held ZnO. Studies in India found that dipping rice seedlings in 2-4% ZnO was sometimes less effective than broadcast application of 11-22 kg ha^{-1} Zn as $ZnSO_4$ or ZnO (Table 2).

Methods of soaking or coating of seeds of field crops in Zn solutions or transplanting from a Zn-enriched nursery were found mostly inferior to soil application on severely Zn deficient soils or just at par on deficient or marginally deficient soil (Takkar et al., 1989; Yoshida and Tanaka, 1969; Yoshida et al., 1970; Castro, 1977; Nayyar et al., 1990). Seed dressing with Zn was insufficient to correct Zn deficiency in soybean (Rasmussen and Boawn, 1969), rice on vertisols in North Nigeria (Kang and Okoro, 1976), or maize on a calcareous, fine-textured soil of Minnesota (Table 3), failing to meet the full Zn requirement of these crops. In contrast, application of Zn as $ZnSO_4$ by seed treatment, foliar spray or soil application proved equally effective for potatoes (Grewel and Trehan, 1979). Soaking of sunflower seed in 0.02% Zn-OEDP complex gave increased yield (Glushchenko et al., 1991). It appears that seed treatments to supply Zn have been successful only in certain crops and lighter soils. New Zn complexes may correct Zn deficiency more effectively (Scott, 1989).

4.2. *Correction of Zn deficiency on a long term basis - the residual effect*

The response to a nutrient by crops following that where it was applied is defined as its residual effectiveness. Zinc has a significant residual effectiveness, so it is not necessary to apply Zn every year. The residual effectiveness of a Zn application is shorter in cropping systems involving flooded rice than in others. Several studies suggest that adequate Zn nutrition persists for between four and seven crops in rice rotations, but

Table 3. Grain Yield Response (t ha^{-1}) of Maize to Rate and Method of Zn Application to Calcareous, Fine-textured Soil of Minnesota, USA.

Method of Zn application	Trial-1[a]	Trial-2[b]	
		Direct	Residual
A. Broadcast and Plowed Under			
5.6 kg ha^{-1}	0.45	2.36	1.62
11.2 kg ha^{-1}	0.63	2.06	1.44
22.4 kg ha^{-1}	0.63	2.96	1.44
B. Coating on seed			
seed alone	0.27	1.08	0.99
seed+11.2 kg Zn ha^{-1} plowed under	0.72	2.79	2.43
Control	6.47	3.86	3.51

a: Slightly Zn deficient soil, b: Severely Zn deficient soil. Source: Gunderson et al., 1965

declines in residual effectiveness can be worsened by a single severe flood (Sakal et al., 1985). Singh and Abrol (1985) describe a rice-wheat rotation grown on an alkali soil (having a shallow watertable and pH 10.45) in which Zn (2.5-27 kg ha^{-1}) was applied either only once initially or with each crop, and irrigated with normal water. After seven crops in double-cropped rotation, yields in plots with continual applications of 2.25 kg Zn ha^{-1} were no different to those with a single initial application of 18 kg Zn ha^{-1}, suggesting that its effectiveness had not declined. Bhardwaj and Prasad (1981) found that the residual effectiveness of an initial application of 5.9 kg Zn ha^{-1} was negligible in the sixth (rice) or seventh (wheat) crops in a rice-wheat system in a silty loam soil. In a third study on a rice-wheat rotation, (Takkar and Nayyar, 1981, 1982; Nayyar et al., 1990) several Zn and gypsum treatments were applied to a highly deteriorated sodic soil (Natraqualf pH 10.4) and irrigated with brackish ground water, the only source for irrigation. After four crops, the plots were split and one portion received a repeat Zn application before transplanting the fifth crop. The increase in yields of the fifth to seventh crops of the repeat Zn treatment over the initial residual indicated that the initially applied Zn (11 to 22 kg ha^{-1}) was inadequate to combat Zn deficiency after the fourth crop in rice-wheat system. Thus, the use of brackish water on highly sodic soil appears to have shortened the residual effectiveness of Zn when compared with the study of Singh and Abrol (1985).

In a 5-year study in 9 countries, 5 kg ha^{-1} Zn as $ZnSO_4$ effectively corrected Zn deficiency in rice for 2 years and though ZnEDTA was a good source, it proved ineffective at low rates (IAEA, 1981). Giordano (1977) reported similar results. The lower yield with ZnEDTA may result from its greater downward movement, beyond the effective root-zone, during flooding as well from its instability in reduced soils (Reddy and Patrick, 1977). In contrast, ZnEDTA was more effective than $ZnSO_4$ in an alkaline alluvial soil (Bansal and Nayyar, 1989).

On a Zn deficient highly sodic soil, Zn-frits and ZnO did not match yield with $ZnSO_4$ or a multi-micronutrient mixture (Nayyar and Takkar, 1980). Again in contrast, a superiority of ZnO over $ZnSO_4$ was observed by Sedberry et al. (1971) and Giordano and

Mortvedt (1973), and in two field experiments on flooded calcareous soil in Egypt (Amer et al., 1980). It appears that effectiveness of ZnO was related to the persistent release of Zn in soil solution of the calcareous soil but not in that of a highly sodic soil.

In studies that did not involve rice or the associated conditions, the residual effectiveness of Zn is longer. Leslie et al. (1973) reported that the residual effectiveness of Zn over five crops on a Waco vertisol, comparing initial single applications with annual applications. They found that the fixation of Zn had "not been a significant problem over the six-year duration of the trial", despite two long-fallow intervals. Takkar et al. (1975) found that an application of 22 kg Zn ha^{-1} would last for at least 7 crops in a wheat-groundnut rotation on a loamy sand, a conservative conclusion reached after observing the decline in extractable Zn over 3 crops. A residual effect from applying 18 kg ha^{-1} Zn was noticed for at least 6 crops of maize in a single cropping rotation on a vertisol (Wier and Holland, 1980). A decline in effectiveness was seen in this study after around 10 years (Holland, pers. com.). In another study on single-cropped maize on a clay soil, Carsky and Reid (pers. com.), found that a single initial application of 9 kg Zn ha^{-1} was just as effective as the same rate applied annually for 7 years.

4.3. Soil management

Studies of the renovation of sodic soils have found that gypsum applications improved rice grain yield (Takkar and Singh, 1978; Sadana and Takkar, 1984). Gypsum interacted with Zn, lowering the Zn rate required for maximum yield, despite also lowering extractable Zn in the soil (Takkar and Singh, 1978; Takkar and Nayyar, 1981). Several other studies have reported beneficial effects of gypsum application on plant Zn uptake (Olsen and Watanabe, 1979; Shukla and Mukhi, 1980; Bowman and Olsen, 1982).

Drainage and drying of wet land has helped in alleviating Zn deficiency (Shiratori, 1977). Also amending soil with organic manures helps cure Zn deficiency. In a three-year field experiment on Zn deficient soil, organic manures applied, at rates of 12 t ha^{-1} farmyard manure, 5.0 t ha^{-1} poultry manure, or 2.5 t ha^{-1} of piggery manure, prior to sowing maize in a maize-wheat system corrected Zn deficiency just as effectively as a broadcast application of 11 kg ha^{-1} Zn (Takkar et al., 1989), although controlling such studies for other nutritional effects seems impossible. Tillage has also been observed to break up mycorrhizal hyphal networks, and in doing so lower the VAM colonisation and Zn concentrations in maize (Evans and Miller, 1990). This could mean that the residual effectiveness of applied Zn may become shortened by tillage, despite its effect on mineralisation.

4.4. Crop management

Differences between and within plant species in their ability to tolerate or avoid Zn deficiency have been observed for a wide range of agricultural plants (Clark, 1982; Takkar, 1991b; Ponnamperuma, 1982). For example, maize, rice (low land, ie. flooded), citrus, and tung tree are more sensitive to Zn deficiency than other species. Among the cereals, rye is well known as a source of genetic efficiency for several micronutrients (Graham, 1984) and efforts to transfer this efficiency to other cereals are described elsewhere in this volume.

Several studies have examined crop cultivars and identified Zn efficient cultivars capable of good yields at low available soil Zn, aiming to mimimise the cost of correcting Zn deficiency (barley: Pathak et al., 1979; Takkar et al., 1983; rice: Dutta et al., 1987; soybean: Rose et al., 1981; wheat: Shukla and Raj, 1974; Best, 1987). The new rice varieties IR 36, IR 46, IR 52 and IR 54 combine efficiency in absorbing Zn with high yield potential, disease-, and insect-resistance to have yield advantages of 2 t ha^{-1} on un-amended soils (Ponnamperuma, 1982).

A distinction should be made between the tolerance of an individual crop and that of a cropping system to Zn deficiency. The latter has been rarely studied but, for example, it has been observed that wheat grown in rotation with rice yielded less than wheat in rotation with maize (Chandi and Takkar, 1982). Wheat in rotation with rice also took up less Zn. Rice itself was observed to take up greater quantities of Zn than eight other species, due in part to higher yields. Subtle effects may be present; a study of the soil chemistry of heavy metal ions under several cropping systems suggested that long term pH changes may be largely responsible for changes in soil solution concentrations of the metal ions (Basta and Tabatabai, 1992). A low cropping frequency has been found to increase the occurrence and severity of Zn deficiency (Duncan, 1967a,b); and related to the decline in mycorrhizae inoculum in the soils.

5. Conclusion

The distribution of Zn deficiency has been unpredictable, being associated with many soil types and crop systems, but persistent problems with flooded soils are very common. Identification and reporting of Zn deficiency is still a sporadic and haphazard process, with the possible exception of India, where a concerted effort has mapped its distribution. Correction of Zn deficiency may be either through soil, seed, or foliar Zn amendments, and no single approach appears most suitable.

References

Alexander A 1986 Foliar Fertilization. Proc. First Int. Symp. Foliar Fertilization, Berlin, March 1985. Developments in Plant and Soil Sciences, Vol. 22. Martinus Nijhoff, Dordrecht. 488 p.

Alexander A and Schroeder M 1987 Modern trends in foliar plant fertilization. J. Plant Nutr. 10, 1391-1399.

Amer F, Rezk A I and Khalid H M 1980 Fertilizer zinc efficiency in flooded calcareous soils. Soil Sci. Soc. Am. J. 44, 1025-1030.

Anderson A J 1970 Trace elements for sheep pastures and fodder crops in Australia. J. Aust. Inst. Agric. Sci. 36, 15-22.

Badillo-Feliciano J and Lugo Lopez M Z 1980 Differential response of corn and sweet potatoes to Zn applications in an Oxisol in northwestern Puerto Rico. Jour Agric Univ Puerto Rico 64, 482-488.

Bansal R L and Nayyar V K 1989 Effect of zinc fertilizers on rice grown on Typic Ustochrepts. IRRN 14, 24.

Basta NT and Tabatabai MA 1992 Effect of cropping systems on adsorption of metals by soils: II. Effect of pH. Soil Sci. 153, 195-204

Bauer A and Lindsay W L 1965 The effect of soil temperature on the availability of indigenous soil zinc. Soil Sci. Soc. Amer Proc. 29, 413-420.

Beeson K C 1957 USDA Year Book Agriculture p. 264

Best E K 1987 Diagnosis and correction of zinc deficiency in Queensland's wheat soils. Final Report to Wheat Res. Comm. Qld. Wheat Res. Inst., Toowoomba. 27p.

Bhardwaj S P and Prasad S N 1981 Response of rice-wheat rotation to zinc under irrigated conditions in Doon Valley. J. Indian Soc. Soil Sci. 29(2), 220-224.

Boawn L C, Viets F Jr and Crawford C L 1957 Plant utilization of zinc from various types of zinc compounds and fertilizer materials. Soil Sci. 83, 219-229.

Bowman R A and Olsen S R 1982 Effect of calcium sulphate on iron and zinc uptake in Sorghum. Agron. J. 74,

923-925.
Brennan, RF 1991 Effectiveness of zinc sulphate and zinc chelate as foliar sprays in alleviating zinc deficiency of wheat grown on zinc-deficient soils in Western Australia. Aust. J. Exp. Agric. 31, 831-834.
Brown A L and Krantz B A 1966 Sources and placement of zinc and phosphorus for corn (*Zea mays* L.) Soil Sci Soc Am. Proc. 30, 86-89
Castro R U 1976 Cost-saving methods of ameliorating zinc deficiency in rice. *In* Proc. Seventh Annu. Meeting Crop Sc. Soc. Philippines, May, 1976, Davao City, Philippines.
Castro R U 1977 Zinc deficiency in Rice. A series of research at the IRRI. IRPS 9, 1-18.
Chandi K S and Takkar P N 1982 Effects of Agricultural cropping systems on micronutrient transformations. 1. Zinc. Plant Soil 69,423-436.
Chandler W H 1937 Botan. Gaz. 98, 625-646.
CIAT 1978 Informe anual. 1977, Cali, Colombia
Clark R B 1982 Plant genotype differences to uptake translocation, accumulation, and use of mineral elements. *In* Genetic Specificity of Mineral Nutrition of Plants. Ed. MR Saric. pp.41-55. Serbian Acad. Sci. Arts, Sci. Assemb. XIII No. 3, Beograd.
Cottenie A, Kang B T, Kiekens L and Sajjapongse A 1981 Micronutrient status. *In* Characterization of Soils. Ed. D J Greenland. pp. 149-163. Clarendon, Oxford
Dantas H da S 1971 Manganese cations permutaveis na unidade. Utiga Pesq. Agropec. Bras. Ser. Agron. 6, 27-30.
Donald C M and Prescott J A 1975 Trace elements in Australian crop and pasture production 1924-74. *In* Trace Elements in soil-plant-animal systems. Eds. D J DNicholas and AR Egan. pp. 7-37. Academic, Sydney.
Duncan O W 1967a Correction of zinc deficiency in wheat on the Darling Downs, Queensland. Qld. J. Agric. Anim. Sci. 24, 287-292.
Duncan O W 1967b Correction of zinc deficiency in linseed on the Darling Downs, Queensland, Qld J. Agric. Anim. Sci. 24, 301-307.
Dutta R K, Muslimuddin and Rahman L 1987 Yield and flowering of rice in relation to fertilizer zinc sulphate. Int. Rice Com. Newsl. 36,16-22.
Egbe N E and Omotoso T I 1972 Nutrient deficiencies of cocoa in Nigeria. Proc 4th Int. Cocoa Res. Conf., Trinidad.
Ellis B G, Davis J F, and Judy W H 1965 Effect of method of incorporation of Zinc in fertilizer on zinc uptake and yield of pea beans (*Phaseolus vulgaris*). Soil Sci. Soc. Am. Proc. 29, 635-636.
Ellis R Jr, Davis J F, and Thurlow D L 1964 Zinc availability in calcareous Michigan soils as influenced by phosphorus level and temperature. Soil Sci. Soc. Am. Proc. 28, 83-86.
Evans D G and Miller M H 1990 The role of the external mycelial network in the effect of soil disturbance upon vesicular-arbuscular mycorrhizal colonization of maize. New Phytol. 114, 65-71.
Fawusi M O A and Ormrod D P 1975 Zinc nutrition and temperature effects on tomato. J. Hort. Sci. 50, 363-371.
Ferrand M, Bachy A and Ollagnier M 1951 Les oligo elements dans la fumure du palmier a huile au moyen-Congo. Oleagineux 11, 629-636.
Finich A H and Kinnison A F 1933 Arizona Agr. Expt. Sta. Tech. Bull. 47, 407-442.
Galrao E Z 1990 Application of micronutrients and chalk on Soybean yield on varzen soils. R. Bras. Ci Solo. 14, 381-384.
Galrao E Z and de M Felho MV 1981 Effect of sources of zinc on the dry matter production of corn in a Cerrado soil. R. Bras. Ci Solo. 5, 167-170.
Gartrell J W and Glencross R N 1969 Copper, zinc and molybdenum fertiliser for new land crops and pastures-1969. J. Agric. West Aust. (4th Ser.) 9, 517-521.
Gartrell J W 1974 *In* Our Land 7:12 CSBP and Farmers, Western Australia.
Giordano P M 1977 Efficiency of zinc fertilization for flooded rice. Plant Soil 48, 673-684.
Giordano P M and Mortvedt J J 1973 Zinc sources and methods of application for rice. Agron. J. 65, 51-53.
Glushchenko L T, Dutchenko Z Y and Mishnev A K 1991 Effectiveness of chelates and micronutrients with sunflower. Khim Sel'sk Khoz. 8, 76-77.
Graham R D 1984 Breeding for nutritional characteristics in cereals. Adv. Plant Nutr. 1,57-102.
Greenwood M and Hayfron R J 1951 Iron and zinc deficiencies in cocoa in the Gold Coast. Emp. J. Exp. Agric. 19, 73-86.
Grewel J S and Trehan S P 1979 Micronutrients for potatoes. Fert. News 24(8), 27-30.
Gunderson O, Bezdicek D and MaxGrejor J 1965 Zinc deficiency of corn in Minnesota. Min. Univ. Agr. Ext. Serv., Ext. Bull. 322 April, 1965

Haque I and Kamara C S 1976 Micronutrient investigations in Sierra Leone FAO/IAEA Coordination meeting on Isotope Aided Micronutrient Studies in Rice Production with Special Reference to Zinc Deficiency Manilla, Philippines May 17-21

Hergert G W, Rehm G W and Wiese R A 1984 Field evaluation of zinc sources band applied in ammonium polyphosphate suspension Soil Sci Soc Am J 48,1190-1193

Hodgson J F, Allaway W H and Lockman R B 1971 Regional plant chemistry as a reflection of environment *In* Environmental Geochemistry on Health and Disease Eds H L Cannon and H C Hopps pp 57-72 Geol Soc Amer Memoir 123

Holden E R and Brown J R 1965 Influences of slowly soluble and chelated zinc on zinc content and yield of alfalfa J Agr Food Chem 13, 180-184

Igue K and Bornemisza E 1967 El problema del Zn en suelos y plantas de regiones tropicales y de zonas templada Fitotecnica Latinoamericana 4, 29-44

IAEA 1981 Zinc fertilization of flooded rice IAEA, Tecdoc 242, Vienna, Austria, 80 p

Iyenger B R V and Raja M E 1988 Response of some vegetable crops to different sources and methods of application of zinc Indian J Agric Sci 58, 565-567

Juo A S R and Uzu F O 1977 Liming and nutrient interactions in two ultisols from southern Nigeria Plant Soil 47, 419-430

Kang B T and Okoro E G 1976 Response to flooded rice grown on vertisol from north Nigeria to zinc source and methods of application Plant Soil 44, 15-25

Kanwar J S and Randhawa N S 1974 Micronutrient Research in Soils and Plants in India (A Review) ICAR Tech Bull (Agric) New Delhi, 185 p

Karim Z, Miah M M U and Hossain S G 1991 Zinc deficiency problems in flood plain agriculture Proc Int Symp Role of S, Mg and Micronutrients in Balance Plant Nutr Ed S Portch pp 337-343 Chengdu PRC

Katyal J C and Friesen D K 1988 Deficiencies of micronutrients and sulfur in wheat *In* Wheat Production Constraints in Tropical Environments Ed A R Klatt pp 99-129 CIMMYT, D F Mexico

Katyal J C and Ponnamperuma F N 1974 Zinc deficiency A widespread nutritional disorder of rice in Agusan del Norte J Philipp Agric 58(3,4) 79-89

Katyal J C and Sharma B D 1979 Role of micronutrients in crop production- a review Fert News 24, 33-50

Kayode G and Agboola A A 1983 Investigations on the use of macro- and micronutrients to improve maize yield in South Western Nigeria Fert Res 4, 211-221

Larez C and Sanchez C 1971 Respuesta del sorgo (*Sorghum vulgare Pers*) a la application de cal y micronutrientos en un suelo ultisol de sabana Oriente Agropecuario 3, 16-23

Leslie J K, Whitehouse M J and Mackenzie J 1973 Zinc fertilizer residual experiment Qld Wheat Res Inst Annu Rep 1972-73, pp 33-35

Lingle J C and Holmberg D M 1957 The response of sweet corn to foliar and soil zinc application on a zinc deficient soil Proc Amer Soc Hort Sci 70, 308-315

Littlemore J, Winston EC, Howitt CJ, O'Farrell P, and Wiffen DC 1991 Improved methods for zinc and boron application to mango (*Mangifera indica* L) cv Kensington Pride in the Mareeba-Dimbulah district of North Queensland Aust J Exp Agric 31, 117-121

Li-shu-fan, Liang Hong, Li Yu-ying and Zang Dong-tie 1991 The effect of sulphur and micronutrients (zinc, copper and iron) on the balance of crop nutrition Proc Int Symp Role of S, Mg and Micronutrients in Balanced Plant Nutr Ed S Portch pp 216-221 Chengdu, PRC

Liu Zheng 1991 Characterization of content and distribution of micronutrients in soils of China Proc Int Symp Role of S, Mg and Micronutrients in Balance Plant Nutr Ed S Portch pp 54-62 Chendu, PRC

Lucas RE and Knezek BD 1972 Climatic and soil conditions promoting micronutrient deficiencies in plants *In* Micronutrient in Agriculture Eds JJ Mortvedt and W L Lindsay pp 265-287 Soil Sci Soc Am , Madison

Macnaeidhe F, Fleming GA and Parle PJ 1986 Zinc deficiency, first time in cereals in Ireland Farm and Food Res pp 57-58

MacNaeidhe FS and Fleming GA 1988 A response in Spring cereals to foliar sprays of zinc in Ireland Irish J Agric Res 27, 91-97

Maftoun M and Karimian N 1989 Relative efficiency of two zinc sources for maize (*Zea mays* L) in two calcareous soils from an arid area of Iran Agronomie 9(8), 771-775

Magar and Babakir 1965 Review of the work done on micronutrients in the clay plain of the Republic of the Sudan Int Symp on Maintenance and Improvement of Soil Fertility, Khartoum OAU/STRC Publications Bureau, Watergate House, York Bldg, London

Malavolta E H P, Mell F A F and Brasil Sobro M O C 1962 On the Mineral nutrition of some tropical crops Int Potash Inst , Berne, Switzerland

Mandal B, Chatterjee J, Hazra GC, and Mandal LN 1992 Effect of preflooding on transformation of applied zinc and its uptake by rice in lateritic soils. Soil Sci. 153, 250-257.

Mann M S, Takkar P N, Bansal R L and Randhawa N S 1978 Micronutrient status of soil and yield of maize and wheat as influenced by micronutrient and farmyard manure application. J. Indian Soc. Soil Sci. 26, 203-214.

Marinho M L and Igue K 1972 Factors affecting zinc absorption by corn from volcanic ash soils. Agron. J. 64, 3-8.

Marschner H and Cakmak I 1989 High light intensity enhances chlorosis and necrosis in leaves of zinc, potassium and magnesium deficient bean (*Phaseolus vulgaris*) plants. J. Plant Physiol. 134, 308-315.

Mikkelsen D S and Shiou K 1977 Zinc fertilization and behaviour in flooded soils. Spec. Publ. No. 5, Comm. Agric. Bur., Farnham Royal. 59 p.

Moraghan JT 1980 Effect of soil temperature on response of flax to phosphorus and zinc fertilizers. Soil Sci. 129, 290-296.

Mordvedt J J 1991 Research techniques with micronutrient fertilizers for use in efficient crop production. Proc. Int. Symp. on Role of S, Mg and micronutrients in balanced plant nutrition. Ed. S. Portch. pp. 30-39. Chengdu, PRC.

Murphy L S and Walsh L M 1972 Correction of micronutrient deficiencies with fertilizers. *In* Micronutrients in Agriculture. Eds. J J Mortvedt, P M Giordano and W L Lindsay. pp. 347-387. Soil Sci. Soc. Amer., Madison USA.

Myhr K 1988 Zinc and manganese fertilization of barley on alkaline soil. Norsk Landbruks Forsking 2, 103-109.

Nambiar E K S 1975 Mobility and plant uptake of micro- nutrients in relation to soil water content. *In* Trace Elements in Soil-Plant-Animal Systems. Eds. D J D Nicholas and AD Egan. pp. 151-163. Academic, London.

Nayyar V K, Singh S P and Takkar P N 1984 Response of sugarcane to zinc and iron sources. J. Res., Punjab Agricultural Univ. 21, 134-136.

Nayyar V K and Takkar P N 1980 Evaluation of various zinc sources for rice grown on alkali soil. Z. Pflanzan. Bodenk. 143, 489-493.

Nayyar V K, Takkar P N, Bansal R L, Singh S P, Kaur N P and Sadana U S 1990 Micronutrients in Soil and Crops of Punjab. Dept. Soil. Res. Bull. 148 p.

Nene Y L 1966 Symptoms, cause and control of Khaira disease of paddy. Bull. Indian Phytopathol. Soc. No. 3,97-101.

Olsen S R and Watanabe F S 1979 Interaction of added gypsum in alkaline soils with uptake of iron, molybdenum, manganese, and zinc by sorghum. Soil Sci. Soc. Amer. J. 43, 125-130.

Ozanne P G 1955 The effect of light on Zn deficiency in subterranean clover. Aust. J. Biol. Sci. 8, 344-355.

Pathak A N, Tiwari K N, Upadhaya R L and Jha A K 1979 Zinc response of barley as influenced by genetic variability. Indian J. Agric. Sci. 49, 25-29.

Peralta F, Bornemisza E and Alvarado A 1981 Zinc adsorption by Andepts from the central plateau of Costa Rica. Commun. Soil Sci. Plant Anal., 12, 669-682.

Pittman H A and Owen R C 1936 Anthracnose and mottle leaf of citrus in Western Australia. J Dept. Agric. West Aust. 13, 137-143.

Ponnamperuma G N 1982 Genotypic adaptability as a substitute for amendments on toxic and nutrient deficient soils. Proc. 9th Int. Plant Nut. Colloq. Warwick, England (August 1982) 467 pp.

Ponomarev V G and Ponomareva T G 1989 Effectiveness of zinc fertilizers on cotton in relation to rates, times and modes of application. Agrokhimiya 5, 90-93.

Prasad B and Sinha M K 1981 The relative efficiency of zinc carriers on growth and zinc nutrition of corn. Plant Soil 62, 45-52.

Randhawa N S and Takkar P N 1975 Micronutrient research in India. The present status and future projection. Fert. News. 20, 11-18.

Rasmussen P E and Boawn L C 1969 Zinc seed treatment as a source of zinc for beans (*Phaseolus vulgaris*) Agron. J. 61, 674-676.

Reddy C N and Patrick W H Jr 1977 Effect of redox potential on the stability of zinc and copper chelates in flooded soils. Soil Sci. Soc. Amer. J. 41, 729-732.

Riceman D S 1948 Mineral deficiency in plants on the soils of the Ninety-mile plain in South Australia. 5. Effect of growth and removal of a crop treated with zinc and copper upon pasture established subsequently on Laffer sand, near Keith. J. Coun. Sci. Ind. Res. Aust. 21, 229-235.

Rose I A, Felton W L and Banks L W 1981 Responses of four soybean varieties to foliar zinc fertilizer. Aust. J. Exp. Agric. Anim. Husb. 21, 236-240.

Ryan P, Lee J and Peebles T F 1967 FAO U N, World Soil Resources Report 31, Rome 1967.

Sadana U S and Takkar P N 1983 Methods of Zinc application to rice on sodic soil. IRRN, 8,21-22.
Sadana U S and Takkar P N 1984 Soil amendments for sodic conditions. IRRN, 9, 10-11.
Sajwan K S and Lindsay W L 1988 Effect of redox, zinc fertilization and incubation time on DTPA-extractable zinc, iron and manganese. Commun. Soil Sci. Plant Anal. 19, 1-11.
Sakal R, Singh A P and Singh B P 1985 Direct and residual effect of zinc and zinc amended organic manures on the zinc nutrition of field crops. Indian J. Agric. Res. 19, 93-97.
Sanchez P A 1977 Manjo de suelos Tropicales en la Amazonia Sudamericana *In* Soc. Colombiane de la Ciencia de Suelo (ed) Memorias del V Congreso Latinoamericao de la Ciencia del Suelo Y IV Coloquio Nacioonal sobre suelos. Suelos Ecuatorials VIII (1), 1-11.
Sanchez P A and Cochrane T T 1980 Soil constraints in relation to major farming systems in tropical America. *In* Soil Related Constraints to Food Production in the Tropics. pp. 109-139. IRRI, Philippines.
Schwartz S M, Welch R M, Grunes D L, Cary E E, Norvell W A, Gilbert M D, Meredith M P and Sanchirico C A 1987 Effect of zinc, phosphorus, and root-zone temperature on nutrient uptake by barley. Soil Sci. Soc. Amer. J. 51, 372-375.
Scott J M 1989 Seed coatings and treatments and their effects on plant establishment. Adv. Agron. 42, 43-83.
Sedberry J E Jr, Peterson F J, Wilson E, Nugent A L, Engler R M and Brapbackor R H 1971 Effects of zinc and other elements on the yield of rice and nutrient content of rice plants. Louisiana Agr. Exp. Sta. Bull. 653.
Servetuik-chaluya G K, Verigina V S, Kornienko V A, Orlov V K Urbisinov Zh K, Kalashmikow I V and Mamutov Z 1991 Hygienic assessment of rice cultivated with the use of zinc salts. Gig. Sanit 5, 42-44.
Sharma B D, Takkar P N and Sadana U S 1982 Evaluation of levels and methods of zinc application to rice in sodic soils. Fert. Res. 3, 161-167.
Sheudzhen A Kh, Bugaevskii V K and Doseeva O A 1991 Effect of zinc fertilization on rice yield. Khim Sel'sk. Khoz. 1, 93-94.
Shiratori K 1977 Proc. Int. Seminar Soil Environ. Fertil. Management Intensive Agric. (SEFMIA) pp. 653-659. Tokyo, Japan.
Shulka U C and Mukhi A K 1980 Ameliorative role of Zn, K, and gypsum on maize growth under alkali soil conditions. Agron. J. 72, 85-88.
Shukla U C and Raj H 1974 Influence of genetic variability on zinc response in wheat. Soil Sci. Soc. Amer. Proc. 38, 477- 479.
Singh M V and Abrol I P 1985 Direct and residual effect of fertilizer zinc application on the yield and chemical composition of rice-wheat crops in an alkali soil. Fert. Res. 8, 179-191.
Singh M V and Abrol I P 1986 Transformation and movement of zinc in an alkali soil and their influence on the yield and uptake of zinc by rice and wheat crops. Plant Soil 94, 445-449.
Stiles W 1961 Trace Elements in Plants. Cambridge Univ Press Third Edition Cambridge, England.
SSSA (Soil Sci. Soc. Am.) Soil Testing Committee 1965 A survey of micronutrient deficiencies in the USA and means of correcting them. Madison, Wisc.
Stephens C G and Donald C M 1958 Australian soils and their responses to fertilizers. Adv. Agron. 10, 167-.
Stratieva S, Sedlarska B and Stoyanov D 1990 Effect of zinc and boron on sugarbeet growth on a leached smonitza chernozem soil. Pochvoznanie i Agrokhimiya 25, 9-14.
Takkar P N 1991a Requirement and response of cultivars to micronutrients in India. Proc. 4th Int. Symp. Genetic Aspects of Plant Mineral Nutrition. Ed Randal, 1991. Plant and Soil (*In Press*).
Takkar P N 1991b Zinc Deficiency in Indian Soils and Crops *In* Zinc in Crop Nutrition. pp. 55-64. Int. Lead Zinc Research Organization, Inc. and Indian Lead Zinc Information Centre, New Delhi.
Takkar P N and Bansal R L 1987 Evaluation of rates, methods and sources of zinc application to wheat. Acta Agronomica, 36, 277-283.
Takkar P N, Chibba I M and Mehta S K 1989 Twenty Years of Coordinated Research of Micronutrients in Soil and Plants (1967-87). Indian Institute of Soil Science, Bhopal. IISS, Bull. I, 314 p.
Takkar P N, Mann M S and Randhawa N S 1974 Methods and time of zinc application. J. Res. Punjab Agric. University 11, 276-283.
Takkar P N, Mann M S and Randhawa N S 1975 Effect of direct and residual available zinc on yield, zinc concentration and its uptake by wheat and groundnut crops. J. Indian Soc. Soil Sci. 23, 91-95.
Takkar P N and Nayyar V K 1981 Effect of gypsum and zinc on rice nutrition on sodic soils. Exptl. Agric. 17, 49-55.
Takkar P N and Randhawa N S 1978 Micronutrients in Indian Agriculture (*Review*). Fert.News 23,5-26.
Takkar P N and Sidhu B S 1979 Kinetics of zinc transformations in submerged alkaline soils in the rice growing tracts of Punjab. J. Agric. Sci. Camb. 93, 4451-447.
Takkar P N and Singh M 1979 Methods of application of micronutrient carriers-soil versus foliar application.

India/FAO/Norway Seminar on Micronutrients in Agriculture, New Delhi. Sept., 1979.

Takkar P N, Singh S P, Bansal R L and Nayyar V K 1983 Tolerance of barley varieties to zinc deficiency. Indian J. Agric. Sci. 53, 971-972.

Takkar P N and Singh T 1978 Zinc nutrition of rice (*Oryza sativa*) as influenced by rates of gypsum and zinc fertilization of alkali soils. Agron. J. 70, 447-450.

Tanner P D and Grant P M 1974 The effectiveness of N, P and K fertilizer and lime as carriers of zinc for maize (*Zea mays* L.) under field conditions. Rhod. J. Agric. Sci. 12, 163-168.

Thorne W 1957 Zinc deficiency and its control. Adv. Agron. 9, 31-65.

Velly J, Celton J and Marquette J 1974 Experimentation sur les oligo-elements en culture seche et en riziere a Madagascar. Comm Reunion d'Agron IRAT, Paris, 26.

Walsh I and Schulte E E 1970 Computer programmed soil test recommendations for field and vegetable crops. Wisconsin Agr. Exp. Sta. Soil Fertility Series Bull. 5.

Welch C D, Gray C, Thomas G W and Anderson W A 1967 Zinc deficiency and fertilization. Texas A and M Univ. Agr. Ext. Serv. Fact Sheet 1-721.

Whitney D A and Murphy L S 1969 Kansas lime and fertilizer recommendations. Kansas Agr. Ext. Circ. 352.

Widdowson J P 1966 Zinc deficiency in the shallow soils of New Zealand J. Agr. Res. 9,44.

Wier R G and Holland J 1980 The residual effects of fertilizer zinc on a black earth from north western New South Wales. Proc. Aust. Agron. Conf., Qld. Agric. Coll., Lawes p.310.

Williams C H and Andrew C S 1970. *In* Australian Grasslands. Ed. RM Moore. p. 321. Australian National Univ. Press, Canberra.

Yoshida S, Ahn J S and Forno D A 1973 Occurrence, diagnosis, and correction of zinc deficiency of lowland rice. Soil Sci. Plant Nutr. 19, 83-93.

Yoshida S, Mclean G W, Shafi M and Mueller K E 1970 Effects of different methods of zinc applications on growth and yields of rice in a calcareous soil, West Pakistan. Soil Sci. Plant Nutr. 16, 147-149.

Yoshida S and Tanaka A 1969 Zinc deficiency of the rice plant in calcareous soils. Soil Sci. Plant Nutr. 15, 75-80.

Chapter 12.

Diagnosis of Zinc Deficiency

R. F. BRENNAN, J. D. ARMOUR and D. J. REUTER

1. Abstract

Zinc deficiency in crops can be diagnosed or predicted successfully using field observations, soil tests and/or plant analysis. Plant species and genotypes differ in their requirements for Zn, so the timing and method of diagnosis is critical. Using whole shoots of plants to diagnose Zn deficiency has been unreliable but specific plant parts (especially young leaves) are good indicators of the Zn status of a plant. Enzyme activities have the potential to be developed into useful diagnostic tests of Zn deficiency. Many chemical extractants have been used to assess the Zn status of soils. These soil tests have often been improved by including measurements of other soil properties (e.g. pH, clay, organic carbon). However, there is a need to standardize the procedures which measure both the quantity and intensity of Zn.

2. Introduction

Detecting nutritional disorders, such as zinc (Zn) deficiency in plants, requires a logical diagnostic framework which integrates local experience (including fertilizer experiments), factors known to promote nutritional deficiencies and observed plant symptoms. The final analysis and interpretation is often confirmed by either soil or plant analysis. Usually, diagnosticians use all relevant information to identify Zn deficiencies.

3. Diagnosis of Zn deficiency by field observations

3.1. Plant symptoms

Symptoms of severe Zn deficiency in plants are characteristic and easy to identify. These distinctive symptoms are useful for both recognizing Zn deficiency and for delineating Zn responsive soils. Numerous publications have described and illustrated symptoms of acute deficiency (for example Asher et al., 1980; Snowball and Robson, 1983; 1986; Grundon, 1987). In the more recent publications, useful systematic keys have been developed for recognizing nutritional disorders in specific plant species.

The most common symptoms of acute Zn deficiency include stunted growth, shortened internodes and petioles and small malformed leaves ('little leaf') which result in the classic 'rosette' symptom in young growth of dicotyledons (for example Snowball and Robson, 1983; 1986) and 'fan shaped stem' in monocotyledons (Grundon, 1987). With careful, frequent observation, symptoms will normally be seen first in young leaves (Zn is considered immobile under conditions of deficiency). These leaves remain small, cup upwards, and develop interveinal chlorosis and on the upper leaf surfaces necrotic spots appear which later coalesce to form brown necrotic and brittle patches. In monocotyledons, chlorotic stripes adjacent to the midrib become necrotic. The necrosis is often more noticeable on middle-aged leaves, which sometimes wilt, bend or collapse. Zn deficiency is typically patchy, even within a single field (Kubota and Allaway, 1972) and symptoms develop rapidly but, depending on the degree of stress, are sometimes transient. In annuals, symptoms are more noticeable during the earlier stages of growth. Seed and fruit formation can be aborted and are severely reduced.

It is now also evident that symptoms are not exhibited by plants suffering moderate to mild levels of Zn stress (for example Reuter et al., 1982a). Indeed, in some studies, dry matter production has been reduced by 40 per cent or more by Zn deficiency in plants which show no visible symptoms (Carroll and Loneragan, 1968). Procedures for diagnosing such 'hidden hunger' are needed for many plant species.

3.2. Climatic conditions

Certain environmental conditions can accentuate Zn deficiency in plants, and often complicate diagnostic interpretations for field-grown plants. For example, in conditions favourable for rapid plant growth, nutrients are sometimes diluted rapidly within the lush growth, and this may induce deficiency symptoms of those nutrients present at a marginally low supply.

Conditions of low temperature, reduced light intensity (or in one case, high light intensity (Marschner and Cakmak, 1989)) and short daylength and their interactions have been implicated in promoting Zn stress in plants (Moraghan and Mascagni, 1991). In the USA, Zn deficiency symptoms in maize are often seen with the onset of cold, wet weather (Lindsay, 1972). In Australia, symptoms of Zn deficiency in annuals are observed during the cool winter months and sometimes disappear with the warmer conditions in spring (for example Rossiter, 1951). These field observations have been attributed to low soil temperatures which restrict the root development of young plants, suppress microbial decomposition of organically bound Zn and reduce the infection of plant roots by mycorrhizal fungi (Moraghan and Mascagni, 1991).

In tropical regions, Zn deficiency is reported to be the most widespread micronutrient disorder affecting wetland rice crops (Lopes, 1980). For dryland crops, soil moisture probably does not have a major direct effect on the incidence of Zn deficiency (Moraghan and Mascagni, 1991).

3.3. Other factors involved in field diagnosis

Diagnosing Zn deficiency in the field can be aided by knowing the soil type. Zn deficiency is most common where soils are sandy, high in organic matter, alkaline or submerged (see Takkar and Walker, this volume). In some studies, sensitive species have been used as 'indicators' of Zn-deficient soils (for example Adams and Piper, 1944).

Several farming practices have a direct impact on the Zn status of plants and so may induce or intensify Zn deficiency. For example, Zn deficiency in crops (Grunes et al., 1961) and pastures (Kleinig and Loveday, 1962) has been caused by removing topsoil. The exposed sub-surface horizon is less enriched with Zn and sometimes has a higher pH and carbonate content than the displaced topsoil. Also, the 'long fallow disorder' (> 12 month cultivated fallow) seen in a range of crops has been associated with declines in viable propagules of vesicular arbuscular mycorrhizas and subsequent development of Zn deficiency (Thompson, 1987).

In addition, knowing the amount, type and regularity of previous Zn fertilizer applications to soil, together with the quantities of Zn removed in farm products, will aid field diagnoses. For example, in Australia, the level of Zn impurity in different macronutrient fertilizers varies widely. Depending on the rate and frequency of fertilizer use, these impurities can supply nutritionally significant amounts of Zn for plants grown on soils of low Zn status (Riley et al., 1992).

Visual diagnosis of Zn deficiency can be aided by knowing the previous applications of lime and phosphatic and nitrogenous fertilizers. At high soil pH, Zn is more strongly absorbed onto soil minerals and oxides (Lindsay, 1991) and Zn availability to plants is correspondingly diminished. Ca-induced restrictions to Zn uptake may also be involved (Chaudhry and Loneragan, 1972). Conversely, the availability of applied zinc sulphate to plants is enhanced by banding Zn with acid-forming fertilizers (Mortvedt and Kelsoe, 1988). Zn deficiency may also be intensified where high rates of phosphatic fertilizers are applied (see Loneragan and Webb, this volume). Increased soil N supply can also modify plant Zn status, either through acidifying the soil and enhancing Zn uptake (Viets et al., 1957) or by promoting growth and increasing the plant's requirement for Zn.

Root systems of Zn-deficient plants are smaller than those of Zn-adequate plants (e.g. Reuter et al., 1982a;b). Visual symptoms of Zn deficiency can be compounded by root disease or by the complicating and exacerbating effects of herbicides, which both reduce Zn uptake by plants. For example, the applications of diclofop-methyl and chlorsulfuron to a Zn-deficient sandy soil intensified Zn deficiency in wheat grown at low Zn supply and induced Zn deficiency on soils of near-adequate status (Robson and Snowball, 1989; 1990). The major effect was to decrease the Zn uptake by plants possibly through effects on root morphology and physiology.

4. Prediction Of Zn deficiency by soil analysis

4.1. Principles

An ideal soil test is one which is rapid, reproducible and correlates reliably with responses in plant yield, plant Zn concentration or Zn uptake.

Soil testing for Zn predicts a potential deficiency of Zn before annual crops and pastures are sown. Generally the surface 0-10 or 0-15 cm of soil is sampled for analysis so soil tests may be misleading for deep-rooted perennials if the subsoils contain markedly different levels of Zn (McLaren et al., 1984). Variations in extractable Zn between soil series can be appreciable in both the surface and subsoil horizons. Subsoils are often more impoverished than surface horizons (for example Follett and Lindsay, 1970; McLaren et al., 1984) which is likely to have important implications for Zn nutrition and root function (Nable and Webb, 1993).

When soil tests are being calibrated for a region or for a specific crop, a series of representative soils, both deficient and adequately supplied in Zn are selected. The growth response to several rates of Zn on each soil is then related to the soil test value from which soils are broadly separated into Zn-deficient and Zn-adequate soils. In some calibration studies, the usefulness of derived criteria have been constrained by a lack of responsive soils (Dolar and Keeney, 1971; Junus and Cox, 1987), by using soils with a limited range of chemical and physical properties (Wear and Evans, 1968) or by omitting yield or uptake data (Sedberry et al., 1979). Calibrations derived from glasshouse studies need to be evaluated under field conditions.

Sampling strategies, involving the pattern and intensity of sample collection in the field, are essential to ensure representative, composite samples are obtained for analysis. Guidelines have been described for macronutrients but are reported only rarely for micronutrients (for example Tiller et al., 1975). Sampling strategies should also account for the variability associated with the placement and lateral movement of previously applied Zn. Reduced tillage minimizes the mechanical mixing of soil and means appropriate sampling techniques are required even where Zn is applied as a contaminant (e.g. superphosphate). This is because Zn movement is limited in soils (Brennan and McGrath, 1988). Special precautions must also be taken during field sampling, sample preparation (drying, grinding, sieving and storage) and analysis to prevent contamination (Soltanpour et al., 1976; Haynes and Swift, 1991).

4.2. *Extractants used for predicting the Zn satus of soils*

Many chemical solutions have been evaluated in the search for a universal extractant to predict Zn availability in soils (Lindsay and Cox, 1985; Sims and Johnson, 1991). These include water, neutral salts, weak and strong acids, and chelating agents such as EDTA (ethylene diamine tetra-acetic acid) and DTPA (diethylene triamine penta-acetic acid) in the presence or absence of inorganic salts. Even so, it is widely acknowledged that soil testing remains empirically based, since during extraction the soil particles are in intimate contact with the extracting solution and considerably more Zn is desorbed from the soil than is accumulated by plants. Moreover, plant roots explore only a small fraction of the soil volume that is sampled.

The test should reflect both the initial concentration of the nutrient in solution (intensity, I) and the ability of the soil to replenish the soil solution (capacity, Q). Recent improvements in instrumentation, particularly flameless atomic absorption and anodic stripping voltammetry, have simplified analysis and allowed the detection limit for measuring Zn in soils, especially in the soil solution and dilute salt extractants, to be markedly decreased. Unfortunately, sensitive analytical instruments and stringent contamination controls are pre-requisites for measuring Zn in the soil solution and the techniques are *presently* unsuited to routine analysis.

However, Tiller et al. (1972) showed that estimates of I were the best predictors of Zn concentration and Zn uptake in clover and wheat plants grown on 25 diverse Australian soils. They found that extractions with 1 mol/L $MgCl_2$, 0.05 mol/L $Ca(NO_3)_2$ and 0.05 mol/L $CaCl_2$ and equilibrium Zn concentrations were all highly correlated with Zn uptake and concentration in the plants, but $CaCl_2$-Zn had the highest correlation coefficient in each case. A significant conclusion was that quantity variables alone do not predict the plant's Zn status unless a limited range of soils is used. The ionic strength of these extractants varied from 0.15 to 3, which is substantially higher than the average ionic strength of about 0.005 for Queensland and Western Australian soils (Gillman and Bell, 1978; Dolling and Ritchie, 1985). McGrath et al. (1985) also found that Zn uptake by clover plants grown on 4 diverse soils was better correlated with Zn concentrations in the soil solution than with DTPA-Zn on all but the most calcareous clay soil. More recently, Armour et al. (1990) found that separate estimates of both I (either Zn in soil solution or extracted with 0.002 *M* $CaCl_2$ or 0.01 *M* $CaCl_2$) and Q (Zn extracted with HCl, EDTA or DTPA) were needed to explain most of the variation in the Zn content of navy beans grown in 13 Australian soils. In this study, soil solution Zn was significantly correlated with the Zn extracted by $CaCl_2$.

Recently, there has been a general trend towards using EDTA (Trierweiler and Lindsay, 1969) and DTPA extractants for simultaneously estimating the micronutrient status of soils. In the USA, in particular, research has concentrated on a "universal" soil test, such as Mehlich (Mehlich, 1984) or ammonium bicarbonate-DTPA tests for Zn, Fe, Cu, Mn, and B (Soltanpour, 1991).

With the development of new soil testing procedures, laboratory procedures must be standardized to ensure that even subtle variations do not confound the interpretation of test results. For example, EDTA and sodium EDTA have been used at concentrations from 0.005 to 0.05 *M*, extractant pH values have varied from 7 to 8.6, and salts such as $(NH_4)_2CO_3$ and $Ca(NO_3)_2$ are sometimes included. Soil/solution ratios (1:2 to 1:10) and extraction periods (0.5 to 1 hour) have also varied widely (Viro, 1955; Trierweiler and Lindsay, 1969; Dolar and Keeney, 1971; Fujii and Corey, 1986). Similar variations have been reported for DTPA (Dolar and Keeney, 1971; Soltanpour and Schwab, 1977). Variations in drying and grinding procedures (Soltanpour et al., 1976; Haynes and Swift, 1991), as well as the intensity and speed of shaking during extraction also affect the results of analyses (Soltanpour et al., 1976; Leggett and Argyle, 1983).

4.3. Soil test criteria

Average critical concentrations for major extractants are 0.5 to 1.0 mg Zn/kg for DTPA, 1.1 mg Zn/mg for Mehlich-1, and 1.0 to 5 mg Zn/kg for 0.1 *M* HCl (Lindsay and Cox, 1985; Cox, 1987). For DTPA alone, the reported range in critical concentrations is 0.1 to 1.0 mg Zn/kg (Cox, 1987; Brennan and Gartrell, 1990).

Critical soil test values vary with the extractants used. For example, in a glasshouse experiment the critical concentration was 0.06 mg Zn/kg for ammonium acetate but 0.59 mg Zn/kg for EDTA (Madziva, 1981). The critical value for a particular extractant also depends on soil type. For example, Brennan and Gartrell (1990) reported a critical DTPA concentration for subterranean clover of 0.13 mg Zn/mg for sands and 0.55 mg Zn/kg for clay soils.

Evaluating other soil properties (Table 1) is commonly used to improve the correlation between a soil test and plant growth and to extend the range of soils over which a test is useful. Although at present it is not possible to specify key parameters, pH, organic carbon and clay have an important influence on the availability of Zn for many plant species and soil types.

An alternative approach is to examine extractants which estimate both I and Q. The selection of 2 'standard' soil tests such as DTPA or EDTA to measure quantity and 0.002 *M* $CaCl_2$ to measure intensity, would reduce the number of Zn soil tests needed and simplify comparisons between regions.

5. Diagnosis of Zn deficiency by plant analysis

As soil tests for Zn are often highly soil and species specific, analysing plant tissue

Table 1. Extractants used and additional soil properties measured to improve the relationship between the crop parameter and the soil test

Crop	Crop parameter	Extractant used	R^2	Soil properties included in multivariate analysis to improve prediction[a]	Revised R^2	Reference
maize, beans, others	yield	0.1 *M* HC1	-	'titratable' alkalinity	-	Nelson et al. (1959)
maize	Zn uptake	EDTA	-	$HCl-H_2SO_4$ Zn and P, pH[b]	-	Alley et al. (1972)
maize	Zn concentration	DTPA	0.36	pH	0.76	Haq and Miller (1972)
oats	Zn uptake	0.1 *M* HC1	0.31	pH, OC, silt and clay	0.70	Osiname et al. (1973)
oats	Zn uptake	EDTA	0.33	OC	0.70	Osiname et al. (1973)
maize	Zn concentration	0.1 M HC1	0.46	Bray-P, pH	0.67	Peaslee (1980)
maize	relative yield	DTPA	0.32	CEC, EC, P, BD, OC	0.87	Singh and Takkar (1981)
maize	relative yield	0.1 *M* HC1	0.38	CEC, EC, P, BD, OC	0.87	Singh and Takkar (1981)
sorghum	Zn uptake	$MgCl_2$	0.15	pH, OC, silt, clay	0.36	Chude and Gabriel (1984)
subterranean clover	critical DTPA concentration	DTPA	0.62	pH, clay	0.93	Brennan and Gartrell (1990)
wheat	critical DTPA concentration	DTPA	0.63	pH, OC, clay	0.88	Brennan (1992)

a OC organic carbon, CEC cation exchange capacity, BD bulk density, EC electrical conductivity
b alternative soil properties instead of EDTA-Zn

is an alternative means of diagnosing Zn deficiency.

Diagnosis estimates the Zn status of the plant at the time of sampling. Plant analysis needs to be related to specific stages of plant development and specific plant organs or tissues. For example, Bates (1971) suggested sampling when symptoms first appear and sampling tissues of a similar physiological age.

There are problems with diagnosis when there are substantial growth differences among the plants sampled. For example, Viets et al. (1954) found little difference in the Zn concentration in leaves of crops growing on Zn-deficient soils (15.4 mg Zn/ha average) and the concentration in the same crops growing on Zn-adequate soils (16.8 mg Zn/kg). Many published values for concentrations of Zn in plant species have been established by comparing deficient and adequately supplied plants.

Another approach is to use the critical concentration; that concentration just deficient for maximum plant growth (Bates, 1971). For most studies, the critical deficiency concentration (CDC; Fig 1) has been defined as that concentration where there is a 10% reduction in the yield of the plant (Ulrich and Hills, 1967).

CDC's are usually determined in glasshouse experiments using Zn-deficient soil, with several rates of applied Zn (all other nutrients non-limiting) and frequent harvests. The relationship between dry matter production and Zn concentrations need to be confirmed in the field.

Principles of plant analysis and sampling procedures have been reviewed elsewhere (Smith, 1986; Jones, 1991). The plant part used for analysis must be easily identifiable, easy to collect, and be related to the mobility and redistribution of the nutrient within the plant (see Chapter 6). The mobility of Zn varies strongly with the adequacy of its supply. This behaviour leads to anomalies in the relationships among Zn supply, Zn concentration in plants and yield. We shall consider the use of Zn concentrations in whole shoots, young tissue and grain as indicators of the Zn status of plants.

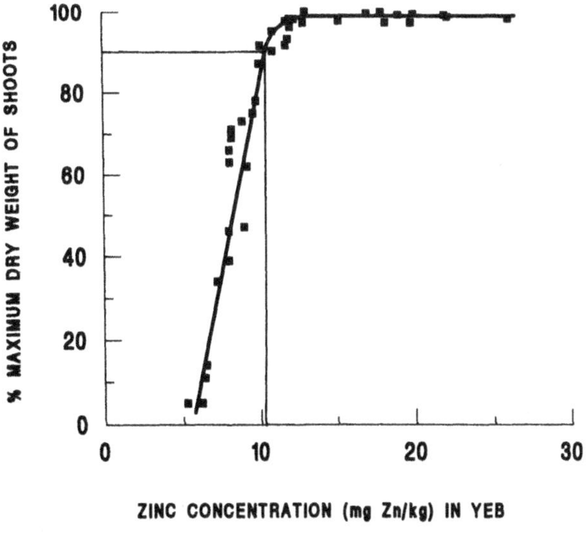

Figure 1. Relationship between the concentration of Zn in the YEB of wheat during tillering and the percentage maximum dry weight of shoots. (adapted from Brennan, 1992).

5.1. *Whole shoots*

Jones (1991) suggested that the average concentration of Zn in plants is 20 mg Zn/kg and that the CDC is about 15 mg/kg. However, plant species differ in their sensitivity to Zn deficiency (for example Jones, 1991; Martens and Westermann, 1991). Therefore critical Zn concentrations need to be established for each plant species. In many situations, Zn concentrations in whole shoots have been unreliable indicators of the Zn status of the plant because of the way Zn moves in plants. Plant age and the anomalous relationship between growth and Zn concentration in the whole tops of very Zn-deficient plants should be considered when using whole tops to diagnose the Zn status of plants.

(i) Plant age: Using whole shoots to diagnose the Zn status of plants is inappropriate because the critical Zn concentrations fall as the plant ages. For example, Reuter et al. (1982a) found that very young, stunted shoots of subterranean clover (which were only 5% of maximum growth) contained 15 mg Zn/kg, but at later harvests, shoots which yielded 95% of maximum yield contained 12 mg Zn/kg. This decline in critical concentration is related to the changing distribution of dry weight and nutrient content among plant parts with time. Stems have lower concentrations of Zn than leaves (Ohki, 1977) so the critical concentration in whole shoots declines when stems comprise a larger proportion of the total dry weight of the plant.

(ii) C-shaped curve: The 'C-shaped' relationship between the concentration of Zn in whole shoots and yield is known as the Piper-Steenbjerg effect (Piper, 1942; Steenbjerg, 1951) and the reasons for this relationship are discussed by Smith (1986). In this portion of the response curve increases in yield are associated with decreases in Zn concentration in very deficient plants. For specific examples of the 'C-shaped' relationship in tropical and subtropical legumes see figure 3 of Andrew et al. (1981). This relationship can confound interpretations of plant analyses.

Although Ulrich and Hills (1967) suggested that problems associated with the (C-shaped) curve for diagnosis may be minimized by sampling when the symptoms *first* appear. Andrew et al. (1981) were reluctant to quote critical concentrations of Zn in whole shoots.

5.2. *Young leaves*

Some of the problems associated with using whole shoots to diagnose Zn deficiency of plants can be overcome by selecting young tissues.

Young growth, often the youngest emerged leaf blade (YEB) of cereals, and the youngest open blade (YOB) of subterranean clover are the usual tissues selected for defining the Zn status of these plants (Table 2). Young, recently matured tissue (for example the YEB or YOB) has been used successfully for assessing the Zn status of a wide range of plants, for example sugarbeet (Rosell and Ulrich, 1964), cotton (Ohki, 1984), soybeans (Ohki, 1977), subterranean clover (Reuter et al., 1982a), pine trees (McGrath and Robson, 1984), apples (Watkins, 1982), peanuts (Bell et al., 1990) and wheat (Brennan, 1992; Riley et al., 1992). There are two reasons for analysing Zn in young tissue. Firstly, symptoms of Zn deficiency characteristically develop in young rather than *old* leaves, reflecting the mobility of Zn within the plant. Second, the concentration of Zn in the youngest leaves is more stable at most stages of their

Table 2. The critical concentrations of Zn in the young tissue of various plant species (after Reuter and Robinson, 1986).

Plant species	Growth stage	Plant part	Critical Zn
Clover (*Trifolium subterraneum*)	21-55 DAE	YOL	12-14
Cassava (*Manihot esculenta*)	63 DAS	YMB	30
Cotton (*Gossypium hirsutum*)	37 DAS	YMB	11
Lupins (*Lupinus angustifolius*)	Veg	YMB	12-14
Peanut (*Arachis hypogaea*)	42 DAS	YMB	20
	EP	YFEL	8-10
Sorghum (*Sorghum vulgare*)		blade 1	10
		blade 5	24
Sugarbeet (*Beta vulgaris*)	83DAS	YMB	9
		YEB	16
Wheat (*Triticum aestivum*)	Tillering	YEB	11
	Post anthesis	YEB	7
Navy bean (*Phaseolus vulgaris*)	31 DAS	YFEL	11*

DAE, days after emergence; DAS, days after seeding; Veg, vegetative growth; EP, early pegging; YOL, youngest open leaf; YMB, youngest mature blade; YFEL, youngest fully emerged leaf; YEB, youngest emerged blade;
* from Armour et al., 1990

development (Reuter et al., 1982a) giving stability to the diagnostic test.

C-shaped relationships between yield and Zn concentrations have also been reported for the *young tissue* (for example sugar beet leaves (Rosell and Ulrich, 1964 and apical growth of subterranean clover(Reuter et al., 1982b)).

For perennial tree crops, the Zn status is often determined by leaf analysis (Leece, 1976; Bould, 1983). In horticultural and forestry crops diagnostic values have often been established by comparing Zn concentrations in unproductive trees, which are usually identified by visual symptoms, with those in adequate or productive trees (Leece, 1976). For tree crops, recently matured leaves associated with an active growing phase of the tree were sampled (Leece, 1976). However, Bould (1983) suggested that analysing mature leaves for immobile nutrients such as Zn is unsatisfactory and suggested that young leaves should be used (for example, in apples Watkins, 1982). Considerable literature on the requirements of temperate woody perennials for Zn and leaf Zn concentrations is available (Robinson, 1986).

5.3. Grain

The concentration of Zn in grain can also be used to assess the adequacy of Zn, but unfortunately it only diagnoses past problems. However, it can be used to identify sites where future crops will respond to Zn fertilizer and regional trends in the Zn status of crops.

Viets (1966) suggested the critical value of Zn in grain is 15 mg Zn/kg. However, recent research showed that the critical concentrations in grain varied with species (for example 10 mg Zn/kg in wheat (Riley et al., 1992) and 43 mg Zn/kg in soybean (Rashid and Fox, 1992)).

5.4. *Biochemical techniques*

Enzyme activities appear more responsive to micronutrient deficiency than the concentration of Zn in young leaves. The activities of 3 enzymes have been used: (i) ribonuclease in rice and maize (Dwivedi and Takkar, 1974) and in orchard trees (Kessler, 1961), (ii) carbonic anhydrase in citrus (Bar-Akiva and Lavon, 1969) and in wheat, maize and mustard leaf tissue (Dwivedi and Randhawa, 1974) and (iii) aldolase in onions (O'Sullivan, 1970). However, Bar-Akiva et al. (1971) failed to establish a specific Zn requirement for the aldolase enzyme in lemon leaves.

Although enzyme activities are reliable as tests and the methods are relatively simple and rapid, they are not used widely for diagnosis. Possible reasons for this include firstly, the complications in enzyme patterns as a result of environmental and growth factors, and second, the problems associated with sampling and preserving tissue from field-grown plants for testing.

5.5. *Nutrient balances*

Ratios of P/Zn, Fe/Zn and Mn/Zn have been suggested as methods of diagnosing Zn deficiency in crops (Millikan, 1963; Nair and Probhat, 1977; Nambiar and Motiramani, 1981; Schropp and Marschner, 1977). However, it appears that the ratios vary widely with species, genotype and probably also with environmental conditions. Other researchers have found no relationship between P/Zn ratios in corn and Zn deficiency (for example Giordano and Mortvedt, 1969; Reuter, 1980).

Zn is thought to help maintain root membrane integrity (Welch et al., 1982) and thus Zn deficiency may cause ions to accumulate in plants. Accumulations of a range of ions in Zn-deficient plants suggest that nutrient/Zn ratios would be unreliable for diagnosing Zn deficiency. Unless the interactions between Zn and other nutrients affect the *utilization* of Zn within the plant it is unlikely that the relationship between Zn concentration in plant tissue and growth will be changed.

Another approach is the concept of an optimal balance between nutrient levels in plant tissues (Bates, 1971). The Diagnostic and Recommendation Integrated System (DRIS, Beaufils, 1973) of interpretation is based on comparing calculated elemental ratio indices with established norms. The DRIS system is not independent of either time of sampling or location (Beverly et al., 1986), hence DRIS is is frequently misleading particularly for Zn which has a small database. Modified DRIS failed to detect 8 of 11 Zn deficiencies of soybean (Hallmark et al., 1989) and it seems doubtful that the DRIS system will be used extensively to diagnose Zn deficiency.

5.6. *Variation between plant species and cultivars*

Differences between plant species in internal nutrient requirement do exist, as observed by the range of critical concentrations of Zn determined for various species (Reuter and Robinson, 1986).

Cultivar differences in the Zn nutrition of different species have been illustrated:

navy bean (Polson and Adams, 1970); corn (Safaya and Gupta, 1979) and wheat (Shukla and Raj, 1974). However, there has been little experimental calibration of critical Zn concentrations for cultivars of the same species.

5.7. *Prognosis of Zn Deficiency*

Prognosis implies that the potential for Zn deficiency can be predicted prior to the appearance of symptoms or loss in plant production. The aim is to predict the absence of a future Zn deficiency. Prognostic values of Zn have not been established for many crop and pasture species.

For example, Riley et al. (1992) have shown in field experiments that Zn deficiency in wheat is unlikely if Zn concentrations in the YEB are 16.5 mg Zn/kg (early tillering) and 7.0 mg Zn/kg at the kernel milk ripe stage. Other attempts at relating Zn concentrations in leaf tissue to subsequent grain yield at maturity have often failed as the critical prognostic values vary, depending on the tissue chosen and growth stage at sampling.

References:

Adams D B and Piper C S 1944 The use of zinc for flax - a progress report for growers. J. Agric. (South Australia) 47, 422-426.

Alley M M, Martens D C, Schnappinger M G and Hawkins G W 1972 Field calibration soil tests for available zinc. Soil Sci. Soc. Am. Proc. 36, 621-624.

Andrew C S, Johnson A D and Haydock K P 1981 The diagnosis of zinc deficiency and effect of zinc on the growth and chemical composition of some tropical and sub-tropical legumes. Commun. Soil Sci. Plant Anal. 12, 1-18.

Armour J D, Ritchie G S P and Robson A D 1989 Changes in time in the availability of soil applied zinc to navy beans and in the chemical extraction of zinc from soil. Aust. J. Soil Res. 27, 699-710.

Armour J D, Robson A D and Ritchie G S P 1990 Prediction of zinc deficiency in navy beans by soil and plant analysis. Aust. J. Exp. Agric. 30, 557-563.

Asher C J, Edwards D G and Howeller R H 1980 Nutritional disorders of cassava (*Manihot esculenta* Crantz.) (Department of Agriculture, University of Queensland, Brisbane.)

Bar-Akiva A and Lavon R 1969 Carbonic anhydrase activity as an indicator of zinc deficiency in citrus leaves. J. Hort. Sci. 44, 359-362.

Bar-Akiva A, Sagiv J and Hasdai D 1971 Effect of mineral deficiencies and other co-factors on the aldolase enzyme activity of citrus leaves. Physiol. Plantarum 25, 386-390.

Bates T E 1971 Factors affecting critical nutrient concentrations in plants and their evaluation. A review. Soil Sci. 112, 116-130.

Beaufils E R 1973 Diagnosis and recommendation integrated system (DRIS). Soil Sci. Bull. 1, Univ. of Natal, Natal, South Africa.

Bell R W, Kirk G, Plaskell D and Loneragan J F 1990 Diagnosis of zinc deficiency in peanut (*Arachis hypogaea* L.) by plant analysis. Commun. Soil Sci. Plant Anal. 21, 273-285.

Beverly R B, Sumner M E, Letzsch W S and Plank C O 1986 Foliar diagnosis of soybean by DRIS. Commun. Soil Sci. Plant Anal. 17, 237-256.

Bould C 1983 Methods of diagnosing nutrient disorders in plants *In* Diagnosis of Mineral Disorders in Plants Vol 1: Principles pp 111-136. Eds. C Bould, E J Hewitt, P Needham HMSO, London.

Brennan R F 1992 The relationship between critical concentration of DTPA - extractable zinc from the soil for wheat production and properties of south western Australian soils responsive to applied zinc. Commun. Soil Sci. Plant Anal. 23, 747-759.

Brennan R F and Gartrell J W 1990 Reactions of zinc with soil affecting its availability to subterranean clover. I. The relationship between critical concentrations of extractable zinc and properties of Australian soils responsive to applied zinc. Aust. J. Soil Res. 28, 293-302

Brennan R F and McGrath J F 1988 The vertical movement of zinc on sandy soils in Southern Western Australia. Aust. J. Soil Res. 26, 211-216.

Carroll M D and Loneragan J F 1968 Response of plant species to concentrations of zinc in solution. I Growth and zinc content of plants. Aust. J. Agric. Res. 19, 859-868.

Chaudhry F M and Loneragan J F 1972 Zinc absorption by wheat seedlings. I Inhibition by macronutrient ions in short term studies and its relevance to long term zinc nutrition. Soil Sci. Soc. Am. Proc. 36, 323-327.

Chude V O and Gabriel O O 1984 A comparison of various extractants for the estimation of extractable zinc in cacao-growing soils of south-western Nigeria. J. Sci. Fd. Agric. 35, 609-612.

Cox F R 1987 Micronutrient soil tests: correlation and calibration. *In* Soil Testing: Sampling, Correlation, Calibration, and Interpretation. Ed. J R Brown. pp 97-117. SSPA Spec. Publ. 21. ASA, CSSA, and SSSA, Madison, USA.

Dolar S G and Keeney D R 1971 Availability of Cu, Zn, and Mn in soils II. Chemical extractability. J. Sci. Fd. Agric. 22, 279-282.

Dolling P J and Ritchie G S P 1985 Estimates of ionic strength and the determination of pH in West Australian soils. Aust. J. Soil Res. 23, 309-314.

Dwivedi R S and Randhawa N S 1974 Evaluation of a rapid test for hidden hunger of zinc in plants Plant Soil 40, 445-451.

Dwivedi R S and Takkar P N 1974 Ribonuclease activity as an index of hidden hunger of zinc in crops. Plant Soil 40, 173-181.

Follett R H and Lindsay W L 1970 Profile distribution of zinc, iron, manganese and copper in Colorado soils. Colorada State Univ. Exp. Stat. Bulletin 110.

Fujii R and Corey R B 1986 Estimation of isotopically exchangeable cadmium and zinc in soils. Soil Sci. Soc. Am. J. 50, 306-308.

Gillman G P and Bell L C 1978 Solution studies on weathered soils from tropical north Queensland. Aust. J. Soil Res. 16, 67-77.

Giordano P M and Mortvedt J J 1969 Response of several corn hybrids to level of water soluble zinc in fertilizers. Soil Sci. Soc. Am. Proc. 33, 145-148.

Grundon N J 1987 Hungry crops: a guide to nutrient deficiencies in field crops. (Queensland Department of Primary Industries, Brisbane, Australia).

Grunes D L, Boawn L C, Carlson C W and Viets F G 1961 Zinc deficiency of corn and potatoes as related to soil and plant analyses. Agron. J. 53, 68-71.

Hallmark W B, Beverly R B, Parker M B, Adams J F, Boswell F C, Ohki K, Shuman L M and Wilson D O 1989 Evaluation of soybean zinc and manganese requirements by the M-DRIS and sufficiency range methods. Agron. J. 81, 770-776.

Haq A U and Miller M H 1972 Prediction of available soil Zn, Cu, and Mn using chemical extractants. Agron. J. 64, 779-782.

Haynes R J and Swift R S 1991 Concentrations of extractable Cu, Zn, Fe and Mn in a group of soils as influenced by air- and oven-drying. Geoderma 49, 319-333.

Jones J 1991 Plant tissue analysis *In* Micronutrients in Agriculture, Second Edition. Eds. J J Mortvedt, F R Cox, L M Shuman and R M Welch. pp 477-521. Soil Sci. Soc. Am. Inc., Madison, Wisconsin, USA.

Junus M A and Cox F R 1987 A zinc soil test calibration based upon Mehlich 3 extractable zinc, pH and cation exchange capacity. Soil Sci. Soc. Am. J. 51, 678-683.

Kessler B 1961 Ribonuclease as a guide for the determination of zinc deficiency in orchard trees. *In* Plant Analysis and Fertilizer Problems. Ed. W Reuther. pp 314-322. Am. Inst. Biol. Sci. Washington DC.

Kleinig C R and Loveday J 1962 Responses of pasture legumes to zinc on calcareous soils in the Riverina, New South Wales. Aust. J. Exp. Agric. Anim. Husb. 2, 228-233.

Kubota J and Allaway W H 1972 Geographic distribution of trace element problems. *In* Micronutrients in Agriculture. Eds. J J Mortvedt, P M Giordano, W L Lindsay. pp 525-554. Soil Sci. Soc. Am., Wisconsin.

Leece D R 1976 Diagnosis of nutritional disorders of fruit trees by leaf and soil analyses and biochemical indices. J. Aust. Inst. Agric. Sci. 42, 7-19.

Leggett G E and Argyle D P 1983 The DTPA-extractable iron, manganese, copper and zinc from neutral and calcareous soils dried under different conditions. Soil Sci. Soc. Am. Proc. 47, 518-522.

Lindsay W L 1972 Zinc in soils and plant nutrition. Adv. Agron. 24, 147-188.

Lindsay W L 1991 Inorganic equilibria affecting micronutrients in soils. *In* Micronutrients in Agriculture, Second Edition. Eds. J J Mortvedt, F R Cox, L M Shuman and R M Welch. pp 89-144. Soil Sci. Soc. Am. Inc., Madison, USA.

Lindsay W L and Cox F R 1985 Micronutrient soil testing for the tropics. Fert. Res. 7, 169-200.

Lopes A S 1980 Micronutrients in soils of the tropics as constraints to food production. *In* Priorities for Alleviating Soil-Related Constraints to Food Production in the Tropics. pp 277-298. Intern. Rice Res. Inst.

and Cornell Univ.

McGrath J F and Robson A D 1984 The distribution of zinc and the diagnosis of zinc deficiency in seedlings of *Pinus radiata*, D. Don. Aust. For. Res. 14, 175-186

McGrath S P, Sanders J R and Adams J M 1985 Comparison of soil solution and chemical extractants to estiamte metal availability to plants. J. Sci. Fd. Agric. 36, 532-533.

McLaren R G, Swift R S and Quin B F 1984 EDTA-extractable copper, zinc, and manganese in soils of the Canterbury Plains. N.Z. J. Agric. Res. 27, 207-217.

Madziva T J T 1981 Methods of measuring available zinc in Zimbabwean soil. Zimbabwe J. Agric. Res. 19, 83-90.

Marschner, H. and Cakmak I 1989 High light intensity enhances chlorosis and necrosis in leaves of zinc, potassium and magnesium deficient bean (*Phaseolus vulgaris*) plants. J. Plant Physiol. 134, 308-315.

Martens D C and Westermann D T 1991 Fertilizer applications for correcting micronutrient deficiencies. In Micronutrients in Agriculture, Second Edition. Eds. J J Mortvedt, F R Cox, L M Shuman and R M Welch. pp 549-592. Soil Sci. Soc. Am. Inc., Madison, Wisconsin, USA.

Mehlich A 1984 Mehlich 3 soil test extractant. A modification of the Mehlich - 2 extractant. Commun. Soil Sci. Plant Anal. 15, 1409-1416.

Millikan C R 1963 Effects of different levels of zinc and phosphorus on the growth of subterranean clover (*Trifolium subterraneum* L.). Aust. J. Agric. Res 14, 180-205.

Moraghan J T and Mascagni H J 1991 Environmental and soil factors affecting micronutrient deficiencies and toxicities. *In* Micronutrients in Agriculture, Second Edition. Eds. J J Mortvedt, F R Cox, L M Shuman and R M Welch. pp 371-425. Soil Sci. Soc. Am. Inc., Madison, USA.

Mortvedt J J and Kelsoe J J 1988 Response of corn to zinc applied with banded acid-type fertilizers and ammonium polyphosphates. J. Fert. Issues 6, 83-88.

Nable RO and Webb M J 1993 New evidence that zinc is required throughout the root zone for optimal development, grain production and water-use in wheat. Plant Soil *(in press)*.

Nair K P P and Probhat G 1977 Differential response of tropical maize genotypes to zinc and manganese nutrition. Plant Soil 47, 149-159.

Nambiar K K M and Motiramani D P 1981 Tissue Fe/Zn ratio as a diagnostic tool for prediction of Zn deficiency in crop plants. I Critical Fe/Zn ratio in maize plants. Plant Soil 60, 357-367.

Nelson J L, Boawn L C and Viets F G 1959 A method of assessing zinc status
of soils using acid extractable zinc and (titratable alkalinity) values. Soil Sci. 88, 275-283.

Ohki K 1977 Critical zinc levels related to early growth and development of determinate soybeans. Agron. J. 69, 969-974.

Ohki K 1984 Lower and upper critical zinc levels in relation to cotton growth and development. Physiol. Plantarum 35, 96-110.

Osiname O A, Schulte E E and Corey R B 1973 Soil tests for available copper and zinc in soils of western Nigeria. J. Sci. Fd. Agric. 24, 1341-1349.

O'Sullivan M 1970 Aldolase activity in plants as an indicator of zinc deficiency. J. Sci. Fd. Agric. 21, 607-609.

Peaslee D E 1980 Effect of extractable zinc, phosphorus, and soil pH values on zinc concentration in leaves of field-grown corn. Commun. Soil Sci. Plant Anal. 11, 417-425.

Piper C S 1942 Investigations on copper deficiency in plants. J. Agric. Sci. 32, 143-183.

Polson D E and Adams M W 1970 Differential response of navy beans (*Phaseolus vulgaris* L.) to zinc. I Differential growth and elemental composition to excessive zinc levels. Agron. J. 62, 557-560.

Rashid A and Fox R L 1992 Evaluating internal zinc requirements of grain crops by seed analysis. Agron. J. 84, 469-474.

Reuter D J 1980 Distribution of copper and zinc in subterranean clover in relation to deficiency diagnosis. PhD thesis, Murdoch Univ., Western Australia.

Reuter D J, Loneragan J F, Robson A D and Plaskett D 1982a Zinc in subterranean clover (*Trifolium subterraneum* L. cv. Seaton Park). I Effects of zinc supply on distribution of zinc and dry weight among plant parts. Aust. J. Agric. Res. 33, 989-999.

Reuter D J, Loneragan J F, Robson A D and Plaskett D 1982b Zinc in subterranean clover (*Trifolium subterraneum* L. cv. Seaton Park). II. Effects of phosphorus supply on the relationship between zinc concentrations in plant parts and yield. Aust. J. Agric. Res. 33, 1001-1008.

Reuter D J and Robinson J B 1986 Plant analysis: An interpretation manual. Inkata Press, Ltd, Melbourne, Australia.

Riley M M, Gartrell J W, Brennan R F, Hamblin J and Coates P 1992 Zinc deficiency in wheat and lupins in

Western Australia is affected by the source of phosphate fertilizers. Aust. J. Exp. Agric. 32, 455-463.

Robinson J B 1986 Fruits, vines and nuts *In* Plant Analysis: An Interpretation Manual. Eds. D J Reuter and J B Robinson. pp 120-147. Inkata Press, Melbourne, Australia.

Robson A D and Snowball K 1989 The effect of 2- (4-2', 4'-dichlorophenoxy-phenoxy) - methyl propanoate on the uptake and utilization of zinc by wheat. Aust. J. Agric. Res. 40, 981-990.

Robson A D and Snowball K 1990 The effect of chlorsulfuron on the uptake and utilization of copper and zinc in wheat. Aust. J. Agric. Res. 41, 19-28.

Rosell R A and Ulrich A 1964 Critical zinc concentrations and leaf minerals of sugar beet leaves. Soil Sci. 97, 152-167.

Rossiter R C 1951 Studies on the nutrition of pasture plants in the South-West of Western Australia. 2. Visual symptoms of mineral deficiencies in the Dwalganup strain of *Trifolium subterraneum* L. Aust. J. Agric Res. 2, 14-23.

Safaya N M and Gupta A P 1979 Differential susceptibility of corn cultivars to zinc deficiency. Soil Sci Soc. Am. J. 40, 719-722.

Schropp A and Marschner H 1977 Effect of high phosphate fertilization on growth rate, zinc content and phosphorus zinc ratio in grapevines *Vitis vinifera*. Z. Planzenernachr. Bodenkd. 140, 515-530.

Sedberry J E, Miller B J and Said M B 1979 An evaluation of chemical methods for extracting zinc from soils. Commun. Soil Sci. Plant Anal. 10, 689-701.

Shukla U C and Raj H 1974 Influence of genetic variability on zinc response in wheat. Soil Sci. Soc. Am. Proc. 38, 477-479.

Sims J T and Johnson G V 1991 Micronutrient soil tests *In* Micronutrients in Agriculture, Second Edition Eds. J J Mortvedt, F R Cox, L M Shuman and R M Welch. pp 427-476.

Soil Sci. Soc. Am. Inc., Madison, USA.

Singh H J and Takkar P N 1981 Evaluation of efficient soil test methods for Zn and their critical values in salt-affected soils for rice. Commun. Soil Sci. Plant Anal. 12, 383-406.

Smith F W 1986 Interpretation of plant analysis concepts and principles. *In* Plant Analysis: An Interpretation Manual. Eds. D J Reuter and J B Robinson. pp 1-12. Inkata Press, Melbourne, Australia.

Snowball K and Robson A D 1983 Symptoms of Nutrient Deficiencies : Subterranean Clover and Wheat. University of Western Australia Press, Nedlands, Western Australia.

Snowball K and Robson A D 1986 Symptoms of Nutrient Deficiencies: Lupins. University of Western Australia Press, Nedlands, Western Australia.

Soltanpour P N 1991 Determination of nutrient availability and elemental toxicity by AB-DTPA soil tests and ICPS. Adv. Soil Sci. 16, 165-190.

Soltanpour P N, Khan A and Lindsay W L 1976 Factors affecting DTPA-extractable Zn, Fe, Mn and Cu from soils. Commun. Soil Sci. Plant Anal. 7, 797-821.

Soltanpour P N and Schwab A P 1977 A new soil test for simultaneous extraction of macro- and micro-nutrients in alkaline soils. Commun. Soil Sci. Plant Anal. 8, 195-207.

Steenbjerg F 1951 Yield curves and chemical plant analyses. Plant Soil 3, 97-107.

Thompson J P 1987 Decline of vesicular-arbuscular mycorrhizae in long fallow disorder of field crops and its expression in phosphorus deficiency of sunflower. Aust. J. Agric. Res. 38, 847-867.

Tiller K G, Honeysett J L and DeVries M P C 1972 Soil zinc and its uptake by plants II. Soil chemistry in relation to prediction of availability. Aust. J. Soil Res. 10, 165-182.

Tiller K G, Cunningham R B, Cartwright B, Clayton P M and Merry R H 1975 Sampling and analytical errors in the estimation of copper, calcium, zinc and lead in surface soils. CSIRO Div. of Soils Report No. 3/1975.

Trierweiler J F and Lindsay W L 1969 EDTA-ammonium carbonate soil test for zinc. Soil Sci. Soc. Am. Proc. 33, 49-53.

Ulrich A and Hills F J 1967 Principles and practices of plant analysis *In* Soil Testing and Plant Analysis Part II - Plant Analysis. Ed. R L Westerman. pp 11-24. Soil Sci. Soc. Am. Inc., Madison, USA.

Viets F G 1966 Zinc deficiency in the soil plant system *In* Zinc Metabolism. Ed. A S Prasad. pp 90-127. C C Thomas, Springfield, Illinois.

Viets F G, Boawn L C and Crawford C L 1954 Zinc contents and deficiency symptoms of 26 crops grown on a deficient soil. Soil Sci. 78, 305-316.

Viets F G, Boawn L C and Crawford C L 1957 The effect of nitrogen and types of nitrogen carrier on plant uptake of indigenous and applied zinc. Soil Sci. Soc. Am. Proc. 21, 197-201.

Viro P J 1955 Use of ethylene diamine tetra-acetic acid in soil analysis. Soil Sci.79, 459-465.

Watkins P A 1982 Copper and zinc nutrition of Granny Smith apple plants. M. Sci. thesis. The University of Western Australia.

Wear J I and Evans C E 1968 Relationship of zinc uptake by corn and sorghum to soil zinc measured by three extractants. Soil Sci. Soc. Am. Proc. 32, 543-546.

Welch R M, Webb M J and Loneragan J F 1982 Zinc in membrane function and its role in P toxicity. pp 710-715. *In* Plant Nutrition 1982. Ed. A Scaife. Proc. Int. Plant Nutr. Coll. 9th Warwick, England. Comm. Agric. Bureaux, Farnham, United Kingdom.

Chapter 13.

Zinc Concentrations and Forms in Plants for Humans and Animals

ROSS M. WELCH

1. Abstract

Foods derived form plants are poorer sources of Zn for humans and monogastric animals when compared to those from animals because they can contain substances which interfere with the absorption and/or utilization (i.e., bioavailability) of Zn. Examples of these substances include phytic acid and certain types of fibre, especially fibre from whole cereal grains. However, these substances are also known to play important roles in either the life cycle of higher plants or possibly, in the prevention of several human diseases including heart disease and certain forms of cancer. Therefore, it is not wise to reduce the content of these substances in food crops or in diets without a more thorough understanding of their role(s) in plant growth or human health. Indeed, current recommended dietary goals for people in the United States advise that citizens double their daily consumption of foods high in fibre by eating twice as much whole cereal grain and legume seed products. Low molecular weight, soluble, anionic, Zn complexes comprise the majority of the naturally occurring forms of Zn in edible portions of food crops. Animal products contain higher concentrations of some substances which promote Zn bioavailability than do plant food products. Thus, it may be more desirable to increase the concentration of promoters of Zn bioavailability in plant foods than to reduce the level of antinutritive substances which interfere with the bioavialability of Zn. The nutritional quality of food crops, with respect to the Zn concentration in edible portions, can be increased significantly by applying available forms of Zn fertilizer to soils at levels in excess of those required for optimum plant growth.

2. Overview

Our current understanding of the importance of Zn to agricultural production systems and to human health and well-being has taken nearly 150 years of scientific investigation to develop. However, much still remains to be learned, especially with respect to the value of Zn in foods of plant origin in meeting the nutritional dietary requirements of people.

Historically, in 1869 Zn was first discovered to be an essential element for the fungus, *Aspergillus niger* by Raulin (1869). Nearly sixty years elapsed when in 1926 Somner and Lipman (1926) reported Zn to be essential for higher plants. In the 1930s, Zn deficiencies in peaches (*Prunus persica* L.), citrus, and other field crops were first identified (Chapman, 1966). Todd and his associates (1934), reported that Zn was essential for higher animals in 1934. Another 20 years passed before Tucker and Salmon (1955) reported that Zn prevented and cured parakeratosis in pigs (*Sus* spp.) and O'Dell and Savage (1957) reported Zn deficiency in poultry. During this same period, Zn deficiency in cattle (*Bos taurus*) under natural conditions was also reported. In the 1960's, Prasad and his co-workers were the first to discover nutritional Zn deficiency in humans (O'Dell, 1984). Since than, acrodermatitis enteropathica has been shown to be a genetic disease of people who lack the biochemical mechanisms required to absorb sufficient quantities of Zn from normal diets, and marginal Zn deficiency has been identified in various human population groups worldwide (Hambidge et al., 1986; Reinbold, 1988; Sandstead, 1984). Additionally, nonexperimental Zn deficiency has been reported in a number of countries in pigs and sheep (*Ovis aries*) (Hambidge et al., 1986; Luecke, 1984).

This review will concentrate on the importance of plant foods as sources of Zn to meet the Zn requirement of humans in developed countries, but much of what is reviewed can also be extended to populations in developing countries. The review focuses on the major forms of Zn in edible plant parts as they relate to the nutritional health of humans and to factors in plant foods that interfere with or promote the absorption of Zn by humans. For information concerning nutritional aspects of plants as sources of Zn for ruminants refer to the chapter on Zn requirements of grazing ruminants by C. L. White in this publication.

While extreme or absolute Zn deficiency is uncommon in animals and humans except under experimental conditions, currently there is convincing evidence that marginal Zn deficiency in various human population groups and in livestock is not a rare occurrence. Indeed, marginal Zn deficiency occurs within free-living human populations within North America and South America, Africa, United Kingdom, Thailand, Sweden, Iran, Turkey, China, and former Yugoslavia (Hambidge et al., 1986; Sandstead, 1984; Sandstead, 1991). Marginal Zn deficiency in humans and animals can result in several undesirable consequences including dermatitis, reduced growth rates of infants and children, impaired immune function, diminished learning ability, delayed male sexual development, prolonged wound healing, taste dysfunction, impaired dark adaptation, anorexia, diarrhoea, depressed mood, abnormal fetal development, fetal growth failure, prematurity, perinatal deaths, and difficulty in parturition (Hambidge et al., 1986; Luecke, 1984; McClain et al., 1985; Sandstead, 1984; Underwood, 1971).

Numerous non-dietary factors, including both genetic and acquired disorders, contribute to the occurrence of Zn deficiency in humans. The primary dietary factors thought to be responsible for Zn deficiency are low dietary intake and, more importantly, various dietary factors which interfere with the absorption and/or utilization of Zn by humans (Hambidge et al., 1986). Most research concerning these negative factors has stressed the importance of certain plant food substances in reducing Zn absorption in humans although some recent work questions many of these assumptions and current dogma written about these plant antinutrients (Hunt et al., 1987).

3. Dietary requirements

Estimations of Zn requirements for humans is beset with uncertainties resulting from the lack of sensitive indicators or tests for human Zn status (National Research Council, 1989; White, 1992) and an incomplete understanding of the factors that affect Zn bioavailability (i.e., absorbable and utilizable Zn) from common diets (Welch and House, 1984). Homeostatic mechanisms control the absorption and excretion of Zn and these mechanisms can maintain an individual in positive Zn balance even though a person's dietary intake is lower than that recommended for a healthy balanced diet (Hambidge et al., 1986; Kirchgessner and Weigand, 1982). However, the long-term effects of suboptimal Zn intake on human health are not known with any certainty, but could result in various undesirable health consequences as listed previously.

In 1989, the Food and Nutrition Board Subcommittee, National Research Council, National Academy of Sciences, USA published new guidelines for the recommended dietary allowance (RDA) of Zn for the U.S. population (National Research Council, 1989). Table 1 lists these RDAs for Zn. They are based on assuming that normal, healthy people absorb and utilize 20% of the Zn in their diets. This 20% bioavailability value for Zn is only a very crude estimate and significant changes in the estimated Zn bioavailability factor could either negate or enhance concern over the adequacy of Zn in U.S. diets. For example, the World Health Organization (WHO) has recommended daily intakes of Zn which differ by a factor of four depending on the amount of food derived form animal sources in the diet (i.e., 22 mg Zn per day for people obtaining less than 10% of their energy needs from animal sources, 11 mg Zn per day if between 10 and 20% and 5.5 mg Zn per day if between 20 and 40% (Mertz,1987)). Apparently, an identical level of dietary Zn can either provide adequate levels of Zn or result in Zn deficiency depending on the amount and types of animal products in the diet. This conclusion is supported by the occurrence of Zn deficiency in Iranian people having Zn intakes much greater than those normally consumed by healthy people in Western nations consuming nutritious diets. Therefore, it is important to determine what factors in plant foods might enhance or inhibit Zn bioavailability from complex diets consumed by people.

Table 1. Recommended dietary allowances (USA) for zinc[a].

Population group	Age range (years)	Recommended dietary allowance (mg day^{-1})
Infants	0-1	5
Children	1-10	10
Adult Males	11-51+	15
Adult Females	11-51+	12
Pregnant Females	all ages	15
Lactating Females	1st 6 mon.	19
	2nd 6 mon.	16

[a]Daily RDAs. Data from Food and Nutrition Board, National Academy of Sciences, USA, National Research Council (National Research Council, 1989).

4. Plant foods as sources of Zn for humans

Approximately 70% of the zinc consumed by most individuals in the United States comes from meat and dairy products which contribute 66% of the total Zn intake (Anonymous, 1988). Grain products contribute 14%, legumes, nuts and soy products contribute about 5%, while vegetables and fruits contribute about 9%. Foods derived from plants are not major contributors to the Zn requirements of most people in the United States, although certain population groups (e.g., vegetarians and infants fed un-supplemented soy-based formula) may receive most of their Zn from plant food sources. It is anticipated that intake of plant foods will increase in developed countries in response to an increasing awareness of the health benefits associated with fruits, vegetables and fibre (Anonymous, 1988).

Unfortunately, some important plant food groups, especially foods derived from whole grain and legume seeds, can contain substances (i.e., antinutrients) which interfere with the bioavailability of Zn to humans (Hambidge et al., 1986). Thus, an increase in plant food consumption at the expense of animal products will lead to increased incidence of marginal Zn deficiency in high risk groups (i.e., pregnant women, lactating women, infants, and children). Improving plant foods as sources of bioavailable Zn for humans would benefit human nutritional health and may become an increasingly important goal for plant scientists as we approach the twenty-first century.

The concentration of Zn in edible plant parts can vary widely depending on a number of complex, dynamic and interacting factors including plant genotype, plant part, and the plant's environment (including soil-Zn availability) during development (Benton-Jones Jr, 1991; Lindsay, 1972; Moraghan and Mascagni, 1991; Welch and House, 1984). Table 2 lists the median and range of Zn concentrations in a number in raw edible parts of food crops grown in major agricultural production regions of the United States. While tables, like Table 2, listing the total concentration of Zn in foods, contain important information that is used by dietitians for developing balanced diets that hopefully embrace adequate Zn levels, bioavailability factors must be considered before judgements can be made on the true nutritive value of a particular plant food in a mixed diet with respect to Zn.

Little is known about most of the antinutritive factors in plant foods that inhibit the bioavailability of Zn to humans or more importantly, their mechanisms of action (Welch and House, 1984). Moreover, there is scant evidence that the chemical nature of naturally occurring forms of Zn in plant tissues plays a critical role in Zn bioavailability under normal dietary regimes (Kirchgessner and Weigand, 1982). Foods of animal origin

Table 2. Concentration of Zn (dry wt. basis) in common food crops grown in major agricultural production regions in the United States[a].

Plant species	Common name and part	Median (mg gdw^{-1})	Range (mg gdw^{-1})
Lactuca sativa L.	Lettuce, washed leaves	45	20-61
Archis hypogaea L.	Peanut, shelled nuts	29	28-56
Soanum tuberosum L.	Potato, peeled tubers	16	9-27
Glycine max L.	Soybean, whole beans	45	36-70
Zea saccharata L.	Sweet corn, whole kernels	25	12-51
Triticum aestivum L.	Wheat, whole kernels	30	13-68
Daucus carota L.	Carrot, scrubbed tap roots	21	6-27
Allium cepa L.	Onion, peeled bulbs	16	8-21
Oryza sativa L.	Rice, whole kernels	16	10-22
Lycopersicon esculentum Mill.	Tomato, fruits with peduncle scars removed	23	18-23
Spinacia oleracea L.	Spinach, washed leaves	42	30-180

[a]Data was summarized from Wolnik et al., 1983; Wolnik et al., 1985.

contain unknown factors that promote Zn bioavailability (Hambidge et al., 1986; Hortin et al., 1993; Welch and House, 1984). Do foods derived from plants lack or contain lower amounts of these factors (i.e., promoters) than do foods from animal products? If so, than perhaps the low level or lack of Zn promoters in plant foods is more important to human Zn status than the presence of antinutritive factors in plant foods. If true, the means of improving crops as sources of Zn for humans might be directed at developing new plant genotypes, through traditional plant breeding or genetic engineering techniques, containing increased amounts of promoters of Zn bioavailability, once these promoters are identified (see discussion below).

5. Plant factors affecting Zn bioavailability to humans

Several references are available which review the literature concerning the factors affecting the bioavialability of Zn in foods from plants (Hambidge et al., 1986; Quarterman, 1973; Welch and House, 1984). Table 3 lists some of the predominant factors studied for major foods of plant origin.

Phytic acid. Phytic acid [myo-inositol hexakis(dihydrogen phosphate)] or its naturally occurring form, phytin (a mixed salt of phytic acid), has received the most attention from nutritional scientists because it forms insoluble precipitates with a number of polyvalent mineral cations (e.g., Ca^{2+}, Fe^{3+}, and Zn^{2+}) *in vitro*, and when added to purified diets, sodium phytate has been shown to decrease Zn absorption in a number of monogastric animal species including humans (Hambidge et al., 1986; Kirchgessner and Weigand, 1982; Welch and House, 1984). High dietary Ca accentuates the effects of phytate on Zn bioavailability and some report that using Zn × phytate/Zn molar ratios in diets is a better predictor of Zn bioavailability than dietary phytate content alone. For humans, Fordyce et al. (1987) reported that ratios above 0.5 *M*/kg dry diet or 200 m*M*/1000 Kcals may be cause of concern.

Phytin is the primary storage form of P in nearly all mature seeds and grains (Welch, 1986). Phytin is accumulated in globoid crystals in membrane bound protein bodies of certain cell types within the developing seed, such as within protein bodies occurring in aleurone cells of the aleurone layer of cereal grains (Lott, 1984; Mazzolini and Legge, 1981; Ogawa et al., 1979). Phytin deposition within globoid crystals of protein bodies is

Table 3. Substances (antinutritives) in plant foods that have been reported to affect the bioavailability of Zn to humans or monogastric animals eating mixed diets under some, but not all circumstances[a].

Substance	Major dietary source
Phytic acid or phytin	Whole legume seeds and cereal grains
Fibre (e.g., cellulose, hemicellulose)	Whole cereal products (e.g., wheat, oat, barely)
Oxalic acid	Spinach leaves
Heavy metals (e.g., Cd, Hg, Pb, Ag)[b]	Plant foods obtained from crops grown on polluted soils (e.g., Cd in rice)

[a]See Hambidge et al., 1986 for a detailed discussion of these factors in relation to Zn bioavailability to humans.

[b]Only studied in animal models, not yet extended to humans.

associated with the accumulation of not only P but also other minerals including K, Mg, Fe, Zn, Cu, Mn, and, in some seed cell types, Ca. As such phytin plays an important role as a storage pool for mineral nutrient reserve required by the developing embryo within the seed during germination, and during early seedling growth and thus contributes to the viability and vigor of the seedling produced (Welch, 1986). Selecting for seed and grain crops with substantially lower phytin content to improve these crops as sources of bioavailable Zn could have very undesirable affects on agricultural production especially in regions having soils of low P status and inadequate mineral fertility.

Despite the great interest in phytin as an antinutritive factor inhibiting Zn absorption by humans, much of the experimental data concerning the potential negative effects of naturally occurring phytin in seeds and grains on Zn bioavailability are confusing and contradictory (Hambidge et al., 1986). Certainly, soluble salts of phytate when added to purified diets reduce Zn bioavailability (Hambidge et al., 1986). Apparently, other dietary factors such as Ca may interact with phytin resulting in the negative effects of phytin on the bioavailability of Zn under some, but not all circumstances. More research is required before an understanding of the role of phytin in the bioavailability of Zn to humans consuming mixed diets is obtained.

Interestingly, recent reports have suggested that phytic acid may have anticarcinogenic activity towards certain types of colon cancer (Messina, 1991; Jariwalla, 1992; Shamsuddin, 1992). Therefore, caution should be used by plant scientists before they try and modify food crops with respect to their phytin content.

Fibre. Usually, products containing significant amounts of natural fibre from major food crops (e.g., whole cereal grain) are associated with depressed Zn bioavailability to humans (Harland, 1989). However, the capability of fibre to reduce Zn bioavailability depends on the chemical nature of the fibre eaten and the chemical composition of other dietary constituents present in the meal. While certain sources of fibre (e.g., those that also contain phytate, such as cereal brans) are associated with reduced Zn bioavailability (Torre et al., 1991), other fibre sources may have no effect (i.e., fiber rich vegetables such as carrots, turnips, cabbage or potatoes) or even enhance Zn bioavailability (Harland, 1989). For example, rats fed semi-synthetic diets supplemented with sugar beet (*Beta vulgaris* L.) fibre resulted in a 39% increase in Zn absorption while diets supplemented with wheat bran reduced Zn absorption by 9% (Fairweather-Tait and Wright, 1990).

Generally, fibre from plant sources is composed of numerous plant cell wall constituents that are not readily hydrolyzed by intestinal digestive enzymes. These substances include not only polysaccharides and lignin but also cutin, suberin, β-glucans and galactomanans (gums), water soluble polysacchorieds (mucilages), polyphenols and other phenolic compounds, and starch and proteins that are resistant to enzymic degradation (Torre et al., 1991). Some of these constituents of dietary fibre have the capacity to tightly bind polyvalent cations. Additionally, large amounts of fibre can dilute the intestinal contents and increase the rate of passage of digesta past the mucosal cell absorption sites. These properties of fibre can reduce Zn bioavailability to humans under some circumstances. Fibre from diverse plant sources is known to have different chemical structures and different divalent cation binding capacities. Because of this complication, it is almost impossible to draw definite general conclusions concerning the relation between plant food fibre and Zn bioavailability. Summarizing, one can only conclude that plant fibre can affect Zn bioavailability under some, but not all circumstances. Clearly, the benefits of dietary fibre outweigh its potential negative effects on Zn bioavailability based on the currently available literature.

Oxalic acid. Oxalic acid forms insoluble precipitates with Ca ions, and has been shown to have negative effects on Ca bioavailability to humans but does not appear to affect Zn bioavailability (Liebman and Doane, 1989; Welch and House, 1984). Thus, at this stage it would appear that oxalic acid is *not* an important plant food antinutrient for Zn in humans.

Promoters of Zn bioavailability. Numerous reports have shown that Zn is more readily absorbed by humans from diets containing meat products than from diets containing only protein of plant origin (Hambidge et al., 1986; Kirchgessner and Weigand, 1982; Shah and Belonje, 1984). Studies using isolated intestinal loops show that histidine and cysteine improve the efficiency of Zn absorption (Kirchgessner and Weigand, 1982; Wapnir et al., 1983). Proteins of animal sources are higher in these amino acids than are important plant food proteins (e.g., soy protein). Furthermore, certain organic acids (e.g., fumaric acid) and some long-chain, saturated fatty acids (e.g., palmitic acid) improve Zn absorption if eaten in sufficient amounts (Wapnir and Lee, 1990). Interest in the effects of fatty acids on Zn bioavailability stems from research showing high bioavailability of Zn in human breast milk (which contains substantial amounts of palmitic acid compared to cows milk) to infants. Most likely, increased absorption of Zn imparted by certain amino acids and some organic acids is the result of competition between organic ligands in the lumen for cation binding and the resulting shift in Zn availability via the formation of soluble Zn complexes. The enhanced absorption of Zn engendered by long-chain fatty acids may be the result of interactions of these hydrophobic diet components with intestinal mucosal cell membranes or with intracellular components of these cells (Wapnir and Lee, 1990). Unfortunately, the major food crops (i.e., cereals and pulses) contain relatively low concentrations of these possible promoters of Zn bioavailability and this may be a factor in the relatively poor bioavailability of Zn in plant foods to humans.

6. Major chemical forms of Zn in plants

As discussed above, very little is know about plant food substances affecting the bioavailability of Zn to humans. What then, is known about the major naturally occurring forms of Zn in plants, and can this information be used to increase our understanding of the nutritional value of food crops with respect to Zn?

An estimate of the subcellular distribution of Zn between different Zn-binding molecules in plant leaf cells is shown in Table 4 (Hewitt, 1983). These calculated values are in close agreement with experimentally determined values. In studies using lettuce (*Lactuca sativa* L.) leaves from plants grown in nutrient solutions radiolabeled with Zn^{65}, Walker and Welch (1987) reported that about 60% of the radioactive Zn could be extracted with 30 m*M* NH_4HCO_3. Of this soluble fraction, 73% of the Zn was associated with molecules ranging in molecular weight from between 1000 to 10,000 Da, 16% was bound to molecules below 1000 Da, and 11% to molecules above 10,000 Da. No free radioactive Zn^{2+} ions were detected using electrophoretic procedures. Thus, about 90% of the soluble Zn in lettuce leaves was associated with compounds of molecular weight below 10,000 Da. Further purification of the major soluble Zn fraction (i.e., the 1000-10,000 Da fraction) revealed that most of the Zn was bound to a negatively charged 1250 Da fraction having several characteristics in common with known Zn-binding peptides of the generic formula, _-glutamylcysteine-*n*-glycine (*n*=3-7) (i.e., phytochelatins). For example, the Zn-binding fraction in lettuce leaves had a similar molecular weight to some phytochelatins and

Table 4. Distribution of Zn between major Zn-binding fractions in leaf cells[a].

Identity of Zn binding molecule	Concn. of molecule in cell (μM)	Concn. of Zn in molecule[b] (μM)	Proportion of Zn bound to molecule (% of cell total)
Superoxide dismutase	0.7	1.4	1.9
Carbonic anhydrase	1.8	11	14.5
RNA polymerase	0.2	1.6	2
Aspartate trans carbamylase	0.2	1.2	1.5
Other Zn binding enzymes	0.1	8	10.1
Other unknown molecules	-	51.8	70

[a]Modified from Hewitt, 1983. See cited reference for assumptions used in calculating values.
[b]Assumes a total Zn concentration in cell of 75μM.

contained reactive sulphydryl groups and ninhydrin-positive components indicating the presence of a-amino groups. Acid hydrolysates contained glutamic acid, glycine and a trace of cysteine and other amino acids. Its UV spectrum was similar to phytochelatins with a UV absorption shoulder at 251 nm (Grill et al., 1985). However, this fraction also contained reducing sugar activity which is not characteristic of phytochelatins.

We studied the possibility that mature pea seeds (containing 1.16% phytic acid) may contain Zn bound to phytin (Welch et al., 1974). When a lyophilized homogenate of mature pea seeds, intrinsically labeled with ^{65}Zn, was extracted with 10 m*M* KH_2PO_4 buffer (pH 6.9), over 60% of the Zn originally present in the homogenate was bound to anionic complexes of less than 1000 Da after centrifugation, ultrafiltration and electrophoresis. There was no correlation between the amount of Zn extracted into various solvents used and phytic acid extracted. Thus, we concluded that it was highly unlikely that Zn was tightly bound to phytin in significant amounts in mature pea seeds.

Table 5 lists some of the major soluble forms of Zn reported for higher plants. Many researchers have reported (see references cited in Table 5) that Zn is primarily associated with low molecular weight (i.e., <10,000 Da) anionic complexes. High molecular weight (i.e., >10,000 Da) compounds containing Zn, are much less abundant in plant cells than these low molecular weight Zn anionic complexes (see Table 4).

What, therefore can be gleaned from our knowledge of naturally occurring major Zn forms in plants and their potential effects on Zn bioavailability? We know that vegan or cereal-based diets contain lower amounts of bioavailable Zn to humans than do diets containing significant amounts of animal proteins. Therefore, either foods derived form plants contain significant amounts of antinutritive substances (capable of tightly binding Zn in the gastrointestinal tract) which interfere with the absorption of Zn, or they lack promoter substances which animal proteins contain, or both. As discussed above, experiments isolating major Zn fractions from a wide variety of plant foods, including mature legume seeds, have shown that the major forms of Zn in plant tissues are soluble, low molecular weight, anionic complexes. If antinutrients were present in the plant tissue homogenates in significant amounts, it is reasonable to assume that they would have been capable of tightly binding Zn under the isolation methods used by various investigators and they should have been detected. Such is not the case. We consider it highly likely that foods from plant

Table 5. Soluble, low molecular weight (<10,000 Da) Zn-binding substances in plants.

Zn binding substance	Example of plant organ or tissue	References
phytosiderophores	cereal roots	Walker and Welch, 1986; Walker and Welch, 1987; Zhang et al, 1989
(e.g., 2'-deoxymugineic acid)	wide range of plant species	Grill et al., 1985; Grill et al., 1991; Steffens, 1990
(γ-glutamylcysteine)-*n*-glycine (*n*=2-7) [also named phytochelatin, cadystin, γ-glutamyl metal binding peptide, or class III metallothionein]	and plant tissues	
polyphenolic complexes (e.g., ellagic acid and catechin)	*Vaccinium myrtillus* L. (blueberry) leaves	Gomah and Davies, 1974
organic acids (e.g., malate, citrate)	Nicotiana tabacum L. (tobacco) cell vacuoles	Krotz et al., 1989
unknown, low-molecular-weight (< 2000 Da), anionic complexes	xylem sap, phloem sap, leaf extracts, and mature legume seeds	Bremner, 1970; Bremner and Knight 1970; Peterson, 1969; Taylor et al., 1988; Tiffin, 1972; Van Goor and Wiersma, 1976; Walker and Welch 1987; Welch et al., 1974

sources are relatively poor sources of dietary Zn compared with foods of animal origin because they contain lower amounts of substances that promote Zn bioavailability.

7. Improving food crops as sources of Zn for humans.

Agronomic practices. Several reviews are available that discuss the effects of agronomic practices on the nutritional quality of major food crops with respect to Zn (Allaway, 1986; Grunes and Allaway, 1985; Quaterman, 1973; Sander et al., 1987; Welch and House, 1984). The most efficient agronomic practice currently available to farmers to increase Zn content of food crops is the application of available forms of Zn fertilizers possibly in excess of those rates required for maximum crop yield (Welch and House, 1984). Indeed, we have shown that increasing the supply of Zn to seed and grain crops increased their mature seed Zn concentration and this increase in Zn did not change the bioavailability of Zn in the foods to monogastric animals (House and Welch, 1989; Welch et al., 1991).

Plant breeding and genetic engineering. As stated previously, animal proteins may promote Zn absorption because they contain relatively high concentrations of certain essential amino acids (cysteine and histidine) or other as yet unidentified factors. These promoters of Zn bioavailability may be stable, soluble, anionic Zn complexes, such as $[Zn(II)\text{-}(cysteine_2)]^{2-}$ (Perrin and Sayce, 1968), which from stable Zn complexes in the pH and E_H range found at major sites of Zn absorption within the small intestine (i.e., the duodenum, ilium and jejunum). These compounds may function by preventing Zn binding to antinutrients, such as phytic acid, thus, ameliorating the negative effects of antinutrients on Zn absorption. If such promoters of Zn bioavailability could be increased in plant foods, then the nutritional quality of these foods with respect to Zn could be greatly enhanced, and there may be no need to reduce the level of antinutrients. Research should be undertaken to study these possibilities.

Interestingly, O'Dell and his associates (O'Dell et al., 1991) have reported that osmotic fragility of rat erythrocyte plasma membranes (a possible index of Zn status; Johanning and O'Dell, 1989; Ruz et al., 1992) induced by inadequate dietary Zn supply can be prevented by supplying elevated levels of S-containing amino acids to rat diets based on soy protein (low in S amino acids), but not to rat diets based on egg white protein (high in S amino acids) containing the same Zn content. In current bioavailability studies in our laboratory, we found that rats maintained on soy protein based diets, low in methionine, absorb significantly less Zn then rats fed the same diet supplemented with methionine at levels typically found in animal protein (unpublished data). If these findings are substantiated in humans (i.e., the low levels of the S-containing amino acids, methionine and cysteine, in major food crops limits Zn bioavailability) then it would be worthwhile to increase the levels of these amino acids in these food crops to overcome the negative effects of antinutritives on Zn bioavailability.

Mutant genotypes of several cereal crops e.g., *floury-2* and *opaque-2* (Nelson and Chang, 1974; Wolf et al., 1972) mutants of *Zea mays* L., which contain substantially higher concentrations of lysine, methionine, and Zn than wild types, are known (Arnold et al., 1977) and could be used in traditional plant breeding programs to increase amino acids and Zn in kernels of these crops (Nelson, 1980; Payne, 1983; Shewry et al., 1981; Sylverter-Bradley and Folkes, 1976). Additionally, currently available genetic engineering techniques could be applied to food crops to increase their S-containing amino acid content if this proves to be a desirable nutritional goal.

8. Summary

Identifying factors in major food crops that inhibit Zn bioavailability to monogastric animals and humans and understanding their mechanisms of action within the gastrointestinal tract are still high priority goals for nutritional and plant scientists undertaking cooperative research efforts. Additionally, identifying factors in animal products that promote Zn bioavailability should receive greater attention among nutritional scientists. Once these promoters are identified, attempts could be made to increase their content in edible parts of major food crops, especially cereal and legume seeds. If successful, there may be no need to reduce the levels of antinutritive substances in plant foods. Reducing the concentration of known antinutritives, such as phytic acid, could lead to undesirable consequences for both plant production, such as reduced seed vigor and viability, and human health, such as decreased content of anti-carcinogens. It is not wise to significantly change the composition of major plant foods without a thorough understanding of the consequences of such actions.

References

Allaway W H 1986 Soilplantanimal and human interrelationships in trace element nutrition. In Trace elements in human and animal nutrition. Ed. W Mertz. 5th Edition. pp 465-488. Academic Press, Orlando, San Diego, New York, Austin, London, Montreal, Sydney, Tokyo, Toronto.

Anonymous 1988 The Surgeon General's Report on Nutrition and Health. 1988. DHHS (PHS) Pub. No. 8850210. U.S. Government Printing Office, Washington, D.C. 727 p.
Arnold J M, Bauman L F and Aycock H S 1977 Interrelations among protein, lysine, oil, certain mineral element concentrations, and physical kernel characteristics in two maize populations. Crop Sci. 17, 421-425.
Benton-Jones J ,Jr. 1991 Plant tissue analysis in micronutrients. In Micronutrients in Agriculture. Eds. J J Mortvedt, F R Cox, L M Shuman and R M Welch. 2nd Edition. pp 477-521. Soil Science Society of America, Madison, WI.
Bremner I 1970 The nature of trace element binding in herbage and gut contents. In Trace Element Metabolism in Animals. Ed. C F Mills. pp 366-369. E. & S. Livingston Pub., Edinburgh, London.
Bremner I and Knight A H 1970 The complexes of zinc, copper and manganese present in ryegrass. Brit. J. Nutr. 24, 279-289.
Chapman H D 1966 Zinc. In Diagnostic Criteria for Plants and Soils. Ed. H D Chapman. pp 484-499. University of California, Div. Agricultural Science, Riverside, CA.
FairweatherTait S J and Wright A J A 1990 The effects of sugarbeet fibre and wheat bran on iron and zinc absorption in rats. Br. J. Nutr. 64, 547-552.
Fordyce E J, Forbes R M, Robbins K R and Erdman J W J 1987 Phytate x calcium/zinc molar ratios: are they predictive of zinc bioavailability?. J. Food Sci. 52, 440-444.
Gomah A M and Davies R I 1974 Identification of the active ligands chelating Zn in some plant water extracts. Plant Soil 40, 119.
Grill E , Winnacker E and Zenk M 1985 Phytochelatins: the principal heavymetal complexing peptides of higher plants. Science 230, 674-676.
Grill E , Winnacker E L and Zenk M H 1991 Phytochelatins. Meth. Enzym. 205, 333-341.
Grunes D L and Allaway W H 1985 Nutritional quality of plants in relation to fertilizer use. In Fertilizer Technology and Use. Ed. O P Engelstad. 3rd Edition. pp 589-619. Soil Science Society of America, Madison, WI.
Hambidge K M, Casey C E and Krebs N F 1986 Zinc. In Trace Elements in Human and Animal Nutrition. Ed. W Mertz. 5th Edition. pp 1137. Academic Press, Inc., New York.
Harland B F 1989 Dietary fibre and mineral bioavailability. Nutr. Res. Rev. 2, 133-147.
Hewitt E J 1983 A perspective on mineral nutrition: essential and functional metals in plants. In Metals and Micronutrients: Uptake and Utilization by Plants. Eds. D A Robb and W S Pierpoint. pp 277-323. Academic Press, New York.
Hortin A E, Oduho G , Han Y , Bechtel P J and Baker D H 1993 Bioavailability of zinc in ground beef. J. Animal Sci. 71, 119-123.
House W A and Welch R M 1989 Bioavailability of and interactions between zinc and selenium in rats fed wheat grain intrinsically labeled with 65Zn and 75Se. J Nutr 119, 916-921.
Hunt J R, Johnson P E and Swan P B 1987 Dietary conditions influencing relative zinc availability from foods to the rat and correlations with in vitro measurements. J. Nutr. 117, 1913-1923.
Jariwalla R J 1992 Anticancer effects of phytate. Am. J. Clin. Nutr. 56, 609
Johanning G L and O'Dell B L 1989 Effect of zinc deficiency and food restriction i rats on erythrocyte membrane zinc, phospholipid and protein content. J. Nutr. 119, 1654-1660.
Kirchgessner M and Weigand E 1982 Zinc absorption and excretion in relation to nutrition. In Metal Ions in Biological Systems. Vol. 15. Zinc and Its Role in Biology and Nutrition. Ed. H Sigel. pp 319-361. Marcel Dekker, Inc., New York and Basel.
Krotz R M, Evangelou B P and Wagner G J 1989 Relationships between cadmium, zinc, Cdpeptide, and organic acid in tobacco suspension cells. Plant Physiol. 91, 780-787.
Liebman M and Doane L 1989 Calcium and zinc balances during consumption of high and low oxalatecontaining vegetables. Nutr. Res. 9, 947-955.
Lindsay W L 1972 Zinc in soils and plant nutrition. Adv. Agron. 24, 147-186.
Lott J N A 1984 Accumulation of seed reserves of phosphorus and other minerals. In Seed Physiology. Vol. I. Development. Ed. D R Murray. pp 139-166. Academic Press, Sydney, New York, London.
Luecke R W 1984 Domestic animals in the elucidation of zinc's role in nutrition. Fed. Proc. 43, 2823-2828.
Mazzolini A P and Legge G J F 1981 The distribution of trace elements in mature wheat seed using the Melbourne proton microprobe. Nuclear. Instr. Meth. 191, 583-589.
McClain C J, Edward M D, Kasarskis J ,Jr. and Allen J J 1985 Functional consequences of zinc deficiency. Prog. Food Nutr. Sci. 9, 185-226.
Mertz W 1987 The practical importance of interactions of trace elements. In Trace Substances in Environmental Health. 21. Proceedings of the University of Missouri's 21st Annual Conference on Trace Substances in Environmental Health, St. Louis, Mo, 2428 May 1987. Eds. Anonymous. pp 526-532. University of

Missouri, Columbia, MO.
Messina M 1991 Phytate's potential role in reducing coloncancer risk. Am. J. Clin. Nutr. 54, 762-763.
Moraghan J T and Mascagni H J 1991 Environmental and soil factors affecting micronutrient deficiencies and toxicities. In Micronutrients in Agriculture. Eds. J J Mortvedt, F R Cox, L M Shuman and R M Welch. 2nd Edition. pp 371-425. Soil Science Society of America, Madison, WI.
National Research Council 1989 Recommended Dietary Allowances. National Academy Press, Washington, D.C. p. 1285.
Nelson O E 1980 Genetic control of polysaccharide and storage protein synthesis in the endosperms of barley, maize, and sorghum. In Advances in Cereal Science and Technology. Vol. III. Ed. Y Pomeranz. pp 4171 American Association of Cereal Chemists, St. Paul, MN.
Nelson O E and Chang M T 1974 Effect of multiple aleurone layers on the protein and amino acid content of maize endosperm. Crop Sci. 14, 374-376.
O'Dell B L 1984 History and status of zinc in nutrition. Fed. Proc. 43, 2821-2822.
O'Dell B L, Browning J D and Reeves P G 1991 Zinc deficiency increases the osmotic fragility of rat erythrocytes. J. Nutr. 117, 1883-1889.
O'Dell B L and Savage J E 1957 Potassium zinc and distillers dried solubles as supplements to a purified diet Poultry Sci. 36, 459-460.
Ogawa M , Tanaka K and Kasai Z 1979 Accumulation of phosphorus, magnesium and potassium in developing rice grains: followed by electron microprobe Xray analysis focusing on the aleurone layer. Plant Cell Physiol. 20, 1927.
Payne P I 1983 Breeding for protein quantity and protein quality in seed crops. In Seed Proteins. Eds. J Daussant, J Mossè and J Vaughan. pp 223-253. Academic Press, London, New York.
Perrin D D and Sayce I G 1968 Complex formation by nickel and zinc with penicillamine and cysteine. J. Am Chem.l Soc. 90, 5357.
Peterson P J 1969 The distribution of Zinc65 in Agrostis tenuis Sibth. nad A. stolonifera L. tissues. J. Exp. Bot. 20, 863-875.
Quarterman J 1973 Factors which influence the amount and availability of trace elements in human food plants. Qual. Plant Pl. Fds. Hum. Nutr. 23, 171-190.
Raulin M J 1869 Études chimiques sur la vègètation. Ann. Sci. Nat. Bot. Biol. Veg. (Ser. 5) 11, 93-299.
Reinhold J G 1988 Problems in mineral nutrition: a global perspective. In Trace Minerals in Foods. Ed. K T Smith. pp 155. Marcel Dekker, Inc., New York and Basel.
Ruz M , Cavan K R, Bettger W J and Gibson R S 1992 Erythrocytes, erythrocyte membranes, neutrophils and platelets as biopsy materials for the assessment of zinc status in humans. Brit. J. Nutr. 68, 515-527.
Sander D H, Allaway W H and Olson R A 1987 Modification of nutritional quality by environment and production practices. In Nutritional Quality of Cereal Grains and Agronomic Improvements. Agronomy Monograph no. 28. Eds. Anonymous. pp 4582. American Society of Agronomy, Madison, WI.
Sandstead H H 1984 Zinc: essentiality for brain development and function. Nutr. Today 19, 2630.
Sandstead H H 1991 Zinc deficiency: a public health problem? Am. J. Dis. Child. 145, 853-859.
Shah B G and Belonje B 1984 Bioavailability of zinc in beef with and without plant protein concentrates. Nutr. Res. 4, 7177.
Shamsuddin A M 1992 Phytate and coloncancer risk. Am. J. Clin. Nutr. 55, 478
Shewry P R, Miflin B J, Forde B G and Bright S W J 1981 Conventional and novel approaches to the improvement of the nutritional quality of cereal and legume seeds. Sci. Prog. 67, 575-600.
Sommer A L and Lipman C B 1926 Evidence on the indispensable nature of zinc and boron for higher plants. Plant Physiol. 1, 231-249.
Steffens J C 1990 The heavy metalbinding peptides of plants. Annu. Rev. Plant Physiol. Plant Mol. Biol. 41, 553-575.
SylverterBradley R and Folkes B F 1976 Cereal grains: their protein components and nutritional quality. Sc. Prog. 63, 241-263.
Taylor K , Albrigo L G and Chase C D 1988 Zinc complexation in the phloem of blightaffected citrus. J. Am. Soc. Hort. Sci. 113, 407-411.
Tiffin L O 1972 Translocation of micronutrients in plants. In Micronutrients in agriculture. Eds. J J Mortvedt, P M Giordano and W H Lindsay. pp 199-229. Soil Science Society of America, Madison, WI.
Todd W R, Elvehjem C A and Hart E B 1934 Zinc in the nutrition of the rat. Am. J. Physiol. 107, 146-156.
Torre M , Rodriguez A R and SauraCalixto F 1991 Effects of dietary fiber and phytic acid on mineral availability. Crit. Rev. Food Sci. Nutr. 1, 122.
Tucker H F and Salmon W D 1955 Parakeratosis or zinc deficiency in the pig. Proc. Soc. Exp. Biol. Med. 88, 613-616.

Underwood E J 1971 Trace Elements in Human and Animal Nutrition. 3rd Edition. Academic Press, New York, London.
Van Goor B J and Wiersma D 1976 Chemical forms of manganese and zinc in phloem exudates. Physiol. Plant 36, 213-216.
Walker C D and Welch R M 1986 Nicotianamine and related phytosiderophores: Their physiological significance and advantages for plant metabolism. J. Plant Nutr. 9(37), 523-534.
Walker C D and Welch R M 1987 Low molecular weight complexes of zinc and other trace metals in lettuce leaf. J. Agric. Food Chem. 35, 721-727.
Wapnir R A, Khani D E, Bayne M A and Lifshitz F 1983 Absorption of zinc by the rat ileum: effects of histidine and other lowmolecularweight ligands. J. Nutr. 113, 1346-1354.
Wapnir R A and Lee S Y 1990 Zinc intestinal absorption: Effect of free fatty acids and triglycerides. J. Trace Elem. Exp. Med. 3, 255-265.
Welch R M, House W A and Allaway W H 1974 Availability of zinc from pea seeds to rats. J. Nutr. 104, 733-740.
Welch R M 1986 Effects of nutrient deficiencies on seed production and quality. Adv. Plant Nutr. 2, 205-247.
Welch R M, Allaway W H, House W A and Kubota J 1991 Geographic distribution of trace element problems. In Micronutrients in Agriculture. Eds. J J Mortvedt, F R Cox, L M Shuman and R M Welch. 2nd Edition. pp 3157. Soil Science Society of America, Madison, WI.
Welch R M and House W A 1984 Factors affecting the bioavailability of mineral nutrients in plant foods. In Crops as Sources of Nutrients for Humans. Eds. R M Welch and W H Gabelman. pp 37-54. American Society of Agronomy, Madison, WI.
White, C L 1992 Zinc deficiency in man and animals: endemic or imaginary? Proc. Nutr. Soc. Aust. 17, 115-124.
Wolf M J, Cutler H C, Zuber M S and Khoo U 1972 Maize with multilayer aleurone of high protein content. Crop Sci. 12, 440-442.
Wolnik K A, Fricke F L, Capar S G, Braude G L, Meyer M W, Satzger R D and Kuennen R W 1983 Elements in major raw agricultural crops in the United States. 2. Other elements in lettuce, peanuts, potatoes, soybeans, sweet corn, and wheat. J. Agric. Food Chem. 31, 1244-1249.
Wolnik K A, Fricke F L, Capar S G, Meyer M W, Satzger R D, Bonnin E and Gaston C M 1985 Elements in major raw agricultural crops in the United States. 3. Cadmium, lead, and eleven other elements in carrots, field corn, onions, rice, spinach, and tomatoes. J. Agric. Food Chem. 33, 807-811.
Zhang F , Romheld V and Marschner H 1989 Effect of zinc deficiency in wheat on the release of zinc and iron mobilizing root exudates. Z. Pflanzenernahr. Bodenk. 152, 205-210.

Chapter 14.

The Zinc Requirements of Grazing Ruminants

C. L. WHITE

1. Introduction

Zinc (Zn) plays an essential role in animal nutrition as a component of a number of critical enzymes. Animals fed diets containing less than 5 mg Zn/kg develop signs of severe Zn deficiency: loss of appetite, reduced growth and immunocompetence, loss of hair and wool and keratotic skin lesions (Underwood, 1981). Signs of severe Zn deficiency are rarely seen under natural conditions because most practical diets contain more than 5 mg Zn/kg. Marginal deficiency, characterised by suboptimum growth, reduced fertility and mild skin disorders, is typically observed in sheep and cattle on pastures containing less than 20 mg Zn/kg, although it has been reported in cattle on pastures containing greater than 30 mg Zn/kg (ARC, 1980). Signs of marginal Zn deficiency cannot easily be distinguished from those caused by a range of other nutritional deficiencies or parasitic and infectious diseases.

Published recommendations on the Zn requirements of grazing animals are based on information derived from field observations, animal house studies and factorial analysis. In many cases the recommended requirements cannot be met by ruminants grazing natural herbage, particularly dry pastures and stubble. This raises the question as to whether published estimates are too generous and of little practical value or whether and to what extent marginal Zn deficiency is endemic.

The aim of this review is to examine critically the evidence underlying published estimates of Zn requirements of sheep and cattle, and suggest possible ways to improve the practical usefulness of these estimates.

2. Methods for estimating Zn requirements.

Published tables of Zn requirements are based on information derived from three sources: experiments using purified diets, field observations of responses to Zn supplements and factorial analysis. Different methods produce different answers and there is a wide range in estimates of Zn requirements for growth and reproduction in sheep and cattle (Table 1). The within-species range reflects differences in the demand for Zn for different physiological processes. For example, animals that are pregnant, lactating or growing rapidly have a greater requirement for Zn than animals at maintenance. Of main concern to the producer or adviser, however, is the wide between-reference range. An examination of the methodology provides some insight into how these differences may arise.

Table 1. Published estimates of Zn requirements for sheep and cattle.

Type of animal	Weight (kg)	Growth rate (kg/day)	Zn requirement (mg/kgDM)			
			ARC (1980)[a]	NRC (1976, 1985)	Towers and Grace (1983)[a]	SCA (1991)[a]
Sheep:						
Growing lamb	20	0.15	32-36	20[b]	17	13[c]
Pregnant ewe	75	twins	20-24	33	23	9
Lactating ewe	75	2 kg/day milk	25-32	33	26	15
Sucking lamb	5	0.15	41	20	-	50
Cattle:						
Growing calf	200	1.0	17-28	20-30[d]	19	12
Pregnant cow	4-500	late pregnancy	13-21		14	9
lactating cow	4-500	20 kg/day milk	19-27		24	14
Sucking calf	40	0.5	28		-	30

a. Factorial assessment.
b. Only two values given. 20 mg Zn/kg for growth and 33 mg/kg for pregnancy and lactation.
c. Growth rate of 0.1 kg/d, critical minimum values.
d. Beef cattle, range only.

2.1. *Purified diets*

The experimental method for determining Zn requirements typically involves feeding animals a semi-purified diet containing a range of Zn concentrations. On the basis of such studies, 7-10 mg Zn/kg dry matter of the diet (mg/kg DM) is considered adequate for normal growth and for the alleviation of clinical symptoms of deficiency in growing lambs and other ruminant species, and approximately 14 mg/kg DM is considered adequate for the maintenance of optimal plasma Zn concentration (Fig. 1 and 2). For example, Figure 1 clearly shows that growth in lambs is sharply reduced when the concentration of Zn in the diet falls below 7 mg/kg DM. Figure 2 shows that clinical signs of deficiency in weaned lambs are evident only when plasma Zn concentration falls below 0.4 mg/L, or about half the normal value.

There is some evidence that dietary Zn requirements are greater for some physiological processes than others. Underwood and Somers (1969) showed that between 17 and 30 mg Zn/kg was necessary for the development of normal spermatogenesis in ram lambs, while less than 17 mg/kg was required for optimal body growth. Martin and White (1992) were unable to confirm this relatively high requirement for Zn for testicular development although there was some indication that testicular growth had a marginally higher requirement for Zn than body growth. Optimum wool growth also required a greater amount of Zn than body growth (Fig. 3) and it was noteworthy that these reductions in testicular and wool growth occurred in the absence of any clinical signs of deficiency.

Critical Zn requirements for pregnancy and lactation in ruminants have not been determined experimentally. For ewes, on the basis of experiments with severely deficient

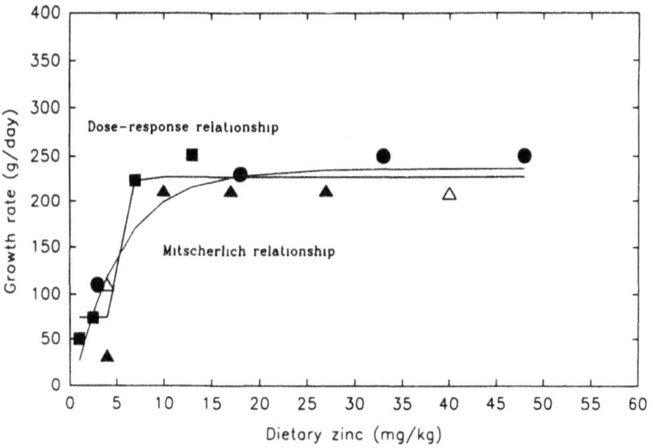

Figure 1.
The relationship between dietary Zn concentration and growth rate of weaned sheep fed semi-purified diets. Symbols represent a mean for several animals: ■ Mills et al., 1967, ● Ott et al., 1965, ▲ Martin and White 1992, △ Martin and White unpublished. An asterisk alongside a symbol represents the appearance of clinical signs. The Mitscherlich relationship is described by Y = 236 - 255exp(-0.19X), r = 0.91. The dose-response relationship is included to show that growth rate is cut off sharply at dietary concentrations below 7 mg/kg; Y = 153/(1+exp(-19(X-7))) + 75, r = 0.91 (Systat software, Evanston, Ill., USA)

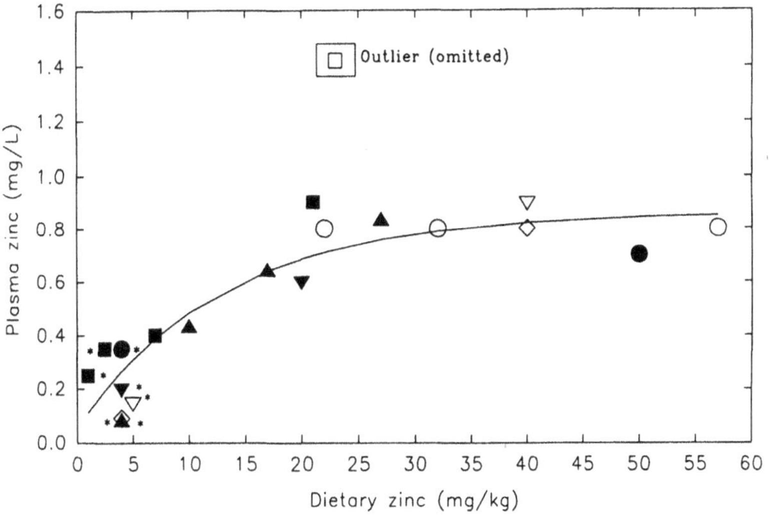

Figure 2.
The relationship between dietary and plasma Zn concentration in young growing sheep fed semi-purified diets. Symbols are: ■ Mills et al., 1967, ● Barge and Mazzocco 1982, ▲ Martin and White 1992, ◇ White and Martin unpublished, ▼ Masters and Moir 1983, ▽ White unpublished, ○ Hatch et al., 1987, □ Pond 1983 (outlier, data omitted from the equations). An asterisk alongside a symbol represents the appearance of clinical signs. The Mitscherlich relationship is described by Y = 0.86 - 0.81exp(-0.08X), r = 0.91. (Systat software, Evanston, Ill., USA).

Table 2. A summary of results from some Zn-responsive experiments using natural diets.

Animal and reference	Diet	Basal dietary Zn (mg/kg)	Plasma Zn (untreated animals) (mg/L)	Clinical signs in untreated animals.	Responses to treatment
Cow/calf[1]	Pasture	18-42	Not measured	SL, HL, D	Cured skin lesions and restored normal hair growth.
Cow/calf[2]	Pasture	28-48	Not measured	SL, HL.	Cured itch and hair licking, increased milk production by 11%.
Cattle[3]	Concentrate	18-29	1.5		Increased growth rate in two experiments (up to 18%), no effects in others.
Ewes[4]	Pasture	16-42	not measured	LD, SL.	Improved lambing rate by up to 20%.
Calves[5]	Concentrates and pasture	32	0.9		Increased growth rate (7%), increased plasma Zn.
Cattle[6]	Concentrate	18-21	1-1.5 (serum)		Increased growth rate (20%) in only 1/7 experiments.
Cow/calf[7]	Pasture and concentrates		12-25	0.81	Increased growth of calves (6%).
Cattle[8]	Winter rations	13-32	0.73-1.1		Improved growth of heifers (50 g/d).
Ewe/lamb[9]	Pasture	16-27	0.23-0.48 ewes	SL, WS, D, A.	
Lambs[10]	Lucerne hay and corn	19-26	1.4		No effect of supplement on growth or feed intake.
Ewe/lamb[11]	Hay, silage and corn		21-34	0.9	No effect on ewe or lamb growth.
Ewes[12]	Pasture	13-24	0.56		Increased lambing rate (14%) and lamb birth weight.
Ewes[13]	Pasture	10-30	0.64-0.85		No effect of Zn on reproductive performance.
Ewe/lamb[14]	Pasture	10-20	0.6 ewe		No effect on growth.

1 Legg and Sears (1960). 2 Haaranen (1963). 3 Perry et al. (1968). 4 Egan, (1972). 5 Spais and Papasteriadis, (1974). 6 Beeson et al. (1977). 7 Mayland et al. (1980). 8 Price and Humphries, (1980). 9 Mahmoud et al. (1983). 10 Pond, (1983). 11 Sword et al. (1984). 12 Masters and Fels, (1980). 13 Masters and Fels, (1985). 14 White et al. (1991).

a. Abbreviations: A = anaemia, D = death, HL = hair loss, LD = leg deformities, SL – skin lesion, WS = wool shedding.

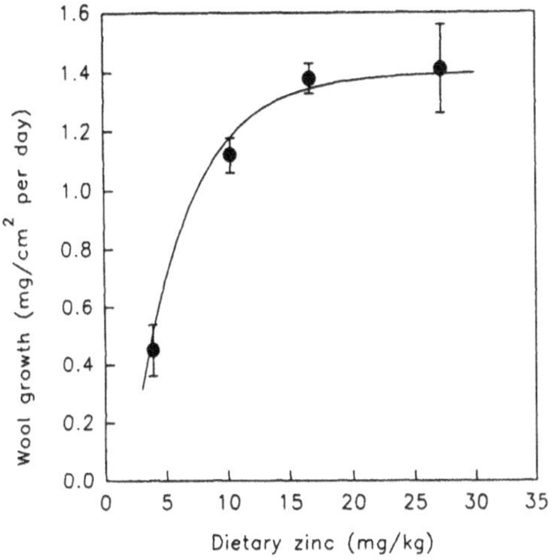

Figure 3
The relationship between dietary Zn concentration and clean wool growth in lambs (White and Martin, unpublished). The fitted curve is described by:
Y = 0.98 - 1.2 exp(-0.076X), r = 0.99.

purified diets, requirements probably lie somewhere between 5 and 40 mg/kg DM (Barge and Mazzocco, 1982; Masters and Moir, 1983; Apgar and Fitzgerald, 1985).

2.2. *Practical diets and field observations*

There are several published reports of Zn either improving growth or reducing the incidence of skin lesions in sheep and cattle consuming forage based diets (Table 2). In many cases the levels of Zn in the diet and plasma were higher than might be expected on the basis of animal house experiments to cause deficiency signs (> 20 mg/kg and > 0.6 mg/L, respectively). In contrast, there are several reports which showed no effects of supplementary Zn on growth or production of grazing animals or those consuming forage based diets when dietary concentrations of Zn were below 20 mg/kg (Beeson et al., 1977; Pond, 1983; Sword et al., 1984: Masters and Fels, 1985; White et al., 1991). This suggests that unknown factors may be involved which influence the availability or demand for Zn.

High calcium intakes have been shown to interfere with normal Zn absorption in ruminants, and it has been suggested that this may contribute to the sporadic occurrence of Zn-responsive diseases (Haaranen, 1963; Perry et al., 1968). However, more recent attempts to reproduce these results have been unsuccessful (Beeson et al., 1977; Pond, 1983) and there remains some doubt whether calcium plays a significant role in affecting Zn availability in ruminants under grazing conditions.

With the exception of certain toxic compounds, there is no evidence of naturally occurring dietary factors reducing Zn availability in ruminants. Parasitic or infectious diseases may increase Zn requirements, but this has not been well quantified.

Producers and advisers are therefore faced with a dilemma. If recommendations of Zn requirements are based on results from field trials then this would mean that a large proportion of the sheep and cattle in Australia would be classified as Zn deficient under

Table 3. Factorial estimates of the critical minimum Zn requirements of sheep and cattle.

Animal	Live weight	Feed intake	Feed quality	Growth or yield	Net Zn required	Minimum dietary Zn requirement		
	(kg)	(kg)	(MJ ME/kg)	(kg/d)	(mg/d)	(mg/d[a])	(mg/kg DM)	(mg/MJ ME)
Growing sheep								
	30	1.4	10	0.180	6.2	16	11	1.1
	30	0.9	6	0.0	2.1	5	6	1.0
Growing cattle								
	200	5.5	10	0.6	25	65	12	1.2
Suckled young								
Lamb	5	0.18 (1 kg milk)		0.2	5.4	9	50	2.0
Calf	40	0.9 (7 kg milk)		0.6	16.3	27	30	1.5

a. Coefficient of absorption is 0.4 for grazing animals and 0.6 for pre-ruminating lambs and calves.

some grazing conditions for at least part of the year. If, however, lower estimates of requirements based on experiments using purified diets are used, then there is a risk of overlooking production losses through marginal Zn deficiency. Unfortunately, there is no simple means of reconciling these differences, although the factorial method described below offers some insight into the extent to which dietary and animal factors can affect requirements.

2.3. *Factorial estimations*

The factorial method for estimating Zn requirements consists of adding up the amount of Zn accreted in tissues and products and subtracting the amount unavoidably lost in urine and faeces. The advantage of this method over empirical methods is the ability to model effects of changes in dietary quality, usually expressed in terms of metabolisable energy (ME).

Net Zn requirements (mg/day) are defined as $Zn_n = Zn_e + Zn_g + Zn_w + Zn_c + Zn_l$. Zn_e is Zn of endogenous origin lost in faeces and urine, and $Zn_{g,w,c,l}$ are Zn utilised for live weight gain, wool, conceptus and milk. The following analysis relates to minimum functional requirements for sheep, and are derived from the literature and unpublished data from this laboratory. It needs to be appreciated that many of these values are subject to wide variation, and in some cases represent little more than an educated guess. Endogenous loss and absorption are estimated at minimum intakes of Zn rather than at average intakes used by the ARC (1980). Assigned values are: Zn_e, 0.055 mg/kg live weight, Zn_g, 24 mg/kg live weight gain; Zn_w, 110 mg/kg clean wool; Zn_c, 1.2 mg/day in late gestation for single foetuses; Zn_l, 5.5 mg/kg milk.

Gross Zn requirements are calculated by divided net requirements by the coefficient of true absorption (TA). Fractional TA values at normal physiological intakes

Table 4. The Zn content of some common Australian feedstuffs and plant material, including estimated critical concentrations for plant growth. The ranges refer to multiple reports.

Species	Reported values[a] (mg/kg DM)	Critical[b] value (mg/kg DM)
Oats:		
Aerial part (green vegetative)	40-60	15-20
grain	26	
straw	8	
Wheat:		
Aerial part (green vegetative)	15-28	14
grain	11-55	
straw	3-16	
Lucerne:		
Aerial part (green vegetative)	23-28	16-20
Lupins: (several varieties)		
Aerial part (green, veg.)	-	12-14
Seeds	20-40	
Straw	12	
Perennial ryegrass (green, veg.)	57-138	10-13
Subterranean clover (green veg)	30-50	12-14

a. AFIC (1987)
b. Smith (1986)

of Zn are around 0.3 (ARC, 1980), while at low Zn intakes they range from 0.6 to 1.0 (Suttle et al., 1982; C.F. Ramberg, pers. comm.). Using a value of 0.4 for TA for non-suckling animals, a 30 kg weaner sheep growing at 180 g/day and grazing a spring pasture of ME value 10 MJ/kg DM would have a critical Zn requirement of 11 mg/kg DM or 1.0 mg Zn/MJ ME (Table 3). For suckling lambs, TA is estimated at 0.6 and predicted zinc requirements are 50 mg/kg dry milk (8 mg/kg fresh), or 2.0 mg/MJ ME (Table 3). Since milk from zinc-supplemented Merino ewes has been reported to contain less than 6 mg/kg (White et al., 1991), it is likely that the factorial estimated dietary zinc requirement for suckling lambs is too high. This discrepancy probably results from underestimating the absorption efficiency.

Factorial estimates of Zn requirements for non-sucking sheep correspond to those derived from studies using purified diets, but are lower than recommendations based on field experimentation. While acknowledging that the factorial method has several weaknesses, it does allow for some useful "what if" predictions. For example, factorial analysis shows that if the pasture or stubble were dry and of poor quality, with an energy value of 6 MJ ME/kg, then the sheep would lose weight, zinc would be redistributed from catabolising tissues and Zn requirements could be met by a diet containing as little as 6 mg Zn/kg. Expressed on an energy basis this is still equivalent to 1.0 mg Zn/MJ ME. If these theoretical considerations apply in practice, then the factorial method predicts that the dietary Zn requirements for growth of sheep grazing pastures in south western Australia ranges from 6 mg/kg on very poor quality dry feed to 11 mg/kg on green pasture (Table 3). This prediction remains to be tested.

Table 5. Effect of season on Zn concentration of annual pastures in SW Australia.

Season	Growth stage	Masters and Somers (1980)[a]		White et al. (1991)[b]
		Minimum	Maximum	
Winter	Slow growth	20	51	-
Spring	Active growth	24	34	20
Summer	Mature, mainly dry	12	39	12
Autumn	Dry, dead	15	27	10

a. Mean of 6 properties.
b. Single site.

3. Relationships between Zn requirements for plant and animal growth.

Most normal grain and green plants have been found to contain between 20 and 60 mg Zn/kg DM (Table 4). On the basis of most published observations, these concentrations of Zn should not be limiting animal production under grazing conditions, provided the plants themselves are not Zn deficient. It is perhaps of evolutionary significance that the critical concentrations of Zn for growth of pasture species such as subterranean clover and perennial ryegrass are similar to the critical requirement for sheep determined experimentally using purified diets (Fig. 1, Table 4).

The main problem faced by the grazing animal is the marked decline in Zn concentration in herbage and stubble as the plant material matures. This is illustrated in Table 4, which shows that cereal and legume straws and stubbles contain a much lower concentration of Zn than the green vegetative material or grain, the amounts being frequently lower than the critical requirement for Zn for growth of animals. This is further demonstrated in pastures sampled in SW Australia where Zn concentration can fall by 50% between spring and summer (Table 5). It is therefore evident that plant material which is adequate in Zn with respect to plant growth can become deficient with respect to animal health and production. To further exacerbate the problem, it is likely that the Zn present in dry herbage is less available (digestible) than that in green feed because of the higher indigestible fibre levels in dry feed. Factorial analysis predicts that the animal, by virtue of a low growth rate on poor quality feed, has a reduced requirement for Zn at this time and thus may not show obvious signs of deficiency. However, pregnancy and lactation increases the requirement for Zn and it is under these conditions that Zn supplements have resulted in live weight and reproductive responses in pregnant and lactating ewes and cows (Egan, 1972; Masters and Fels, 1980; Mayland, 1980).

4. Conclusion

The Zn requirements of grazing animals are poorly understood and recent data suggest that requirements have been overestimated in the past. Part of the difficulty lies in the fact that different methods of determining requirements provide different answers, and none of the published estimates of requirements are entirely satisfactory. Estimates of critical requirement range from 10 to over 30 mg Zn/kg DM, leaving producers with the dilemma of not knowing if their livestock are currently, or likely to become, deficient in Zn or not. Some inconsistencies disappear when needs are expressed on an ME basis, but this approach has not been applied under field conditions. There is no simple cost-effective means of supplying extra Zn to animals raised under extensive grazing systems, and producers need to be reasonably certain of a response before embarking on a supplementation program. Despite the obvious difficulties, future research should concentrate on studying zinc metabolism under practical grazing conditions in order to determine practically useful estimates of requirements.

References

ARC (Agricultural Research Council) 1980 *In* The nutrient requirements of ruminant livestock. pp 256-262. Commonwealth Agricultural Bureaux, Slough, UK.

Apgar J and Fitzgerald J A 1985 Effect on the ewe and lamb of low zinc intake throughout pregnancy. J. Anim. Sci. 60, 1530-1538.

AFIC (Australian Feeds Information Centre) 1987 Australian Feed Composition Tables. AFIC-CSIRO, Sydney.

Barge M T and Mazzocco P 1982 The functional activity of zinc in the feeding of ruminants of economic importance. 2nd part. Experimental deficiency by full-grown ewes. Z. Tierphysiol. Tierernahrg. Futtermittelkde. 48, 36-46.

Beeson W M, Perry T W and Zurcker T D 1977 Effect of supplemental zinc on growth and on hair and blood serum levels of beef cattle. J. Anim. Sci. 45, 160-165.

Dynna O and Havre G N 1963 Interrelationship of zinc and copper in the nutrition of cattle: a complex zinc-copper deficiency. Acta Vet. Scand. 4, 197-208.

Egan A R 1972 Reproductive responses to supplemental zinc and manganese in grazing Dorset Horn ewes. Aust. J. Exp. Agric. Anim. Husb. 12, 131-135.

Haaranen S 1963 Some observations on the zinc requirement of cattle for the prevention of itch and hair licking at different calcium levels in the feed. Nord. Vet. Med. 15, 536-542.

Hatch P A, Spears J W, Goode L and Johnson B H 1987 Influence of dietary zinc on growth and testicular development in ram lambs fed a high fibre diet. Nutr. Rep. Internat. 35, 1175-1183.

Legg S P and Sears L 1960 Zinc sulphate treatment of parakeratosis in cattle. Nature 186, 1061-1062.

Mahmoud O M, Samani F E, Bakheit A O and Hassan M A 1983 Zinc deficiency in Sudanese desert sheep. J. Comp. Path. 93, 591-595.

Martin G B and White C L 1992 Effects of zinc deficiency on gonadotrophin secretion and testicular growth in young male sheep. J. Reprod. Fertil. 96, 497-507.

Masters D G and Fels H E 1980 Effect of zinc supplementation on the reproductive performance of grazing merino ewes. Biol. Trace Element Res. 2, 281-290.

Masters D G and Fels H E 1985 Zinc supplements and reproduction in grazing ewes. Biol. Trace Element Res. 7, 89-93.

Masters D G and Moir R J 1983 Effect of zinc deficiency on the pregnant ewe and developing foetus. Br. J. Nutr. 49, 365-372.

Masters D G and Somers M 1980 Zinc status of grazing sheep: seasonal changes in zinc concentrations in plasma, wool and pasture. Aust. J. Exp. Agric. Anim. Husb. 20, 20-24.

Mayland H F, Rosenau R C and Florence A R 1980 Grazing cow and calf responses to zinc supplementation. J. Anim. Sci. 51, 966-974.

Mills C F, Dalgarno A C, Williams R B and Quarterman J 1967 Zinc deficiency and the zinc requirements of calves and lambs. Br J. Nutr. 21, 751-768.

NRC (National Research Council) 1985 Nutrient requirements of sheep. 6th edition. pp 19-20. National Academy Press, Washington DC.

NRC (National Research Council) 1976 Nutrient requirements of beef cattle. 5th edition. p 10. National Academy Press, Washington DC.

Ott E A, Smith W H, Stob M, Parker H E, Harrington R B and Beeson W M 1965 Zinc requirement of the growing lamb fed a purified diet. J. Nutr. 87, 459-463.

Perry T W, Beeson W M, Smith W H and Mohler M T 1968 Value of zinc supplementation of natural rations for fattening beef cattle. J. Anim. Sci. 27, 1674-1677.

Pond W G 1983 Effect of dietary calcium and zinc levels on weight gain and blood and tissue mineral concentrations of growing Columbia - and Suffolk-sired lambs. J. Anim. Sci. 56, 952-959.

Price J and Humphries W R 1980 Investigation of the effect of supplementary zinc on growth rate of beef cattle on farms in N Scotland. J. Agric. Sci. Camb. 95, 135-39.

SCA (Standing Committee on Agriculture) 1991. Feeding Standards for Australian Livestock, Ruminants. pp 175-181. CSIRO Publications, Melbourne.

Smith F W 1986. Pasture species. *In* Plant Analysis. Eds. DJ Reuter and JB Robinson. pp 103. Inkata Press, Melbourne.

Spais A G and Papasteriadis A A 1974 Zinc deficiency in cattle under Greek conditions. *In* Trace element metabolism in Animals - 2. Eds. WG Hoekstra, JW Suttie, HE Ganther, W Mertz. pp 628-631. University Park Press, Baltimore.

Suttle N F, Lloyd-Davies, H and Field A C 1982 A model for zinc metabolism in sheep given a diet of hay. Br. J. Nutr. 47, 105-109.

Sword J T, Ataja A M, Pope A L and Hoekstra W G 1984 Effect of calcium phosphates and zinc in salt-mineral mixtures on ad libitum salt-mix intake and on zinc and selenium status of sheep. J. Anim. Sci. 59, 1594-1600.

Towers N R and Grace N D 1983 Zinc *In* The mineral requirements of grazing ruminants. pp 84-91. New Zealand Society of Animal Production Occasional; Publication No 9.

Underwood E J and Somers M 1969. Studies of zinc nutrition in sheep 1. The relation of zinc to growth, testicular development and spermatogenesis in young rams. Aust. J. Agric. Res. 20, 889-897.

Underwood E J 1981 "The Mineral Nutrition of Livestock" 2nd Edition. Commonwealth Agricultural Bureau, Slough. 180 p.

White C L, Chandler B S and Peter D W 1991 Zinc supplementation of lactating ewes and weaned lambs grazing improved mediterranean pastures. Aust. J. Exp. Agric. 31, 183-9.

Developments in Plant and Soil Sciences

1 J. Monteith and C. Webb (eds.): *Soil Water and Nitrogen in Mediterranean-type Environments*. 1981 ISBN 90-247-2406-6
2 J.C. Brogan (ed.): *Nitrogen Losses and Surface Run-off from Landspreading of Manures*. 1981 ISBN 90-247-2471-6
3 J.D. Bewley (ed.): *Nitrogen and Carbon Metabolism*. 1981 ISBN 90-247-2472-4
4 R. Brouwer, I. Gasparíková, J. Kolek and B.C. Loughman (eds.): *Structure and Function of Plant Roots*. 1981 ISBN 90-247-2510-0
5 Y.R. Dommergues and H.G. Diem (eds.): *Microbiology of Tropical Soils and Plant Productivity*. 1982 ISBN 90-247-2624-7
6 G.P. Robertson, R. Herrara and T. Rosswall (eds.): *Nitrogen Cycling in Ecosystems of Latin America and the Caribbean*. 1982 ISBN 90-247-2719-7
7 D. Atkinson, K.K.S. Bhat, M.P. Coutts, P.A. Mason and D.J. Read (eds.): *Tree Root Systems and Their Mycorrhizas*. 1983 ISBN 90-247-2821-5
8 M.R. Saric and B.C. Loughman (eds.): *Genetic Aspects of Plant Nutrition*. 1983 ISBN 90-247-2822-3
9 J.R. Frency and J.R. Simpson (eds.): *Gaseous Loss of Nitrogen from Plant-Soil Systems*. 1983 ISBN 90-247-2820-7
10 United Nations Economic Commission for Europe (ed.): *Efficient Use of Fertilizers in Agriculture*. 1983 ISBN 90-247-2866-5
11 J. Tinsley and J.F. Darbyshire (eds.): *Biological Processes and Soil Fertility*. 1984 ISBN 90-247-2902-5
12 A.D.L. Akkermans, D. Baker, K. Huss-Danell and J.D. Tjepkema (eds.): Frankia *Symbioses*. 1984 ISBN 90-247-2967-X
13 W.S. Silver and E.C. Schroder (eds.): *Practical Application of Azolla for Rice Production*. 1984 ISBN 90-247-3068-6
14 P.G.L. Vlek (ed.): *Micronutrients in Tropical Food Crop Production*. 1985 ISBN 90-247-3085-6
15 T.P. Hignett (ed.): *Fertilizer Manual*. 1985 ISBN 90-247-3122-4
16 D. Vaughan and R.E. Malcolm (eds.): *Soil Organic Matter and Biological Activity*. 1985 ISBN 90-247-3154-2
17 D. Pasternak and A. San Pietro (eds.): *Biosalinity in Action*. Bioproduction with Saline Water. 1985 ISBN 90-247-3159-3
18 M. Lalonde, C. Camiré and J.O. Dawson (eds.): Frankia *and Actinorhizal Plants*. 1985 ISBN 90-247-3214-X
19 H. Lambers, J.J. Neeteson and I. Stulen (eds.): *Fundamental, Ecological and Agricultural Aspects of Nitrogen Metabolism in Higher Plants*. 1986 ISBN 90-247-3258-1
20 M.B. Jackson (ed.): *New Root Formation in Plants and Cuttings*. 1986 ISBN 90-247-3260-3
21 F.A. Skinner and P. Uomala (eds.): *Nitrogen Fixation with Non-Legumes*. (Proceedings of the 3rd Symposium, Helsinki, 1984) 1986 ISBN 90-247-3283-2
22 A. Alexander (ed.): *Foliar Fertilization*. 1986 ISBN 90-247-3288-3
23 H.G. v.d. Meer, J.C. Ryden and G.C. Ennik (eds.): *Nitrogen Fluxes in Intensive Grassland Systems*. 1986 ISBN 90-247-3309-X
24 A.U. Mokwunye and P.L.G. Vlek (eds.): *Management of Nitrogen and Phosphorus Fertilizers in Sub-Saharan Africa*. 1986 ISBN 90-247-3312-X
25 Y. Chen and Y. Avnimelech (eds.): *The Role of Organic Matter in Modern Agriculture*. 1986 ISBN 90-247-3360-X
26 S.K. De Datta and W.H. Patrick Jr. (eds.): *Nitrogen Economy of Flooded Rice Soils*, 1986 ISBN 90-247-3361-8
27 W.H. Gabelman and B.C. Loughman (eds.): *Genetic Aspects of Plant Mineral Nutrition*. 1987 ISBN 90-247-3494-0
28 A. van Diest (ed.): *Plant and Soil: Interfaces and Interactions*. 1987 ISBN 90-247-3535-1
29 United Nations Economic Commission for Europe and FAO (eds.): *The Utilization of Secondary and Trace Elements in Agriculture*. 1987 ISBN 90-247-3546-7
30 H.G. v.d. Meer, R.J. Unwin, T.A. van Dijk and G.C. Ennik (eds.): *Animal Manure on Grassland and Fodder Crops*. Fertilizer or Waste? 1987 ISBN 90-247-3568-8
31 N.J. Barrow: *Reactions with Variable-Charge Soils*. 1987 ISBN 90-247-3589-0
32 D.P. Beck and L.A. Materon (eds.): *Nitrogen Fixation by Legumes in Mediterranean Agriculture*. 1988 ISBN 90-247-3624-2
33 R.D. Graham, R.J. Hannam and N.C. Uren (eds.): *Manganese in Soils and Plants*. 1988 ISBN 90-247-3758-3
34 J.G. Torrey and L.J. Winship (eds.): *Applications of Continuous and Steady-State Methods to Root Biology*. 1989 ISBN 0-7923-0024-6
35 F.A. Skinner, R.M. Boddey and I. Fendrik (eds.): *Nitrogen Fixation with Non-Legumes*. (Proceedings of the 4th Symposium, Rio de Janeiro, 1987) 1989 ISBN 0-7923-0059-9
36 B.C. Loughman, O. Gasparíková and J. Kolek (eds.): *Structural and Functional Aspects of Transport in Roots*. 1989 ISBN 0-7923-0060-2, Pb 0-7923-0061-0
37 P. Plancquaert and R. Haggar (eds.): *Legumes in Farming Systems*. 1990 ISBN 0-7923-0134-X
38 A.E. Osman, M.M. Ibrahim and M.A. Jones (eds.): *The Role of Legumes in the Farming Systems of the Mediterranean Areas*. 1990 ISBN 0-7923-0419-5
39 M. Clarholm and L. Bergstrom (eds.): *Ecology of Arable Land. Perspectives and Challenges*. 1989 ISBN 0-7923-0424-1
40 J. Vos, C.D. van Loon and G.J. Bollen (eds.): *Effects of Crop Rotation on Potato Production in the Temperate Zones*. 1989 ISBN 0-7923-0495-0
41 M.L. van Beusichem (ed.): *Plant Nutrition - Physiology and Applications*. 1990 ISBN 0-7923-0740-2
42 N. El Bassam, M. Dambroth and B.C. Loughman (eds.): *Genetic Aspects of Plant Mineral Nutrition*. 1990 ISBN 0-7923-0785-2

Developments in Plant and Soil Sciences

43. Y. Chen and Y. Hadar (eds.). *Iron Nutrition and Interactions in Plants* 1991 ISBN 0-7923-1095-0
44 J.J.R. Groot, P. de Willigen and E.L.J Verberne (eds.). *Nitrogen Turnover in the Soil-Crop System* 1991 ISBN 0-7923-1107-8
45 R.J. Wright, V.C. Baligar and R.P. Murrmann (eds.): *Plant-Soil Interactions at Low pH*. 1991 ISBN 0-7923-1105-1
46. J. Kolek and V. Kozinka (eds.) *Physiology of the Plant Root System 1992* ISBN 0-7923-1205-8
47. A.U. Mokwunye (ed.): *Alleviating Soil Fertility Constraints to Increased Crop Production in West Africa*. 1991 ISBN 0-7923-1221-X; Pb 0-7923-1222-8
48. M. Polsinelli, R. Materassi and M. Vincenzini (eds): *Nitrogen Fixation* (Proceedings of the 5th Symposium, Florence, 1990). 1991 ISBN 0-7923-1410-7
49 J.K Ladha, T. George and B B Bohlool (eds) *Biological Nitrogen Fixation for Sustainable Agriculture* 1992 ISBN 0-7923-1774-2
50 P.J Randall, E. Delhaze, R.A. Richards and R Munns (eds.)· *Genetic Aspects of Plant Mineral Nutrition* 1993 ISBN 0-7923-2118-9
51. K.S Kumarasinghe and D.L. Eskew (eds.)· *Isotopic Studies of Azolla and Nitrogen Fertilization of Rice*. 1993 ISBN 0-7923-2274-6
52. F A. Bliss and G. Hardarson (eds.) *Enhancement of Biological Nitrogen Fixation of Common Baen in Latin America*. 1993 ISBN 0-7923-2451-X
53 M.A.C Fragoso and M.L. van Beusichem (eds): *Optimization of Plant Nutrition*. 1993 ISBN 0-7923-2519-2
54 N J. Barrow (ed). *Plant Nutrition - From Genetic Engineering to Field Practice*. 1993 ISBN 0-7923-2540-0
55 A.D Robson (ed)· *Zinc in Soils and Plants* 1993 ISBN 0-7923-2631-8

GPSR Compliance

The European Union's (EU) General Product Safety Regulation (GPSR) is a set of rules that requires consumer products to be safe and our obligations to ensure this.

If you have any concerns about our products, you can contact us on ProductSafety@springernature.com

In case Publisher is established outside the EU, the EU authorized representative is:

Springer Nature Customer Service Center GmbH
Europaplatz 3
69115 Heidelberg, Germany

Batch number: 10809441

Printed by Printforce, the Netherlands